Industrial Applications of Soil Microbes

(Volume 2)

Edited by

Shampi Jain

Ashutosh Gupta

&

Neeraj Verma
Department of Agriculture Science, AKS University, Satna, Madhya Pradesh
- 485001, India

Industrial Applications of Soil Microbes

(Volume 2)

Editors: Shampi Jain, Ashutosh Gupta & Neeraj Verma

ISSN (Online): 2811-0773

ISSN (Print): 2811-0765

ISBN (Online): 978-981-5050-26-4

ISBN (Print): 978-981-5050-27-1

ISBN (Paperback): 978-981-5050-28-8

© 2023, Bentham Books imprint.

Published by Bentham Science Publishers Pte. Ltd. Singapore. All Rights Reserved.

First published in 2023.

BENTHAM SCIENCE PUBLISHERS LTD.
End User License Agreement (for non-institutional, personal use)

This is an agreement between you and Bentham Science Publishers Ltd. Please read this License Agreement carefully before using the ebook/echapter/ejournal (**"Work"**). Your use of the Work constitutes your agreement to the terms and conditions set forth in this License Agreement. If you do not agree to these terms and conditions then you should not use the Work.

Bentham Science Publishers agrees to grant you a non-exclusive, non-transferable limited license to use the Work subject to and in accordance with the following terms and conditions. This License Agreement is for non-library, personal use only. For a library / institutional / multi user license in respect of the Work, please contact: permission@benthamscience.net.

Usage Rules:

1. All rights reserved: The Work is the subject of copyright and Bentham Science Publishers either owns the Work (and the copyright in it) or is licensed to distribute the Work. You shall not copy, reproduce, modify, remove, delete, augment, add to, publish, transmit, sell, resell, create derivative works from, or in any way exploit the Work or make the Work available for others to do any of the same, in any form or by any means, in whole or in part, in each case without the prior written permission of Bentham Science Publishers, unless stated otherwise in this License Agreement.
2. You may download a copy of the Work on one occasion to one personal computer (including tablet, laptop, desktop, or other such devices). You may make one back-up copy of the Work to avoid losing it.
3. The unauthorised use or distribution of copyrighted or other proprietary content is illegal and could subject you to liability for substantial money damages. You will be liable for any damage resulting from your misuse of the Work or any violation of this License Agreement, including any infringement by you of copyrights or proprietary rights.

Disclaimer:

Bentham Science Publishers does not guarantee that the information in the Work is error-free, or warrant that it will meet your requirements or that access to the Work will be uninterrupted or error-free. The Work is provided "as is" without warranty of any kind, either express or implied or statutory, including, without limitation, implied warranties of merchantability and fitness for a particular purpose. The entire risk as to the results and performance of the Work is assumed by you. No responsibility is assumed by Bentham Science Publishers, its staff, editors and/or authors for any injury and/or damage to persons or property as a matter of products liability, negligence or otherwise, or from any use or operation of any methods, products instruction, advertisements or ideas contained in the Work.

Limitation of Liability:

In no event will Bentham Science Publishers, its staff, editors and/or authors, be liable for any damages, including, without limitation, special, incidental and/or consequential damages and/or damages for lost data and/or profits arising out of (whether directly or indirectly) the use or inability to use the Work. The entire liability of Bentham Science Publishers shall be limited to the amount actually paid by you for the Work.

General:

1. Any dispute or claim arising out of or in connection with this License Agreement or the Work (including non-contractual disputes or claims) will be governed by and construed in accordance with the laws of Singapore. Each party agrees that the courts of the state of Singapore shall have exclusive jurisdiction to settle any dispute or claim arising out of or in connection with this License Agreement or the Work (including non-contractual disputes or claims).
2. Your rights under this License Agreement will automatically terminate without notice and without the

need for a court order if at any point you breach any terms of this License Agreement. In no event will any delay or failure by Bentham Science Publishers in enforcing your compliance with this License Agreement constitute a waiver of any of its rights.
3. You acknowledge that you have read this License Agreement, and agree to be bound by its terms and conditions. To the extent that any other terms and conditions presented on any website of Bentham Science Publishers conflict with, or are inconsistent with, the terms and conditions set out in this License Agreement, you acknowledge that the terms and conditions set out in this License Agreement shall prevail.

Bentham Science Publishers Pte. Ltd.
80 Robinson Road #02-00
Singapore 068898
Singapore
Email: subscriptions@benthamscience.net

CONTENTS

FOREWORD i
PREFACE ii
LIST OF CONTRIBUTORS iii
CHAPTER 1 MICROORGANISMS AND THEIR INDUSTRIAL USES 1
Meenakhi Prusty, Ashish Kumar Dash, Suman G. Sahu and *Neeraj Verma*
    INTRODUCTION 1
    ANTIBIOTICS 2
    ENZYMES 3
        Proteases 3
            Uses 3
        Lipases 3
            Uses 4
        Carbohydrases 4
            α-Amylase 4
            Uses 4
            β-Amylase 4
            Uses 4
            Amyl Glucosidase (Glucoamylase) 4
            Uses 4
            β-Galactosidase (Lactase) 4
            Uses 5
            Glucose Isomerase 5
            Uses 5
            Glucose Oxidase 5
            Uses 5
            Inulinase 5
            Uses 5
            Invertase 5
            Uses 5
            Pullulanase 5
            Uses 6
            Pectinases 6
            Uses 6
        Hemicellulases and β-Glucanases/Cellulases 6
            Uses 6
            Hemicellulases 6
            Uses 6
        Miscellaneous Enzymes and Their Uses 6
            Catalase 6
            Phytase 7
            Urease 7
    ENZYMES IN MISCELLANEOUS BIOTRANSFORMATION PROCESSES 7
        Penicillin and Cephalosporin Acylases (Amidases) 7
        L-Amino Acid Acylases 7
    NATURAL FOOD PRESERVATIVES 7
    VITAMINS 8
    FERMENTATION PRODUCTS 8
    AMINO ACIDS 9

AGRICULTURE APPLICATION	10
CONCLUDING REMARKS	11
REFERENCES	11

CHAPTER 2 SOIL INHABITANT BACTERIA: MORPHOLOGY, LIFE CYCLE AND IMPORTANCE IN AGRICULTURE AND OTHER INDUSTRIES 12
Safi Ur Rehman Qamar and *Mayer L. Calma*

INTRODUCTION	12
ACTINOBACTERIA	13
Morphological Characteristics and Life Cycle	13
Industrial Importance and Applications of Actinobacteria	14
CYANOBACTERIA	15
Morphological Characteristics and Life Cycle	15
Industrial Importance and Applications of Cyanobacteria	16
NITROGEN-FIXING BACTERIA	17
Azotobacter	19
Azospirillum	19
Rhizobium	19
INDUSTRIAL IMPORTANCE AND APPLICATIONS OF COMMON NITROGEN-FIXING BACTERIA	20
CONCLUDING REMARKS	20
REFERENCES	21

CHAPTER 3 ROLE OF SOIL MICROBES IN THE SUSTAINABLE DEVELOPMENT: AGRICULTURE, RECOVERY OF METALS AND BIOFUEL PRODUCTION 26
Anurag Singh, Priya Bhatia, Shreya Kapoor, Simran Preet Kaur, Sanjay Gupta, Nidhi S. Chandra and *Vandana Gupta*

INTRODUCTION	27
SOIL MICROBES AS BIOFERTILIZERS	28
Nitrogen-Fixing Microorganisms	29
Phosphate Solubilizing Microorganisms	30
Plant Growth Promoter	30
Other Potential Bioinoculum	32
Future Aspects	33
SOIL MICROBES AS BIOCONTROL AGENT	34
Microbial Biological Control Agents (MBCA)	34
Mechanism of Action of Microbial Biocontrol Agents	36
Direct Interaction with the Microbes	36
Indirect Interaction with the Microbes	37
Induced Resistance	37
Future Aspects	38
SOIL MICROBES IN BIOLEACHING	38
Mesophilic Bacteria	38
Acidithiobacillus	38
Leptospirillum	39
Thermophilic Bacteria	39
Heterotrophic Microorganisms	40
Bioleaching Mechanisms	40
Bacterial or Chemical Oxidation of Metal Sulfides to Sulfate	40
Precipitation/ Metal Recovery	41
Industrial Applications	42
Future Aspects	42

SOIL MICROORGANISMS IN BIOFUEL PRODUCTION	43
Microbial Communities Involved in Biofuel Production	44
Bacteria	44
Fungi	44
Algae	46
Alcohol Production	46
Pre-Treatment and Hydrolysis	46
Fermentation	47
Distillation	48
Future Aspects	48
CONCLUDING REMARKS	48
ACKNOWLEDGEMENTS	50
REFERENCES	50
CHAPTER 4 INDUSTRIAL APPLICATIONS OF SOIL MICROBES: PRODUCTION OF ENZYMES, ORGANIC ACIDS AND BIOPIGMENTS	56
Simran Preet Kaur, Tanya Srivastava, Anushka Sharma, Sanjay Gupta, Nidhi S. Chandra and Vandana Gupta	
INTRODUCTION	56
SOIL MICROORGANISMS IN ENZYME PRODUCTION	58
SOME IMPORTANT INDUSTRIAL ENZYMES	59
ENZYMES PRODUCED BY COMMON SOIL MICROBES	61
Bacteria	61
Bacillus	61
Actinomycetes	61
Pseudomonas	62
Staphylococcus	62
Arthrobacter	62
Xanthomonas	63
Corynebacterium	63
Fungi	63
Aspergillus	63
Penicillium	63
Mucor	64
Yeasts	64
Others	64
ENZYMES FROM ACIDOPHILES	65
ENZYMES FROM ALKALIPHILES	67
ENZYMES FROM THERMOPHILES	68
FUTURE PERSPECTIVES	70
INDUSTRIAL PRODUCTION OF ORGANIC ACIDS	71
WHY ARE SOIL MICROORGANISMS USED FOR ACID PRODUCTION?	72
Important Organic Acids Derived from Soil Microbes	73
Alpha Ketoglutaric Acid	73
Citric Acid	73
Fumaric Acid	74
Gluconic Acid	74
Malic Acid	74
Itaconic Acid	74
Lactic Acid	74
Butyric Acid	74

Propionic Acid	74
Future Aspects	76
BIOPIGMENTS	77
Future Aspects	79
CONCLUDING REMARKS	79
ACKNOWLEDGEMENT	80
REFERENCES	80

CHAPTER 5 APPLICATIONS OF MICROBIAL BIOPESTICIDES 85
Poonam Meena, Neelam Poonar, Sampat Nehra and *P.C. Trivedi*

INTRODUCTION	85
Plant-Incorporated Protectants (PIPs)	86
Biochemical Pesticides	86
BACTERIAL BIOPESTICIDES	87
Sporulating Bacteria	88
Bacillus thuringiensis (Bt)	88
Bacillus sphaericus (Bs)	89
Paenibacillus Species	89
Non Sporulating Bacteria	90
Pseudomonas aeruginosa (Schroeter) Migula	90
Serratia marcescens	90
Clostridium Species	90
Gammaproteobacteria	91
Photorhabdus and Xenorhabdus	91
Serratia Species	91
Yersinia entomophaga	91
Pseudomonas entomophila	91
Betaproteobacteria	91
Burkholderia Species	91
Chromobacterium Species	91
Actinobacteria	92
Streptomyces Species	92
Saccharopolyspora spinosa	92
FUNGAL BIOPESTICIDES	92
VIRAL PESTICIDES	94
PROTOZOANS	95
NEMATODES	95
ENDOPHYTES	96
ROLE OF MICROBIAL INSECTICIDES (MIS)	96
LIMITATION OF MICROBIAL INSECTICIDES	97
CONCLUDING REMARKS	97
REFERENCES	97

CHAPTER 6 SOIL MICROFLORA - A POTENTIAL SOURCE OF ANTIBIOTICS 102
S. Bharathi and *Prajeesh P. Nair*

INTRODUCTION	102
ACTINOMYCETES - A POTENTIAL SOURCE OF ANTIBIOTICS	104
ANTIBIOTIC PRODUCING BACTERIA	105
FUNGI AS A SOURCE OF ANTIBIOTICS	106
CLASSIFICATION OF ANTIBIOTICS	107
Classification Based on Their Structure	107
Penicillin-Derived Antibiotics	107

Amino Glycosidase Antibiotics	107
Macrolide Antibiotics	107
Tetracycline Antibiotics	108
Peptide Antibiotics	108
Antifungal Antibiotics	108
Chloramphenicol Antibiotics	108
Other Groups of Antibiotics	108
Classification Based on Their Mechanism of Action	108
Inhibition of Cell Wall Synthesis	108
Protein Synthesis Inhibitors	108
Inhibitors of Nucleic Acid Synthesis	109
Other Biomolecule Inhibitors	109
ISOLATION OF ANTIBIOTIC-PRODUCING ORGANISMS FROM SOIL	109
Crowded Plate Technique	109
A GENERALIZED PROTOCOL FOR THE PRODUCTION OF ANTIBIOTICS	111
CONCLUDING REMARKS	112
REFERENCES	112

CHAPTER 7 ROLE OF NONPATHOGENIC STRAINS IN RHIZOSPHERE ... 113
Rana Muhammad Sabir Tariq, Maheen Tariq, Sarah Ali, Shahan Aziz and *Jam Ghulam Mustafa*

INTRODUCTION	113
IMPORTANCE OF NON-PATHOGENIC STRAINS OF RHIZOSPHERE	114
Rhizobiaceae	114
Trichoderma	115
Arbuscular Mycorrhizal Fungi (AMF)	115
BENEFICIAL ROLE OF NON-PATHOGENIC STRAINS IN RHIZOSPHERE	115
EFFECTS OF NON-PATHOGENIC RHIZOSPHERE STRAINS ON PLANTS' HEALTH	117
INDUSTRIALIZATION OF NON-PATHOGENIC STRAINS OF RHIZOSPHERE	121
CONCLUDING REMARKS	121
REFERENCES	121

CHAPTER 8 BIOREMEDIATION INDUSTRY: A MICROBIAL PERSPECTIVE ... 129
Pooja Singh

INTRODUCTION	129
BIOREMEDIATION	130
POTENTIAL POLLUTANTS FOR BIOREMEDIATION	130
Organic Pollutants	130
Inorganic Pollutants	131
MYCOREMEDIATION	132
BIOLOGICAL PROCESSES INVOLVED IN BIOREMEDIATION	134
USES OF BIOREMEDIATION TECHNOLOGY	135
In Situ Bioremediation	136
Biostimulation	136
Bioaugmentation	136
Bioventing	137
Biosparaging	138
Ex Situ Bioremediation	138
Land Farming	139
Composting	139
Biopiles	139
Bioreactors	140

BENEFITS OF BIOREMEDIATION	140
BIOREMEDIATION BY GENETICALLY ENGINEERED MICROORGANISMS	141
CONCLUDING REMARKS	142
REFERENCES	142

CHAPTER 9 ALLEVIATION OF SALINITY STRESS BY MICROBES 145
Sampat Nehra, Raj Kumar Gothwal, Alok Kumar Varshney, Pooran Singh Solanki, Poonam Meena, P.C. Trivedi and P. Ghosh

INTRODUCTION	145
Salt Stress Damage to Plants	146
SALT TOLERANT BACTERIA	148
Characterization of Salt Tolerant Bacteria	149
Soil Tolerant Bacteria in Salt Stress Alleviation of Plants	151
MECHANISM OF SALT STRESS ALLEVIATION	153
Phytohormones Synthesis	153
IAA Production	153
ABA Production	154
Cytokinin Production	155
Ethylene Production	155
Protective Compounds	156
Antioxidative Enzymes	157
Exopolysaccharides	157
Ion Homeostasis	158
ENHANCEMENT OF NUTRIENT UP-TAKE	160
Nitrogen	160
Phosphorus	161
ROOT HYDRAULIC CONDUCTANCE	162
CONCLUDING REMARKS	162
REFERENCES	162

CHAPTER 10 LIGNOCELLULOSE DEGRADING BACTERIA IN SOIL 175
Archana Rawat, Parul Bhatt Kotiyal, Soni Singh and Neeraj Verma

INTRODUCTION	175
FACTORS AFFECTING THE BIODEGRADATION PROCESS	176
BIODEGRADATION VS. COMPOSTING	177
LIGNIN-MODIFYING ENZYMES (LMES)	179
Bacterial Lignin Modifying Enzymes	179
Lignocellulosic Biomass	180
ENZYMES RESPONSIBLE FOR LIGNIN DEGRADATION	183
Important Groups of LMEs	184
Lignin Peroxidases (LiP)	185
Manganese Dependent Peroxidase (MnP)	185
Laccases	185
Glyoxal Oxidase (GLOX)	186
Aryl Alcohol Oxidase (AAO)	186
Heme-Thiolate Haloperoxidases	186
Flavin adenine dinucleotide (FAD)-dependent glucose dehydrogenase (GDH)	187
Pyranose 2-Oxidase (POX)	187
Cellobiose Dehydrogenase (CDH)	187
CONCLUDING REMARKS	187
REFERENCES	188

CHAPTER 11 AN OVERVIEW OF DIVERSE YEAST SPECIES PERFORMING THE BIOCONTROL FUNCTION IN AGRICULTURE 193
Abhishek Sinha
- INTRODUCTION 193
- BENEFITS AND DRAWBACKS OF USING YEASTS AS BIOCONTROL AGENTS 194
- DIFFERENT YEASTS FOR BIOCONTROL AND THEIR MODES OF ACTION 195
- ENVIRONMENTAL CONDITIONS AFFECTING BIOCONTROL BY YEASTS 197
  - Heat Stress 198
  - Desiccation Shock 198
  - Oxidative Stresses 199
  - Other Environmental and Production Condition Stresses 200
- CONCLUDING REMARKS 201
- ACKNOWLEDGEMENT 201
- REFERENCES 201

CHAPTER 12 *MACROPHOMINA PHASEOLINA*: AN AGRICULTURALLY DESTRUCTIVE SOIL MICROBE 203
Ramesh Nath Gupta, Kishor Chand Kumhar and *J.N. Srivastava*
- INTRODUCTION 204
- SURVIVAL OF THE PATHOGEN 205
- THE DISEASE 205
- DISEASE SYMPTOMS 206
- INFLUENCE OF WEATHER CONDITIONS ON DISEASE DEVELOPMENT 207
- MANAGEMENT OF DISEASE 207
  - Management With Non-Conventional Chemicals 207
    - *Salicylic Acid (SA)* 208
    - *Acetylsalicylic Acid (ASA)* 208
    - *Indole-3-Acetic Acid (IAA)* 208
    - *Indole Butyric Acid (IBA)* 209
    - *Riboflavin* 209
    - *Thiamine* 209
  - Impact of Non-Conventional Chemicals on Phenol Content 210
    - *Peroxidase Content* 210
    - *Polyphenol Oxidase Content* 211
    - *Phenylalanine Ammonia Lyase Content* 211
    - *Catalase Content* 211
  - Application of Phytoextracts and Fungicides 212
  - Use of Resistant Varieties 212
- CONCLUDING REMARKS 213
- ACKNOWLEDGEMENTS 213
- REFERENCES 213

CHAPTER 13 *BEAUVERIA BASSIANA*: AN ECOFRIENDLY ENTOMOPATHOGENIC FUNGI FOR AGRICULTURE AND ENVIRONMENTAL SUSTAINABILITY 219
Purnima Singh Sikarwar and *Balaji Vikram*
- INTRODUCTION 219
  - Taxonomy 220
- MORPHOLOGY 221
- MODE OF ACTION AGAINST INSECTS 221
  - Adhesion 221
  - Germination and Differentiation 221

    Penetration 222
    Dissemination within and to Another Host 222
  APPLICATIONS OF *B. BASSIANA* IN INSECT PEST MANAGEMENT 223
    Efficacy of *B. bassiana* Against Thrips 223
    Efficacy of *B. bassiana* Against Insect Pests of Stored Grains 224
    Efficacy of *B. bassiana* Against Crop Pests 224
  CONSTRAINTS TO USE *B. BASSIANA* 225
    Environmental Conditions 225
    Temperature 225
    Humidity 225
    Other Constraints 226
  ADVANTAGES OF USING *B. BASSIANA* 227
    Improved Plant Nutrient Uptake and Plant Growth 228
    Using *B. Bassiana* For Insect Management 229
    Using *B. Bassiana* as an Insect Management of Field Crops 230
    To Control Termites and Other Soil Insects 230
      Drip System 230
      Preparation of Spray Solution 230
  DIFFICULTIES FACED IN FORMULATION OF BIOPESTICIDE BASED ON *B. BASSIANA* 231
  CONCLUDING REMARKS 231
  REFERENCES 232

CHAPTER 14 ERGOT, ERGOTISM AND ITS PHARMACEUTICAL USE 234
Doomar Singh
  INTRODUCTION 234
    History 235
  PRODUCTION OF ERGOT ALKALOIDS 236
  ERGOT ALKALOIDS 239
  SYMPTOMS OF ERGOTISM IN HUMANS AND ANIMALS 239
  PHARMACEUTICAL USE OF ERGOT ALKALOIDS 243
  CONCLUDING REMARKS 245
  ACKNOWLEDGMENTS 246
  REFERENCES 246

CHAPTER 15 IDENTIFICATION OF FUNGI ON RHIZOPLANE AND IN RHIZOSPHERE OF LEGUMINOUS CROP BY ADOPTING DIFFERENT TECHNIQUES 248
Deepali Chaturvedi
  INTRODUCTION 248
  SOIL SAMPLING 249
  ISOLATION OF FUNGAL SPECIES FROM RHIZOSPHERE AND RHIZOPLANE 250
    Methods 250
      Direct Inoculation 250
      The Soil Plate Method 250
      Soil Dilution and Plate Count 251
      Immersion Techniques 251
  ISOLATION OF FUNGAL HYPHAE FROM SOIL 252
    Physical Dissection 252
  RESULTS 252
  DISCUSSION 254
  CONCLUDING REMARKS 254
  REFERENCES 255

CHAPTER 16 *TRICHODERMA*: A POTENTIAL ARSENAL FOR INDUSTRIES 256
P.B. Khaire, S.S. Mane and S.V. Pawar
   INTRODUCTION 256
   BIOTECHNOLOGICAL APPLICATIONS 257
      Production of Secondary Metabolites 257
      Production of Antibiotics and Bioactive Compounds 259
      Production of Hydrolytic Enzymes 260
   ROLE OF *TRICHODERMA* IN WINEMAKING AND BREWERY INDUSTRY 261
   ROLE OF *TRICHODERMA* IN BIOREMEDIATION 261
   ROLE OF *TRICHODERMA* IN AGRICULTURE 262
   MASS PRODUCTION, FORMULATION AND ITS SCOPE FOR THE MANAGEMENT OF PLANT DISEASES 263
      Characteristics of *Trichoderma* for Formulation Development 263
      Characteristics of an Ideal Formulation 264
      Model Methods for the Mass Multiplication of *Trichoderma* 264
         Formulation Based on Talc 264
         Formulation Based on Vermiculite and Wheat Bran 265
         Formulation Based on Pesta Granules 265
         Formulation Reliant on Alginate Prills 265
         Formulation Based on Press Mud 265
         Formulation Based on Coffee Husks 265
         Formulations Based on Oil 266
         Formulations Focused on Banana Waste 266
      Shelf Life of *Trichoderma* Formulations 266
   ***TRICHODERMA* DELIVERY FOR DISEASE MANAGEMENT** 267
      Treatment of *Trichoderma* to seeds 267
      Treatment of *Trichoderma* as biopriming seeds 268
      Treatment of *Trichoderma* to the soil 268
      Treatment of *Trichoderma* to the Root 269
      Spraying the Leaves/Dressing for Wounds 269
   MASS PRODUCTION AND FORMULATION DEVELOPMENT 270
      Fermentation 270
         Liquid Fermentation 270
         Solid Fermentation 270
   CONSTRAINTS TO COMMERCIALIZATION 271
   CONCLUDING REMARKS 272
   REFERENCES 272

SUBJECT INDEX 279

FOREWORD

Microbes, or microscopic organisms, are widely used in large-scale industrial processes. They are crucial for the production of a variety of metabolites, such as ethanol, butanol, lactic acid, and riboflavin, as well as the transformation of chemicals that help in reducing the environmental pollution. For instance, microbes can be used to create biofertilizers or to reduce metal pollutants. Certain non-microbial products can also be used to produce certain microbial products, such as the diabetes medication insulin.

The immense use of chemical fertilizers, pesticides, fungicides, and weedicides has diminished the soil texture, quality, and ecosystem of the soil. It also leads to an unfavourable environment for favourable soil microbes. It may also lead to the development of resistant pathogenic strains, which may have devastating effects on the plants. Therefore, it seems that it is the best time to reduce the use of chemicals in agriculture and farmers should opt for biologically derived chemicals in their fields. The beneficial soil microbes, upon interaction with the plants, enhance productivity. They directly or indirectly take part in various enzymatic activities, leading to various byproducts beneficial to plants and ultimately animals, including human beings. Therefore, the use of natural bio-molecules will not only increase soil productivity but will also be environmentally eco-friendly.

This book, which deals with "INDUSTRIAL APPLICATION OF SOIL MICROBES" Vol. 2, is a timely, appreciable, and praise-worthy contribution to the field of Industrial Microbiology. The different chapters of this book, written by various experts in their fields, have made this book more valuable and highly scientific. A team of a galaxy of experts has explained the dynamics of soil microbes and the knowledge about their direct or indirect use in agriculture and industry. This book will help to launch a programme to make every family farm a bio-fortified farm in the days to come. I, therefore, hope that this book will be widely read by students, teachers, and researchers and will serve the purpose of being an important reference book.

I congratulate all the contributors and members of the editorial board for preparing this script of microbiological wisdom.

Prof. K.R. Maurya
M.Sc. (Ag) Kanpur, D.Phil. (Allahabad)
F.H.S.I., Former Vice-Chancellor - RAU, Pusa
Samastipur (Bihar) and M.J.R.P. University
Jaipur (Rajasthan), Director (Horticulture)
AKS University, Satna
Madhya Pradesh, India

PREFACE

Modern agricultural practices largely rely on high inputs of mineral fertilizers to high yields and involve the application of chemical pesticides to protect crops from diseases and pests. These practices are now being reevaluated and coming under scrutiny as our awareness of potential health and environmental consequences of excessive mineral fertilizers and chemical pesticide usage improves. It is widely recognized that the application of mineral fertilizers (especially nitrogen) can result in soil sickness and groundwater contamination by nitrates leaching through the soil profiles. To preserve our most precious natural resources like soil and water, this is high time to go for organic agriculture.

Thus, the present condition of soil and groundwater again recalls the scientists and researchers to maintain its productivity as well as quality. As the world population is increasing enormously, there is a need to sustain agriculture and increase the yield of agricultural products by using biochemicals instead of synthetic chemicals to reduce the negative impact on the soil and water environment. Using these biochemicals not only the productivity of soil will increase due to microorganism activity, but also the ecosystem could be saved.

For the use of soil microorganisms as biochemicals or the bioproducts obtained from them, one should know about the soil microorganisms effectively and their utility, which has been established over the years of research. Various chemicals like enzymes, vitamins, minerals, organic acids, biopigments, antibiotics, *etc.* have been successfully isolated from these microorganisms and are being exploited in various industrial fields like pharmaceuticals, Agriculture (biofertilizers, biopesticides, etc), food industries, beverages, and dairy industries.

Several studies have been conducted to exploit these microorganisms to produce biochemical helpful for not only humans but also for animals and the environment. However, these research findings are in scattered form. Although, the subject of microbes and their use in agriculture is very important there is a dearth of research-based books. Hence in this text, we have tried to collect the research findings of several eminent scientists and compile them in the form of a book. Therefore, this book will provide exhaustive knowledge about microbes and their uses for the welfare of mankind. The industrialists, professors, extension workers, scientists, and students will get most of the information about these soil microbes and their uses in one place.

The editors are highly thankful to the publisher of this text. Since this is the first edition, it may contain some errors that crept in inadvertently in spite of all editorial care. So, we are always interested to hear comments and suggestions from the learned readers.

Shampi Jain

Ashutosh Gupta

&

Neeraj Verma
Department of Agriculture Science, AKS University,
Satna, Madhya Pradesh
- 485001, India

List of Contributors

Abhishek Sinha	Department of Microbiology, Swami Vivekanand University, N.H. 26 Narsinghpur road, Sagar, Madhya Pradesh, India
Alok Kumar Varshney	Department of Bioengineering, Birla Institute of Technology, Mesra, Ranchi, Jaipur Campus, Jaipur, Rajasthan, India, Statue Circle, Jaipur, Rajasthan, India
Anurag Singh	Department of Microbiology, Ram Lal Anand College, University of Delhi, Benito Juarez Road, New Delhi 110021, India
Anushka Sharma	Department of Microbiology, Ram Lal Anand College, University of Delhi, Benito Juarez Road, New Delhi 110021, India
Archana Rawat	Forest Ecology and Climate Change Division, Forest Research Institute, Dehradun, India
Ashish Kumar Dash	Department of Soil Science and Agricultural Chemistry, OUAT, Bhubaneswar, India
Balaji Vikram	Department of Horticulture, AKS University, Satna (M.P.), India
S. Bharathi	Department of Microbiology, The Oxford College of Science, Bangalore, India
Deepali Chaturvedi	Department of Botany, Lucknow University, U.P., India
Doomar Singh	Department of Plant Pathology, Faculty of Agriculture Science & Technology, AKS University, Satna 485001, India
Jam Ghulam Mustafa	Department of Agriculture, Agribusiness Management, University of Karachi, Karachi, Pakistan
J.N. Srivastava	Chairman, Department of Plant Pathology, Bihar Agricultural University, Sabour - 813210, India
Kishor Chand Kumhar	Assistant Director-Plant Protection, Deen Dayal Upadhyay Centre of Excellence for Organic Farming, CCS HAU, Hisar - 125004, India
Maheen Tariq	Punjab College Sambrial Wazirabad Road, University Road, Sambrial, Sialkot, Pakistan
Mayer L. Calma	Department of Biology, College of Science, University of the Philippines Baguio, Baguio City, 2600, Benguet, Philippines
Meenakhi Prusty	RRTTS, Dhenkanal, Odisha University of Agriculture and Technology, Bhubaneswa, India
Neelam Poonar	Department of Botany, University of Rajasthan, Jaipur, India
Neeraj Verma	Department of Agriculture Science, AKS University, Satna, MP, India
Nidhi S. Chandra	Department of Microbiology, Ram Lal Anand College, University of Delhi, Benito Juarez Road, New Delhi 110021, India
P.B. Khaire	Department of Plant Pathology and Agril. , Microbiology, PGI, MPKV, Rahuri, India
Poonam Meena	Department of Botany, University of Rajasthan, Jaipur, India
Pooja Singh	Department of Botany, DDU Gorakhpur University, Gorakhpur, UP, India

P. Ghosh	Department of Bioengineering, Birla Institute of Scientific Research, Statue Circle, Jaipur, Rajasthan, India
Pooran Singh Solanki	Department of Bioengineering, Birla Institute of Technology, Mesra, Ranchi, Jaipur Campus, Jaipur, Rajasthan, India
P.C. Trivedi	Deptartment of Botany, University of Rajasthan, Jaipur, Rajasthan, India
Parul Bhatt Kotiyal	Forest Ecology and Climate Change Division, Forest Research Institute, Dehradun, India
Priya Bhatia	Department of Microbiology, Ram Lal Anand College, University of Delhi, Benito Juarez Road, New Delhi 110021, India
Prajeesh P. Nair	Department of Microbiology, The Oxford College of Science, Bangalore, India
Purnima Singh Sikarwar	Department of Horticulture, AKS University, Satna (M.P.), India
Raj Kumar Gothwal	Department of Bioengineering, Birla Institute of Technology, Mesra, Ranchi, Jaipur Campus, Jaipur, Rajasthan, India
Rana Muhammad Sabir Tariq	Punjab College Sambrial, Wazirabad Road, University Road, Sambrial, Sialkot, Pakistan
Ramesh Nath Gupta	Assistant Professor-cum-Jr. Scientist, Department of Plant Pathology, Bihar Agricultural University, Sabour - 813210, India
S.S. Mane	Department of Plant Pathology and Agril Microbiology, PGI, MPKV, Rahuri, India
S.V. Pawar	Department of Plant Pathology, Sumitomo Chemical India Limited, Mumbai, India
Suman G. Sahu	Department of Soil Science and Agricultural Chemistry, OUAT, Bhubaneswar, India
Safi Ur Rehman Qamar	Chulabhorn Graduate Institute, 54 Kamphaeng Phet 6 Road, Lak Si, 5 Bangkok 10210, Thailand
Shreya Kapoor	Department of Microbiology, Ram Lal Anand College, University of Delhi, Benito Juarez Road, New Delhi 110021, India
Simran Preet Kaur	Department of Microbiology, Ram Lal Anand College, University of Delhi, Benito Juarez Road, New Delhi 110021, India
Sanjay Gupta	Independent Scholar (Former Head, Department of Biotechnology, Jaypee Institute of Information Technology, Sector 62, Noida, UP - 201309, India
Sampat Nehra	Department of Biotechnology, Birla Institute of Scientific Research, Statue Circle, Jaipur, Rajasthan, India
Sarah Ali	Department of Agriculture, Agribusiness Management, University of Karachi, Karachi, Pakistan
Shahan Aziz	Department of Agriculture, Agribusiness Management, University of Karachi, Karachi, Pakistan
Soni Singh	Forest Ecology and Climate Change Division, Forest Research Institute, Dehradun, India
Tanya Srivastava	Department of Microbiology, Ram Lal Anand College, University of Delhi, Benito Juarez Road, New Delhi 110021, India

Vandana Gupta Department of Microbiology, Ram Lal Anand College, University of Delhi, Benito Juarez Road, New Delhi 110021, India

CHAPTER 1

Microorganisms and their Industrial Uses

Meenakhi Prusty[1,*], **Ashish Kumar Dash**[2], **Suman G. Sahu**[2] and **Neeraj Verma**[3]

[1] *RRTTS, Dhenkanal, Odisha University of Agriculture and Technology, Bhubaneswar, India*
[2] *Department of Soil Science and Agricultural Chemistry, OUAT, Bhubaneswar, India*
[3] *Department of Agriculture Science, AKS University, Satna, MP, India*

Abstract: For human beings, the diversity of microorganisms is still an undiscovered aspect. For the well-being of society, the huge microbial population performs many vital activities. Microorganisms play an important role in sustainable agriculture, environmental protection, and human and animal health. Microorganisms have a major contribution to agricultural issues like crop productivity, plant health protection, soil health maintenance, and environmental issues like bioremediation of soil and water from many pollutants. In addition to these activities, microorganisms also produce many products, either directly or through industrial processes, which are essential for human survival. In this chapter, we deal with various industrial products that are produced by microorganisms through different reactions, like antibiotics, enzymes, natural food preservatives, vitamins, fermentation products, amino acids, and agricultural products.

Keywords: Antibiotics, Biopesticides, Enzymes, Fermentation, Microbial Biotransformation.

INTRODUCTION

Microorganisms are mainly used to provide a large number of products and services. Because of the ease of their mass cultivation, speed of growth, use of cheap substrates (which are mainly wastes), and the diversity of potential products, they are considered very beneficial to human beings. As for the ease of their genetic manipulation, they have limitless possibilities for new products and services from the fermentation industry. The branch of biotechnology and microbiology that mainly deals with the study of various microorganisms and their applications in industrial processes is referred to as industrial microbiology. Even before the existence of microorganisms was known, they were used in industrial processes. Microorganisms play multiple roles in the industry. There are

---

* **Corresponding author Meenakhi Prusty:** RRTTS, Dhenkanal, Odisha University of Agriculture and Technology, Bhubaneswar, India; E-mail: meenakhi.prusty@gmail.com

Shampi Jain, Ashutosh Gupta and Neeraj Verma
All rights reserved-© 2023 Bentham Science Publishers

many ways to manipulate microorganisms so that product yields will be increased. The microorganisms are manipulated to produce many products like antibiotics, vitamins, enzymes, amino acids, solvents, alcohol, and dairy products. The discovery of microorganisms with their multiplicity of highly specific biochemical activities has stimulated a steady growth of industrial fermentation processes.

Microbes are widely used to synthesize several products valuable to human beings in industrial processes. Numerous industrial products have been derived from microbes such as:

- Antibiotics
- Enzymes
- Natural food preservatives
- Vitamins
- Fermentation products
- Amino acids
- Agricultural products

ANTIBIOTICS

For medical applications, the production of new drugs synthesized by a specific organism for medical purposes is the main focus of industrial microbiology. Antibiotic treatment is essential for many bacterial infections. Many naturally occurring antibiotics and precursors are produced through the fermentation process. As the microorganisms grow in a liquid medium, the population size can be controlled to maximize the amount of the product. During the production of antibiotics, the nutrient environment, pH, temperature, and oxygen should be controlled to produce a maximum number of cells without any mortality.

Certain microbes produce antibiotics, which function either by killing or retarding the growth of harmful microbes without affecting the host cells. Since 1928, when Alexander Fleming discovered the first antibiotic penicillin from the fungus *Penicillium notatum*, several microorganisms (fungal and bacterial) have been reported to produce many other antibiotics. Antibiotics such as streptomycin, tetracycline, chloramphenicol, erythromycin, vancomycin, and neomycin have been isolated from various *Streptomyces* spp. (to treat many bacterial infections), while antifungals such as Amphotericin B and Nystatin have also been isolated from *Streptomyces* spp. Bacitracin is an antibiotic that has been isolated from *Bacillus subtilis* [1].

Most of the antibiotics are produced by fungi or actinomycetes, whereas bacteria produce subtilin and bacitracin. But the latter is used very limitedly because of its toxic effect.

Steroids can be produced by microbial biotransformation. Steroids can be consumed either orally or by injection. Arthritis is mainly controlled by the use of steroids. Cortisone is an anti-inflammatory drug that fights against arthritis as well as several skin diseases. Testosterone is also a steroid, which is produced from dehydroepiandrosterone by using *Corynebacterium* sp.

ENZYMES

The biological catalysts that are mainly used to control certain biochemical reactions in the living system are enzymes. Enzymes have a wide range of applications both in the medical and non-medical fields. The enzymes, which are obtained from certain microbes, are referred to as microbial enzymes. Industrial enzymes are mainly produced by microorganisms through safe gene transfer methods. In the year 1896, from fungal amylase, the first industrially produced microbial enzymes were obtained and were used for indigestion and other digestive disorders. Some enzymes synthesized by microorganisms with their uses are described as follows [1]:

Proteases

These are obtained mostly as neutral proteases or zinc metalloprotease from *Aspergillus* spp. and *Bacillus* spp., along with some alkaline serine proteases.

Uses

a. Biological detergents: The alkaline proteases are used in this, which are produced by *Bacillus licheniformis* and *B. subtilis*.
b. Baking: dough modification/gluten reduction and flavour enhancement
c. Beer brewing: used to chill proof beer to remove protein haze.
d. Tendering and baiting with leather
e. Cheese processing: coagulation of milk protein, accelerated ripening, and flavoring.
f. Tenderization of meat and removal of meat from bones
g. Flavoring control and food product production
h. Waste management: Silver recovery from used photographic films

Lipases

These are mainly produced by various species of *Bacillus*, *Aspergillus*, *Rhizopus*, and *Rhodotorula*.

Uses

These are used as biological detergents, in leather processing (for fat removal), in the production of flavour compounds and to accelerate ripening in dairy and meat products.

Carbohydrases

These are also mostly obtained from *Aspergillus* spp. and *Bacillus* spp. Some are mentioned below:

α-Amylase

It can be obtained from *Aspergillus* and *Bacillus* spp.

Uses

It is used in starch processing, baking, brewing, textile manufacturing, and as biological detergents.

β-Amylase

It is obtained from *Bacillus* spp., *Streptomyces* spp., and *Rhizopus* spp.

Uses

It is used in maltose syrup production and in the brewing industry (to increase wort fermentability).

Amyl Glucosidase (Glucoamylase)

It is obtained from *Aspergillus niger* and *Rhizopus niveus*.

Uses

It is generally used in glucose syrup production (for complete starch saccharification), baking (to improve bread crust colour), in the brewing industry (for the production of low-carbohydrate beer), and to remove starch from wine and fruit juice.

β-Galactosidase (Lactase)

It is obtained from various species of *Bacillus*, *Kluyveromyces*, and *Candida*.

Uses

It has various important industrial uses, such as in whey syrup production (to give greater sweetness), in the processing of milk and dairy products to reduce lactose, and in ice-cream manufacturing.

Glucose Isomerase

It is obtained from species of *Actinoplanes*, *Arthrobacter*, and *Streptomyces*.

Uses

It is used in the manufacture of high-fructose syrups.

Glucose Oxidase

It is mainly produced by *Aspergillus niger* and *Penicillium notatum*.

Uses

It can be used as an antioxidant (used along with catalase to remove oxygen from various food products).

Inulinase

It is obtained from *Aspergillus niger*, *Candida* spp., and *Kluyveromyces* spp.

Uses

It is used in the processing of Jerusalem artichoke tubers (for hydrolysis of polyfructans and levans).

Invertase

It is obtained from *Saccharomyces* spp. and *Kluyveromyces* spp.

Uses

It is used in the sweets and confectionery industries (for the liquefaction of sucrose) and to invert sugar production, *i.e.*, sucrose conversion to glucose and fructose.

Pullulanase

This is obtained mainly from *Klebsiella pneumoniae* and *B. acidopullulyticus*.

Uses

It is used in starch processing (for debranching of starch in sugar syrup manufacture and brewing).

Pectinases

These are a mixture of enzymes such as polygalacturonase, produced by *Aspergillus niger* and *Penicillium* spp.

Uses

These are used in the extraction and clarification of plant juice, oil, fruit and wine clarification, and coffee bean fermentation.

Hemicellulases and β-Glucanases/Cellulases

β-Glucanases (cellulases) are obtained from *Trichoderma reesei, Penicillium funiculosum, Aureobasidium pullulans*, and *Bacillus* spp.

Uses

These are used in the production and processing of fruit juice and olive oil, wine and beer, malting (to speed modification of grains) in textile industries, and in wood pulp processing.

Hemicellulases

These are obtained from *Cryptococcus* and *Trichosporon*.

Uses

These are used in baking, brewing, animal feedstuffs, nutraceuticals, and wood pulp processing.

Miscellaneous Enzymes and Their Uses

Catalase

It is obtained from *Aspergillus niger, Corynebacterium glutamicum*, and *Micrococcus lysodeikticus*. It is used in textiles for bleaching and cheese manufacturing.

Phytase

It is used in feeds for monogastric animals (to release phosphate from phytic acid).

Urease

It is obtained from *Lactobacillus fermentum*. It is used in wine production and in ceramics manufacturing.

ENZYMES IN MISCELLANEOUS BIOTRANSFORMATION PROCESSES

Penicillin and Cephalosporin Acylases (Amidases)

Bacillus megaterium and *Escherichia coli* produce penicillin acylase, and *Pseudomonas* species produce cephalosporin acylase. These are used for antibiotic conversion to remove the side groups of penicillins or cephalosporins to form 6-aminopenicillanic acid or 7-aminocephalosporanic acid, respectively.

L-Amino Acid Acylases

These are obtained from *Aspergillus* spp. and used for the resolution of L-amino acids from acylatedracemic mixtures of D, L-amino acid mixtures.

NATURAL FOOD PRESERVATIVES

Many organic acids, enzymes, and antibiotics are produced by microorganisms (Fig. 1). Organic acids, mainly lactic acid, are extensively used in food preservation, and bacteriolytic microbial enzymes similar to lysozyme exhibit preservative potential. Antibiotics, mainly those used in chemotherapy, cannot be used as food preservatives due to fears of their toxicity and allergenicity. Several compounds originating from microorganisms are widely used as food additives and preservatives. A vast list of them is there and some of them are worth mentioning, such as acidulants (*e.g.*, citric acid from *Aspergillus niger*, *Candida guilliermondii*, and *Candida lipolytica*; lactic acid from *Lactobacillus delbrueckii* and *Lactobacillus bulgaricus*) and amino acids (*e.g.*, lysine from *Corynebacterium glutamicum* and *Brevibacterium flavum*; tryptophan from *Klebsiella aerogenes*).

Some microorganisms have been used as colouring agents to produce colour, such as β-carotene (from *Blakeslea trispora* and *Dunaliella salina*), lycopene (from *Blakeslea trispora* and *Streptomyces chrestomyceticus*), zeaxanthin (from

Flavobacterium species), lutein (from *Spongiococcum excentricum* and *Chlorella pyrenoidosa*), etc.

Various products used as flavours and flavour enhancers have been isolated from microorganisms, *e.g.*, vanillin (from *Saccharomyces* spp.), monosodium glutamate (from *Corynebacterium glutamicum* and *Brevibacterium flavum*), etc.

VITAMINS

Organic compounds like vitamins perform many life-sustaining functions inside the human body. Just like essential micronutrients, the requirement for vitamins is in small quantities for the metabolism of the body. As our body cannot synthesize these vitamins, they need to be supplied through the diet. Apart from plant and animal sources, microbes are also capable of synthesizing vitamins. A group of microbes living in the digestive tracts of both humans and other animals called the gut microbiota help in synthesizing these vitamins. Vitamin K is synthesized by these microbes. Other examples of microbial vitamins include thiamin (vitamin B1), riboflavin (vitamin B2), pyridoxine (vitamin B6), cobalamin (vitamin B12), biotin, folic acid, l-ascorbic acid (vitamin C), β-carotene (provitamin A), ergosterol (provitamin D2), and pantothenic acid. Some vitamins can be produced by direct fermentation processes, whereas many are produced through biotransformations or combined chemical and microbiological processes [1].

Vitamins have also been isolated from various microorganisms, such as β-carotene (from *Blakeslea trispora* and *Dunaliella salina*), lycopene (from *B. trispora* and *Streptomyces chrestomyceticus*), zeaxanthin (from *Flavobacterium* spp.), cantaxanthin (*Cantharellus cinnabarinus* and *Rhodococcus maris*) and astaxanthin (from *Agrobacterium aurantiacum*, *Haematococcus pluvialis*, and *Mycobacterium lacticola*).

FERMENTATION PRODUCTS

In the fermentation process, sugar can be converted into gas, alcohol, or acids. The fermentation reaction occurs in the absence of oxygen, *i.e.* anaerobically. Many products are produced by yeasts and bacteria. One of the oldest known fermentation processes is acetic acid (vinegar) fermentation, which occurs naturally as unwanted spoilage of wine. Acetic acid is the active ingredient in vinegar, which is produced through the oxidization process of an alcohol containing fruit juice by acetic acid bacteria (*Acetobacter* sp.). Production of vinegar occurs in two stages. In the first stage, yeast such as *Saccharomyces cerevisiae* produces ethanol from plant sugar, and in the second stage, acetic acid bacteria like *Acetobacter* and *Gluconobacter* are involved in the aerobic fermentation process. Alcohol, which is consumed as ethanol, is also used in powering automobiles as a fuel source.

From natural sugars like glucose, drinking alcohol is produced, and during the reaction, carbon dioxide is produced as a side product, which can be used for making bread and carbonated beverages (*Saccharomyces cerevisiae*). Alcoholic beverages like beer and wine are fermented anaerobically by microorganisms. Acetone-butanol production by *Clostridium acetobutylicum* is the most famous industrial fermentation [2]. During the First World War, acetone, sorely needed for the manufacture of cordite, was produced by the fermentation process. Nowadays, microorganisms are involved in the production of fuels and chemical feedstocks. Some of those worth mentioning are acetone (from *Clostridium* sp.), butanol (from *Clostridium acetobutylicum*), ethanol (from *Zymomonas mobilis* and *Saccharomyces cerevisiae*), methane (from methanogenic archaea), *etc*.

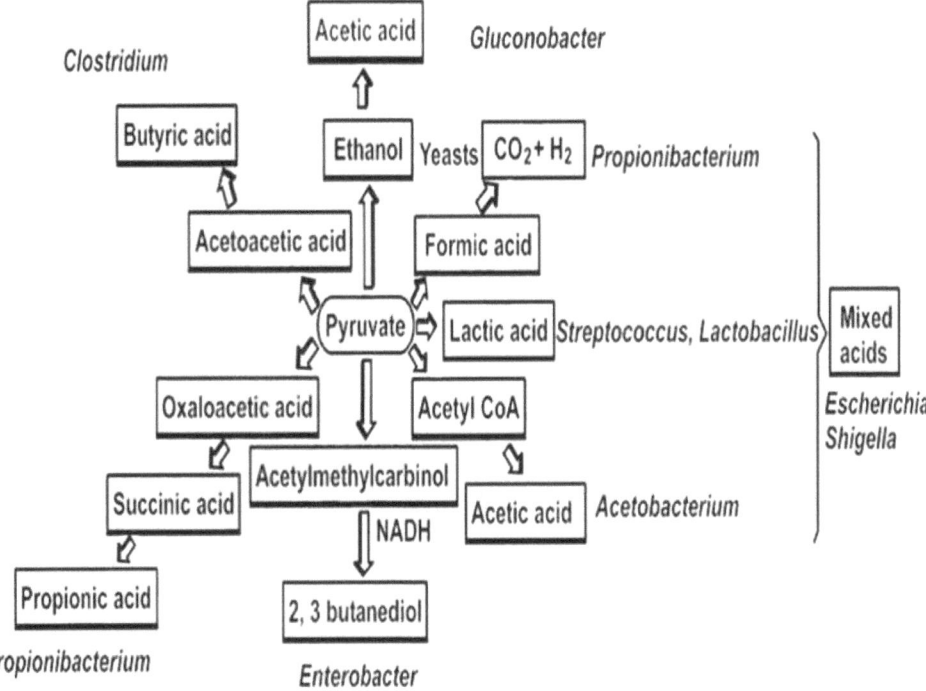

Fig. (1). Schematic representation of production of various organic acids by different bacteria [3].

AMINO ACIDS

Amino acids and organic solvents are synthesized by microbes. The essential amino acids such as L-methionine, L-lysine, L-tryptophan and the non-essential amino acid L-glutamic acid are used mainly for feed, food, and pharmaceutical industries. These amino acids are produced by *Corynebacterium glutamicum* through the fermentation process. *C. glutamicum* produces L-lysine and L-glutamic acid. L-glutamic acid is used for the production of monosodium glutamate (MSG), which is used as a food flavouring agent. Previously, in *E. coli*, diaminopimelic acid (DAP) was produced from which L-lysine was synthesized but when *C. glutamicum* was discovered for the production of L-glutamic acid, this organism and other autotrophs were used to yield other amino acids such as lysine, aspartate, methionine, isoleucine, and threonine. L-lysine is used for the feeding of pigs and chickens as well as to treat nutrient deficiency and viral infections. *Corynebacterium* and *E. coli* produce L-tryptophan through the fermentation process. Though the production is not as large as other amino acids, it is produced for pharmaceutical purposes since it can be converted and used to produce neurotransmitters.

AGRICULTURE APPLICATION

The increasing need for various fertilizers and pesticides enhances the demand for agricultural products. The overuse of chemical fertilizers and pesticides has long-term effects on soil and the environment. The soil becomes infertile for growing crops due to the excessive use of chemical fertilizers and pesticides. To alleviate the bad effects of chemicals, biofertilizers, biopesticides, and organic farming are the only alternatives.

From these naturally occurring substances, biochemical pesticides can be produced, which can control pest populations in a non-toxic manner. Garlic and pepper are used as biochemical pesticides, which can repel insects from the desired location. As a microbial pesticide, primarily a virus, bacterium, or fungus is used, which can control pest populations more specifically. The most commonly used microbe for the production of microbial biopesticides is *Bacillus thuringiensis*, also known as Bt. The delta-endotoxins produced by this spore-forming bacterium cause the insect or pest to stop feeding on the crop or plant as the endotoxin destroys the lining of the digestive system.

Several microbes that live in association with plants, like *Rhizobium*, mycorrhiza, and free-living fungi, are there. They have many functions in the promotion of plant growth by different mechanisms. Phosphate solubilization is one such example, and the organisms involved in this activity are known as phosphate solubilizers. Many other bacteria that promote the growth of plants are termed PGPR (plant growth promoting rhizobacteria), including species of *Azotobacter*, *Azospirillum*, *Acetobacter*, *Burkholderia*, and *Bacillus*. Free-living non-deleterious bacteria which could promote plant growth directly or indirectly may be considered as PGPR [4].

Of the fungi species, around 12,000 species can be considered mushrooms, of which at least 2,000 species have different degrees of edibility. From the wild species, around 200 species of mushroom have been collected and used for medical purposes. Commercially, 35 mushroom species have been cultivated till now, out of which 20 are cultivated on an industrial scale. Small-scale mushroom production becomes an opportunity and option for landless farmers. The organic wastes of different agricultural and food processing by-products can be used as growing media for edible fungi.

CONCLUDING REMARKS

Resource utilization and environmental perturbation have increased rapidly with the rise of the world's population. The exploitation of unexplored microbial diversity and available culturable microbes is the major challenge of the twenty-first century. For industrial technology, it is very useful to manipulate the genetic potential of various microorganisms. Applications of microbes for future research and extension will have to go a long way towards the improvement of industrial products, environmental quality, agricultural productivity, human health, and novel uses such as global climate change. Our future efforts should be directed towards (1) Exploration of various unexplored habitats of microbial resources; (2) Exploitation of plant health, plant genome promotion, and bioremediation research; (3) The role of microbes in global climate change; and (4) The role of microbes in new drug development and transgenic development.

REFERENCES

[1] Waites MJ, Morgan NL, Rockey JS, et al. Industrial Microbiology: An Introduction. London, UK: Blackwell Science Ltd 2001 p. 302.

[2] Okafor N, Ed. Modern Industrial Microbiology and Biotechnology. Enfield (NH), USA: Science Publishers 2007; p. 530.

[3] Bisen PS, Ed. Microbes in Practice. New Delhi, India: I K International Publishing House Pvt. Ltd 2014; 1158-224.

[4] Jacoby R, Peukert M, Succurro A, et al. The role of soil microorganisms in plant mineral nutrition - current knowledge and future directions. Front Plant Sci 2017; 8: 1617.
[http://dx.doi.org/10.3389/fpls.2017.01617] [PMID: 28974956]

CHAPTER 2

Soil Inhabitant Bacteria: Morphology, Life Cycle and Importance in Agriculture and Other Industries

Safi Ur Rehman Qamar[1,*] **and Mayer L. Calma**[1,2]

[1] *Chulabhorn Graduate Institute, 54 Kamphaeng Phet 6 Road, Lak Si, 5 Bangkok 10210, Thailand*

[2] *Department of Biology, College of Science, University of the Philippines Baguio, Baguio City, 2600, Benguet, Philippines*

> **Abstract:** There are many bacteria in the soil, but they have less biomass because of their small size. Soil-inhabitant bacteria are an essential source of nutrients for plants. Some studies highlighted their industrial importance, like in the pharmaceutical industry, perfume manufacturing, and agriculture product scale-up production, including biofertilizers. Most of the studies have been carried out on Actinobacteria and *Nitrobacter* because of their potential to produce biofertilizers and chemical constituents on a large scale. This chapter discussed their taxonomic and morphological characteristics and gathered details about their practical applications from limited studies carried out in this field.

Keywords: Actinobacteria, *Azotobacter*, *Azospirillum*, Cyanobacteria, *Rhizobium*.

INTRODUCTION

Soil harbours a variety of bacterial communities. One gram of soil contains 10^3 to 10^6 unique species of bacteria. Most of these bacteria have not been fully characterized. However, through recent molecular advancements, we have gained in-depth knowledge of these bacteria from cellular to molecular level [1]. This data provided crucial information about the ecology of these bacteria and their importance to the soil ecosystem. Several taxa of these bacteria have been well studied in terms of their ecology, morphology, life cycle and importance in the industry. These are as follows:

---

[*] **Corresponding author Safi Ur Rehman Qamar:** Chulabhorn Graduate Institute, 54 Kamphaeng Phet 6 Road, Lak Si, 5 Bangkok 10210, Thailand; E-mail: ranasafi73@gmail.com
Contribution: Both authors contributed equally.

Shampi Jain, Ashutosh Gupta and Neeraj Verma
All rights reserved-© 2023 Bentham Science Publishers

1. **Actinobacteria** act as an essential nutrient (calcium, phosphorus and sodium) regulator in the soil [2].
2. *Streptomyces* act as a catabolic organism and increase soil fertility by converting xylan, lignin, cellulose and lignocellulose into organic matter [3].
3. *Azospirillum* acts as an influencer of growth due to its ability to fix nitrogen and stimulate auxin production [4].
4. *Azotobacter* acts as a phytohormone accelerators (*e.g.*, indole 3 acetic acid), metabolizes heavy metals, and pesticides and degrades oil globules [5].
5. *Rhizobium* acts as a nitrogen fixer [6].
6. *Nitrobacter* acts as a nitrite oxidizer by oxidizing nitrite into nitrate, which is a primary source of inorganic nitrogen [7].

This chapter discusses the physiological characteristics of soil-inhabiting bacteria and their importance in various industries.

ACTINOBACTERIA

Actinobacteria are freely distributed in both aquatic and terrestrial environments. These are Gram-positive filamentous bacteria. The feeding nature of these bacteria is saprophytic, which means they feed on organic matter. These bacteria are found abundantly in alkaline soil containing high organic matter content, whereas they are less abundant in water or air. These bacteria inhibit the upper surface and deep ground up to 1.5 m of soil [8].

Morphological Characteristics and Life Cycle

The characteristics of Actinobacteria mainly relate to cell wall composition, phospholipids, type of menaquinone, and sugar content of cells. Fragmented mycelium is regarded as a unique form of vegetative reproduction. However, these bacteria usually reproduce by asexual spores. Moreover, various morphologies have been observed in these bacteria, mainly the absence or presence of aerial and substrate mycelium. Besides, mycelial colouring and diffusible melanoid pigment production are also considered as one the criteria to distinguish between actinobacteria species [9]. It reproduces from both types of mycelia depending on the conditions, *i.e.*, it reproduces from aerial hyphae to form spores on a solid surface. On the other hand, substrate mycelia develop from germinating spores (Fig. **1**).

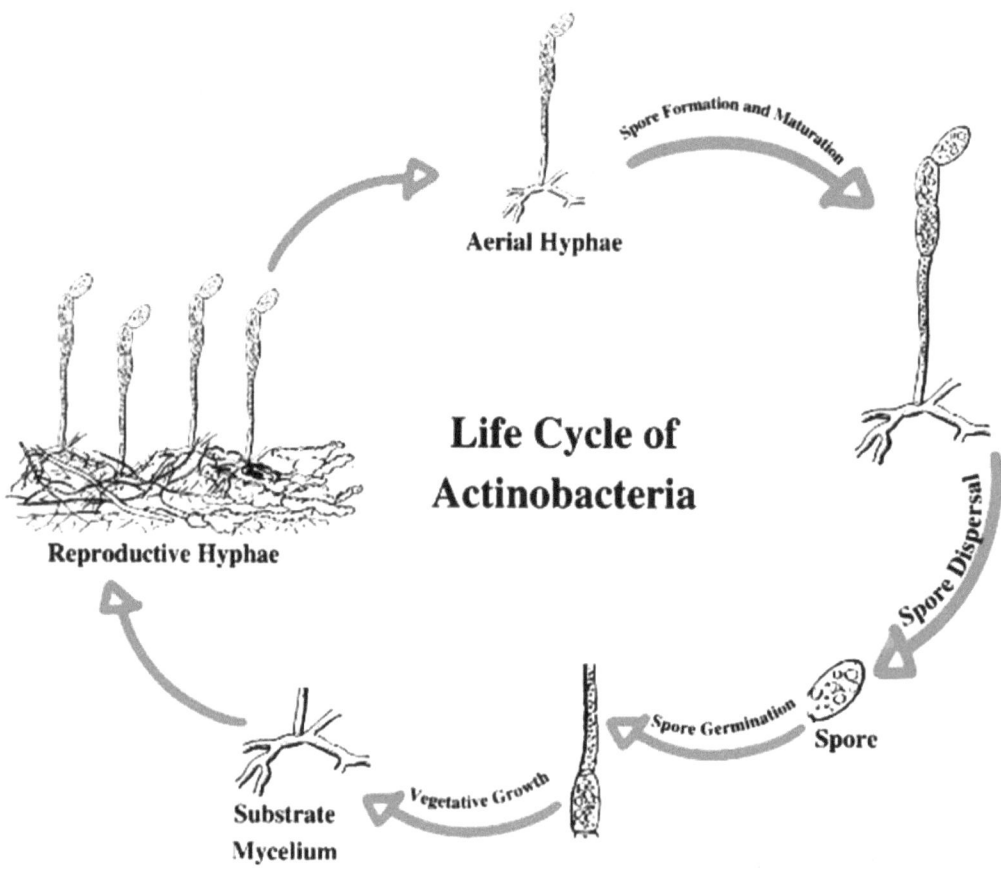

Fig. (1). Schematic illustration of the life cycle of typical actinobacteria.

Industrial Importance and Applications of Actinobacteria

Several genera of actinobacteria have been described for their industrial importance. The main industrial applications of these bacteria deal with biomedical science and biotechnology. Through recent advancements in bioinformatics tools and DNA sequencing, we can understand metaproteomics (proteomics for large-scale production of microbial enzymes) [10]. Actinobacteria have been deeply explored for their potential to produce amylases, proteases, cellulases, pectinases, chitinases, and xylanases. With their applications in paper, pulp, waste management, detergents, and agriculture, some of these industrially essential enzymes are listed in Table **1**.

Table 1. Industrially important enzymes synthesized using actinobacteria.

Enzyme	Bacterial Strain	Optimum Conditions	Industrial Application	Reference
Amylase	*Thermobifida fusca*	60°C & 6 pH	Pulp, paper, and textile	[11]
	Streptomyces erumpens	45°C & 9 pH	Detergent and baking	[12]
Pectinase	*Streptomyces lydicus*	45°C & 6.5 pH	Textile and beverage	[13]
Cellulase	*Thermobifida halotolerans*	45°C & 7 pH	Pulp, paper, and textile	[14]
	Streptomyces ruber	37°C & 6 pH	Detergent	[15]
Chitinase	*Nocardiopsis prasina*	55°C & 7 pH	Leather	[16]
	Streptomyces thermoviolaceus	60°C & 6 pH	Textile	[17]
Protease	*Streptomyces thermoviolaceus*	65°C & 6.5 pH	Detergent, food, and brewing	[18]
	Streptomyces pactum	40°C & 7 pH	Leather and pharmaceutical	[19]
Lipase	*Nocardiopsis alba*	30°C & 7 pH	Cosmetics and detergent	[20]
	Streptomyces exfoliates	37°C & 6 pH	Pulp and paper	[21]
Keratinase	*Actinomadura keratinilytica*	70°C & 10 pH	Leather	[22]
Xylanase	*Actinomadura* sp.	70°C & 4 pH	Animal feed	[23]
	Streptomyces spp.	50°C & 9 pH	Pulp and paper	[24]

CYANOBACTERIA

Cyanobacteria are Gram-negative filamentous bacteria. They inhibit all types of environments, *i.e.*, terrestrial, marine, and freshwater. Because of this fact, they can fix nitrogen and are photosynthetic. Their ecological importance can be assessed because the "red sea" is named after the cyanobacteria (*Trichodesmium erythraeum*) that produce red-colour [25].

Morphological Characteristics and Life Cycle

Cyanobacteria are filamentous, unicellular, and colonial. The filament consists of a mucilage sheet, and multiple strands are called trichomes (Fig. **2**). These filaments are subdivided into two types: undifferentiated homo-cysts and differentiated hetero-cysts [26]. Their prokaryotic cells have a peptidoglycan wall that surrounds 70S ribosomes, DNA, and photosynthetic unit thylakoid. They increase their progeny by asexual reproduction, including fragmentation, monocytes, binary fission, exospore, endospore, and akinetes (Fig. **2**). In comparison, sexual reproduction is absent in these bacteria [27].

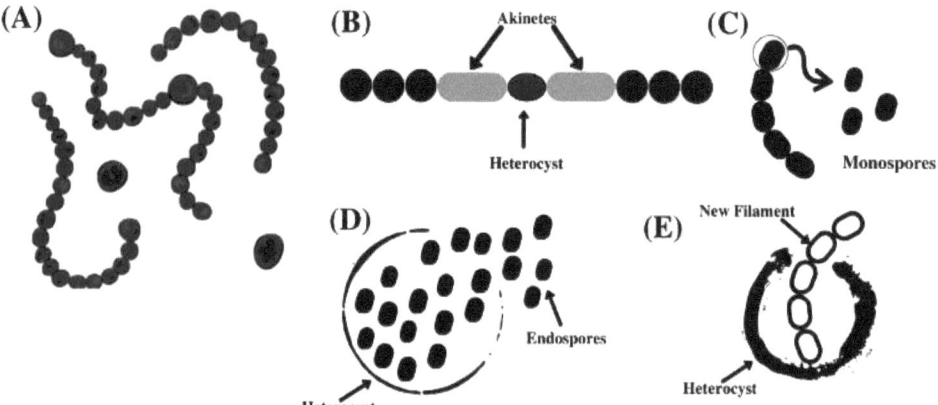

Fig. (2). (A) Morphological illustration of cyanobacteria trichome and **(B-E)** various types of asexual reproduction.

Industrial Importance and Applications of Cyanobacteria

Cyanobacteria have gained massive attention in recent years due to their possible applications in industry and agriculture. These bacteria are considered a potential source of biologically active complexes with antibacterial, antifungal, antiviral, and anticancer activities [28]. Some strains of these bacteria accumulate polyhydroxyalkanoates (PHA), which are used as a supernumerary for photochemical-based non-biodegradable plastics [29]. Recent studies showed that cyanobacteria were found in abundance in the oil spill area because of their ability to degrade oil components. These bacteria help other oil-degrading bacteria by providing them with the required fixed nitrogen and oxygen [30]. Hydrogen produced by cyanobacteria is considered a potential alternative source of energy. In addition to these essential applications, cyanobacteria are also being used in wastewater treatment, food, aquaculture, fertilizers, pharmaceuticals, and the manufacturing of secondary metabolites such as vitamins, toxins, and exopolysaccharides. Some of the industrially important strains of cyanobacteria and their applications are given in Table **2**.

Table 2. Industrially important strains of cyanobacteria and their applications.

Biologically Active Compounds Producing Cyanobacteria			
Species	Biologically Active Compound	Application	Reference
Lyngbya lagerheimii	Sulpholipid	Antiviral	[31]

(Table 2) cont.....

Biologically Active Compounds Producing Cyanobacteria			
Species	**Biologically Active Compound**	**Application**	**Reference**
Lyngbya majuscule	Cyclic polypeptide, lyngbyatoxins, malyngolide, and curacin A	Antiviral and antibacterial	[32]
Phormidium tenue	Galactosyl diacylglycerols	Antiviral and antialgal	[33]
Nostoc spp.	Nostocyclamide	Antifungal	[34]
Synechocystis spp.	Naienones A-C	Anticancer	[35]
PHA Producing Cyanobacteria			
Chlorogloea fritschii	**PHA Type**	Additive material for biochemically produced non-biodegradable plastic	[36]
Trichodesmium thiebautii	P (3HB)		[37]
Synechocystis spp. PCC6803			[38]
Oscillatoria limosa	P (3HV)		[39]
Hydrogen Producing Cyanobacteria			
Anabaena cylindrical	**Optimum Conditions**	Hydrogen production	[40]
Microcystis PCC 7820	Air with photon fluence rate of 20 µE m^{-2} s^{-1}		[41]
Synechocystis PCC 6714			[42]
Biofertilizer Producing Cyanobacteria			
Nostoc commune TUBT 05	**Target crop**	Biofertilizer	[43]
	Oryza sativa L.		
Anabaena spp.	*Zea mays* L. hybrids	N-supplement	[44]
Aulosira prolifica, Stigonema dendroideum, Nostoc muscorum, Nostoc calcicola, Anabaena oryzae, Scytonema varium, Tolypothrix tenuis, and *Schizothrix vaginata*	*Zea mays, Triticum aestivum,* and *Hordeum vulgare*	Growth promoter	[45]

NITROGEN-FIXING BACTERIA

Biological nitrogen fixation (BNF) is the process of converting nitrogen from the environment into a more complex form that plants can absorb. It plays a vital role in nitrogen cycling and is performed by specific prokaryotic groups with an enzyme called nitrogenase [46]. This ATP-dependent and oxygen-sensitive enzyme catalyzes the reduction of nitrogen gas (N_2) into ammonia (NH_3) coupled with the reduction of protons into hydrogen gas (H_2) [46, 47]. BNF is a highly regulated process using different nitrogen fixation genes (*nif*) in response to varying levels of fixed nitrogen, carbon, ATP, and oxygen [46]. The ubiquitous soil-inhabiting bacteria capable of BNF are termed diazotrophs. They are a physiologically diverse group of organisms with a complex phylogenetic relationship (Fig. 3) [48]. Traditionally, they can be classified into three groups based on their way of life [49]. Free-living diazotrophs such as *Azotobacter* spp. do not have direct interaction with plants. Associative diazotrophs such as *Azospirillum* spp. form close interactions with some groups of plants in a system known as the rhizosphere. As for endosymbiotic diazotrophs, they rely on a host plant for nutrients while they fix nitrogen into a usable form for their host in a structure called a rhizosphere [49]. One of the most studied and well-understood rhizospheres of endosymbiotic diazotrophs is the root nodules of leguminous plants, usually colonized by *Rhizobium* spp. or *Bradyrhizobium* spp [50]. Using molecular techniques, the subsets of diazotroph communities in the soil can be established using the *nifH* marker gene as a fingerprinting tool [51, 52]. In this section, some common diazotrophs will be described with their morphology, physiology, plant-microbe interaction, and commercial applications. Moreover, some methods of culturing common diazotrophs under laboratory conditions are briefly described.

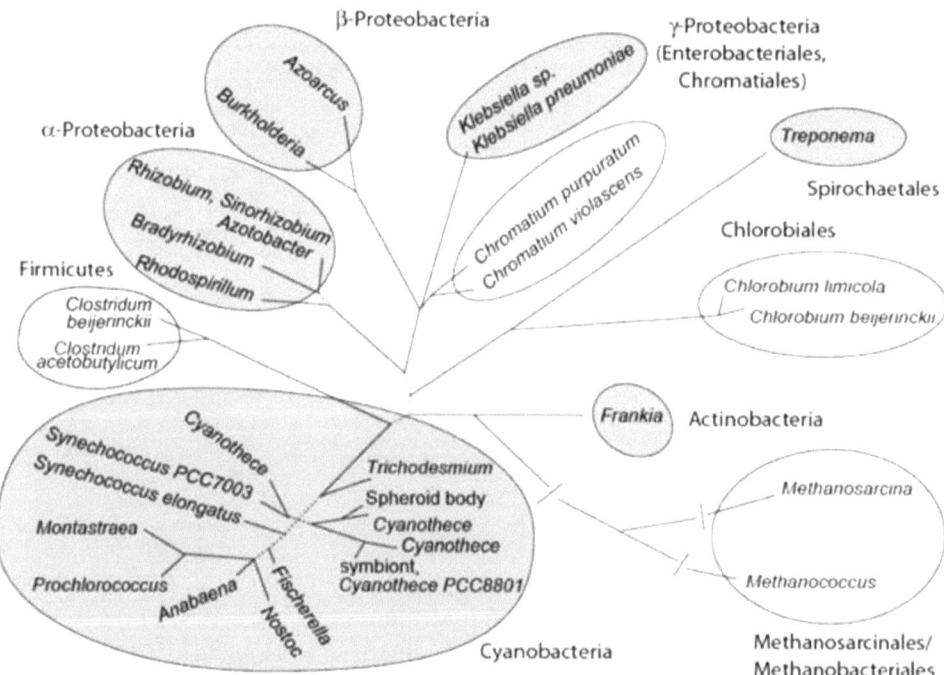

Fig. (3). Phylogenetic relationship of non-associative and associative diazotrophs [48].

Azotobacter

Azotobacter is a genus of Gram-negative ovoid or spheroid gamma-proteobacteria capable of forming thick-walled cysts under stressful conditions. It is comprised of free-living diazotrophs, with six currently reported species [5]. BNF is carried out by three metal-dependent nitrogenases (molybdenum, vanadium, and iron only), which are oxygen-sensitive [5, 53]. The cells are capable of forming a thick wall made of alginate exopolysaccharide to prevent the diffusion of oxygen in the cytoplasm that can disrupt the activity of nitrogenases [54]. Although considered as free-living, *Azotobacter* spp. can be found in the rhizosphere soil portion [52, 55]. A culture-based method such as the use of selective medium (Ashby's nitrogen-free agar) coupled with 16S rDNA for molecular identification or a non-culture-based method by using *nifH* gene sequencing can reveal the diversity of *Azotobacter* spp. in soils [52].

Azospirillum

Azospirillum is a genus of Gram-negative, twisted, rod-shaped, non-spore-forming alpha-proteobacteria capable of BNF. Members of this genus usually form an association in the soil component of the rhizosphere of non-leguminous plants such as maize, rice, sugarcane, and wheat [49]. BNF is carried out by an oxygen-sensitive nitrogenase complex with two components [4]. Cells from soil samples can be grown in Nfb (nitrogen-free bromothymol blue) semisolid medium. *Azospirillum* isolates can be grown in Nfb solid medium for morphological and biochemical analysis [56].

Rhizobium

Rhizobium is a genus of rod-shaped Gram-negative alpha-proteobacteria with an endosymbiotic relationship with leguminous plants (Fabaceae) in structures known as root nodules. Nitrogen fixation is catalyzed by an oxygen-sensitive nitrogenase complex similar to *Azospirillum* [57]. The bacterial cells known as bacteroids are found inside the host plant's root cells, where the mutualistic relationship occurs. The *Rhizobium*-legume symbiosis involves the interplay of plant-microbe factors. The process starts when the root cells of the host plant release flavonoids and isoflavonoids [58], which are signalling molecules for the expression of nodulation (*nod*) genes of the bacteria. Once expressed, the bacterial cells release lipo-chito-oligosaccharide signalling molecules that trigger root hair coiling and cortical cell division. The invasion of plant cells by bacterial cells eventually results in the formation of root nodules used for nitrogen fixation [57]. Root nodule-inhabiting diazotrophs can be isolated from the nodules and cultured in YEMA (yeast extract, mannitol, and agar) medium containing bromothymol blue [59].

INDUSTRIAL IMPORTANCE AND APPLICATIONS OF COMMON NITROGEN-FIXING BACTERIA

The productivity of agriculture relies on the availability of essential nutrients such as nitrogen, phosphorus, and potassium. Globally, these nutrients are mainly supplemented in the form of chemical fertilizers. However, the excessive use of these fertilizers results in ecological perturbations and eutrophication. Thus, sustainable methods in agriculture should be practiced [58]. One approach that is environment-friendly involves the utilization of BNF performed by diazotrophs. BNF is most productive in endosymbiotic diazotrophs, but associative and free-living species also contribute significantly, accounting for around 30% of global BNF [58]. Legumes comprise 14% of the total cultivated crops globally [50], while the remaining crops are non-legumes that produce the bulk of human food [49]. As an alternative to chemical fertilizers, biofertilizers are commercially produced to increase the absorbable nitrogen for plants through BNF. *Rhizobium*, *Bradyrhizobium*, *Azotobacter*, and *Azospirillum* are common diazotrophs [58]. Some strains of *Rhizobium*, including *Agrobacterium rhizogenes*, are considered necessary. These stains produce biologically active compounds such as protocatechuate, *p*-coumarate, cholinium lysinate, 1-ethyl-3-methylimidazolium acetate, α-pinene, and d-limonene. These compounds have been reported to have industrial importance as bio-jet-fuel candidates, fragrance candidates, and cleaning product components [43, 60]. *Azotobacter* is considered a unique biofertilizer for maintaining the nitrogen level in soil and promoting plant growth by affecting gibberellins and indole acetic acid release. Due to their ability to enhance plant growth, these bacteria can be used universally to promote agricultural yield [61]. Several studies have highlighted the importance of *Azospirillum* as a biofertilizer in agriculture [62, 63]. *Azospirillum* has been shown in studies to provide sulphate and phosphorous to plants through the oxidation of organic compounds such as phytates, resulting in plant growth [63, 64].

CONCLUDING REMARKS

The soil ecosystem is the habitat of many microbial species that are beneficial for plants because they provide essential nutrients such as nitrogen and phosphorous. Simultaneously, many bacterial species produce industrially important chemicals such as α-pinene and d-limonene for perfume production. Mass scale culturing of these bacteria can be a step towards sustainable and environmental-friendly green industry and agriculture approaches. Although there are benefits to these soil-inhabiting bacteria, few publications address the importance of soil bacteria, especially industrial importance. There is a need to focus on important environmental-friendly industries to sustain the economy and achieve a greener and waste-free ecosystem.

REFERENCES

[1]   Tringe SG, von Mering C, Kobayashi A, *et al.* Comparative metagenomics of microbial communities. Science 2005; 308: 554-7.
[http://dx.doi.org/10.1126/science.1107851] [PMID: 15845853]

[2]   Olanrewaju OS, Babalola OO. *Streptomyces*: Implications and interactions in plant growth promotion. Appl Microbiol Biotechnol 2019; 103: 1179-88.
[http://dx.doi.org/10.1007/s00253-018-09577-y] [PMID: 30594952]

[3]   Zhang B, Wu X, Tai X, *et al.* Variation in actinobacterial community composition and potential function in different soil ecosystems belonging to the Arid Heihe river basin of northwest China. Front Microbiol 2019; 10: 2209.
[http://dx.doi.org/10.3389/fmicb.2019.02209] [PMID: 31608036]

[4]   Steenhoudt O, Vanderleyden J. *Azospirillum*, a free-living nitrogen-fixing bacterium closely associated with grasses: Genetic, biochemical and ecological aspects. FEMS Microbiol Rev 2000; 24: 487-506.
[http://dx.doi.org/10.1111/j.1574-6976.2000.tb00552.x] [PMID: 10978548]

[5]   Sumbul A, Ansari RA, Rizvi R, *et al. Azotobacter*: A potential bio-fertilizer for soil and plant health management. Saudi J Biol Sci 2020; 27: 3634-40.
[http://dx.doi.org/10.1016/j.sjbs.2020.08.004] [PMID: 33304174]

[6]   Zahran HH. *Rhizobium*-legume symbiosis and nitrogen fixation under severe conditions and in an arid climate. Microbiol Mol Biol Rev 1999; 63: 968-89.
[http://dx.doi.org/10.1128/MMBR.63.4.968-989.1999] [PMID: 10585971]

[7]   Han S, Li X, Luo X, *et al.* Nitrite-oxidizing bacteria community composition and diversity are influenced by fertilizer regimes, but are independent of the soil aggregate in acidic subtropical red soil. Front Microbiol 2018; 9: 885.
[http://dx.doi.org/10.3389/fmicb.2018.00885] [PMID: 29867799]

[8]   Goodfellow M, Williams ST. Ecology of actinomycetes. Annu Rev Microbiol 1983; 37: 189-216.
[http://dx.doi.org/10.1146/annurev.mi.37.100183.001201] [PMID: 6357051]

[9]   van Dissel D, Claessen D, van Wezel GP. Morphogenesis of *Streptomyces* in submerged cultures. Adv Appl Microbiol 2014; 89: 1-45.
[http://dx.doi.org/10.1016/B978-0-12-800259-9.00001-9] [PMID: 25131399]

[10] Pieper R, Huang S-T, Suh M-J. Proteomics and Metaproteomics. In: Nelson KE, Ed. Encyclopedia of Metagenomics. New York, NY: Springer 2013; pp. 1-11.
[http://dx.doi.org/10.1007/978-1-4614-6418-1_690-9]

[11] Zhang F, Chen JJ, Ren WZ, *et al.* Cloning, expression and characterization of an alkaline thermostable GH9 endoglucanase from *Thermobifida halotolerans* YIM 90462T. Bioresour Technol 2011; 102: 10143-6.
[http://dx.doi.org/10.1016/j.biortech.2011.08.019] [PMID: 21907577]

[12] El-Sersy NA, Abd-Elnaby H, Abou-Elela GM, *et al.* Optimization, economization and characterization of cellulase produced by marine *Streptomyces ruber*. Afr J Biotechnol 2010; 9: 6355-64.
[http://www.academicjournals.org/AJB]

[13] Jacob N, Ashapoorna C, Prema P. Purification and partial characterization of polygalacturonase from *Streptomyces lydicus*. Bioresour Technol 2008; 99: 6697-701.
[http://dx.doi.org/10.1016/j.biortech.2007.10.002] [PMID: 17996445]

[14] Yang CH, Liu WH. Purification and properties of a maltotriose-producing α-amylase from *Thermobifida fusca*. Enzyme Microb Technol 2004; 35: 254-60.

[http://dx.doi.org/10.1016/j.enzmictec.2004.05.004]

[15] Kar S, Ray RC. Statistical optimization of alpha-amylase production by *Streptomyces erumpens* MTCC 7317 cells in calcium alginate beads using response surface methodology. Pol J Microbiol 2008; 57: 49-57. [http://www.pjmonline.org/wp-content/uploads/2015/12/vol5712008049.pdf] [PMID: 18610656]

[16] Horikoshi K. Alkaliphiles: Some applications of their products for biotechnology. Microbiol Mol Biol Rev 1999; 63: 735-50.
[http://dx.doi.org/10.1128/MMBR.63.4.735-750.1999] [PMID: 10585964]

[17] Bhattacharya D, Nagpure A, Gupta RK. Bacterial chitinases: Properties and potential. Crit Rev Biotechnol 2007; 27: 21-8.
[http://dx.doi.org/10.1080/07388550601168223] [PMID: 17364687]

[18] Bentley SD, Chater KF, Cerdeño-Tárraga AM, *et al.* Complete genome sequence of the model actinomycete *Streptomyces coelicolor* A3(2). Nature 2002; 417: 141-7.
[http://dx.doi.org/10.1038/417141a] [PMID: 12000953]

[19] Wietzorrek A, Bibb M. A novel family of proteins that regulates antibiotic production in streptomycetes appears to contain an OmpR-like DNA-binding fold. Mol Microbiol 1997; 25: 1181-4.
[http://dx.doi.org/10.1046/j.1365-2958.1997.5421903.x] [PMID: 9350875]

[20] Gandhimathi R, Seghal Kiran G, Hema TA, *et al.* Production and characterization of lipopeptide biosurfactant by a sponge-associated marine actinomycetes *Nocardiopsis alba* MSA10. Bioprocess Biosyst Eng 2009; 32: 825-35.
[http://dx.doi.org/10.1007/s00449-009-0309-x] [PMID: 19288138]

[21] Magda MM, Tork S, Al-Garni SM, *et al.* Production of lipase from genetically improved *Streptomyces exfoliates* LP10 isolated from oil-contaminated soil. Afr J Microbiol Res 2012; 6: 1125-37.
[http://dx.doi.org/10.5897/AJMR11.1123]

[22] Habbeche A, Saoudi B, Jaouadi B, *et al.* Purification and biochemical characterization of a detergent-stable keratinase from a newly thermophilic actinomycete *Actinomadura keratinilytica* strain Cpt29 isolated from poultry compost. J Biosci Bioeng 2014; 117: 413-21.
[http://dx.doi.org/10.1016/j.jbiosc.2013.09.006] [PMID: 24140106]

[23] Brzezinski R, Déry CV, Beaulieu C. Thermostable xylanase DNA, protein and methods of use. US Patent 1999. 5871730.

[24] Priya BS, Stalin T, Selvam K. Efficient utilization of xylanase and lipase producing thermophilic marine actinomycetes (*Streptomyces albus* and *Streptomyces hygroscopicus*) in the production of ecofriendly alternative energy from waste. Afr J Biotechnol 2012; 11: 14320-5.
[http://www.academicjournals.org/AJB] [DOI: 10.5897/AJB12.835]

[25] Bergman B, Sandh G, Lin S, *et al. Trichodesmium*-a widespread marine cyanobacterium with unusual nitrogen fixation properties. FEMS Microbiol Rev 2013; 37: 286-302.
[http://dx.doi.org/10.1111/j.1574-6976.2012.00352.x] [PMID: 22928644]

[26] Kumar K, Mella-Herrera RA, Golden JW. Cyanobacterial heterocysts. Cold Spring Harb Perspect Biol 2010; 2: a000315.
[http://dx.doi.org/10.1101/cshperspect.a000315] [PMID: 20452939]

[27] Lau NS, Matsui M, Abdullah AAA. Cyanobacteria: Photoautotrophic microbial factories for the sustainable synthesis of industrial products. BioMed Res Int 2015; 2015: 754934.
[http://dx.doi.org/10.1155/2015/754934] [PMID: 26199945]

[28] Singh RK, Tiwari SP, Rai AK, *et al.* Cyanobacteria: An emerging source for drug discovery. J Antibiot (Tokyo) 2011; 64: 401-12.
[http://dx.doi.org/10.1038/ja.2011.21] [PMID: 21468079]

[29] Johnston B, Radecka I, Hill D, *et al.* The microbial production of polyhydroxyalkanoates from waste polystyrene fragments attained using oxidative degradation. Polymers (Basel) 2018; 10: 957.

[http://dx.doi.org/10.3390/polym10090957] [PMID: 30960882]

[30] Xu X, Liu W, Tian S, *et al*. Petroleum hydrocarbon-degrading bacteria for the remediation of oil pollution under aerobic conditions: A perspective analysis. Front Microbiol 2018; 9: 2885.
[http://dx.doi.org/10.3389/fmicb.2018.02885] [PMID: 30559725]

[31] Reshef V, Mizrachi E, Maretzki T, *et al*. New acylated sulfoglycolipids and digalactolipids and related known glycolipids from cyanobacteria with a potential to inhibit the reverse transcriptase of HIV-1. J Nat Prod 1997; 60: 1251-60.
[http://dx.doi.org/10.1021/np970327m] [PMID: 9428159]

[32] Jha R, Zi-rong X. Biomedical compounds from marine organisms. Mar Drugs 2004; 2: 123-46.
[http://dx.doi.org/10.3390/md203123]

[33] Murakami N, Morimoto T, Imamura H, *et al*. Studies on glycolipids. III. Glyceroglycolipids from an axenically cultured cyanobacterium, *Phormidium tenue*. Chem Pharm Bull (Tokyo) 1991; 39: 2277-81.
[http://dx.doi.org/10.1248/cpb.39.2277] [PMID: 1804541]

[34] Bhadury P, Wright P. Exploitation of marine algae: Biogenic compounds for potential antifouling applications. Planta 2004; 219: 561-78.
[http://dx.doi.org/10.1007/s00425-004-1307-5] [PMID: 15221382]

[35] Nagle DG, Gerwick WH. Nakienones A-C and nakitriol, new cytotoxic cyclic C11 metabolites from an okinawan cyanobacterial (*Synechocystis* sp.) overgrowth of coral. Tetrahedron Lett 1995; 36: 849-52.
[http://dx.doi.org/10.1016/0040-4039(94)02397-T]

[36] Jensen TE, Sicko LM. Fine structure of poly-β-hydroxybutyric acid granules in a blue-green alga, *Chlorogloea fritschii*. J Bacteriol 1971; 106: 683-6.
[http://dx.doi.org/10.1128/jb.106.2.683-686.1971] [PMID: 4102335]

[37] Siddiqui P, Bergman B, Björkman P-O, *et al*. Ultrastructural and chemical assessment of poly-β-hydroxybutyric acid in the marine cyanobacterium *Trichodesmium thiebautii*. FEMS Microbiol Lett 1992; 94: 143-8.
[http://dx.doi.org/10.1016/0378-1097(92)90598-I] [PMID: 1325936]

[38] Panda B, Mallick N. Enhanced poly-β-hydroxybutyrate accumulation in a unicellular cyanobacterium, *Synechocystis* sp. PCC 6803. Lett Appl Microbiol 2007; 44: 194-8.
[http://dx.doi.org/10.1111/j.1472-765X.2006.02048.x] [PMID: 17257260]

[39] Stal LJ. Poly(hydroxyalkanoate) in cyanobacteria: An overview. FEMS Microbiol Lett 1992; 103: 169-80.
[http://dx.doi.org/10.1111/j.1574-6968.1992.tb05835.x]

[40] Masukawa H, Nakamura K, Mochimaru M, *et al*. Photobiological Hydrogen Production and Nitrogenase Activity in Some Heterocystous Cyanobacteria. In: Miyake J, Matsunaga T, Pietro AS, Eds. Biohydrogen II-an Approach to Environmentally Acceptable Technology. Pergamon 2001. pp. 63-6.

[41] Moezelaar R, Bijvank SM, Stal LJ. Fermentation and sulfur reduction in the mat-building cyanobacterium *Microcoleus chthonoplastes*. Appl Environ Microbiol 1996; 62: 1752-8.
[http://dx.doi.org/10.1128/aem.62.5.1752-1758.1996] [PMID: 16535319]

[42] Howarth DC, Codd GA. The uptake and production of molecular hydrogen by unicellular cyanobacteria. Microbiology (Reading) 1985; 131: 1561-9.
[http://dx.doi.org/10.1099/00221287-131-7-1561]

[43] Chittapun S, Limbipichai S, Amnuaysin N, *et al*. Effects of using cyanobacteria and fertilizer on growth and yield of rice, Pathum Thani I: A pot experiment. J Appl Phycol 2018; 30: 79-85.
[http://dx.doi.org/10.1007/s10811-017-1138-y]

[44] Prasanna R, Adak A, Verma S, *et al*. Cyanobacterial inoculation in rice grown under flooded and SRI

modes of cultivation elicits differential effects on plant growth and nutrient dynamics. Ecol Eng 2015; 84: 532-41.
[http://dx.doi.org/10.1016/j.ecoleng.2015.09.033]

[45]   Mohan A, Kumar B. Growth performance and yield potential of cereal crops (wheat, maize and barley) in association with cyanobacteria. Int J Curr Microbiol Appl Sci 2017; 6: 744-58.
[http://dx.doi.org/10.20546/ijcmas.2017.610.091]

[46]   Dixon R, Kahn D. Genetic regulation of biological nitrogen fixation. Nat Rev Microbiol 2004; 2: 621-31.
[http://dx.doi.org/10.1038/nrmicro954] [PMID: 15263897]

[47]   Badalyan A, Yang ZY, Seefeldt LC. A voltammetric study of nitrogenase catalysis using electron transfer mediators. ACS Catal 2019; 9: 1366-72.
[http://dx.doi.org/10.1021/acscatal.8b04290]

[48]   Kneip C, Lockhart P, Voß C, et al. Nitrogen fixation in eukaryotes – New models for symbiosis. BMC Evol Biol 2007; 7: 55.
[http://dx.doi.org/10.1186/1471-2148-7-55] [PMID: 17408485]

[49]   Mus F, Crook MB, Garcia K, et al. Symbiotic nitrogen fixation and the challenges to its extension to nonlegumes. Appl Environ Microbiol 2016; 82: 3698-710.
[http://dx.doi.org/10.1128/AEM.01055-16] [PMID: 27084023]

[50]   Raza A, Zahra N, Hafeez MB, et al. Nitrogen Fixation of Legumes: Biology and Physiology. In: Hasanuzzaman M, Araújo S, Gill SS, Eds. The Plant Family Fabaceae - Biology and Physiological Responses to Environmental Stresses 2020; pp. 43-74.
[http://dx.doi.org/10.1007/978-981-15-4752-2_3]

[51]   Bürgmann H, Widmer F, Von Sigler W, et al. New molecular screening tools for analysis of free-living diazotrophs in soil. Appl Environ Microbiol 2004; 70: 240-7.
[http://dx.doi.org/10.1128/AEM.70.1.240-247.2004] [PMID: 14711647]

[52]   Zhengtao Z, Wenge H, Di Y, Yuan H, et al. Diversity of *Azotobacter* in relation to soil environment in Ebinur Lake wetland. Biotechnol Biotechnol Equip 2019; 33: 1280-90.
[http://dx.doi.org/10.1080/13102818.2019.1659181]

[53]   Natzke J, Noar J, Bruno-Bárcena JM. *Azotobacter vinelandii* nitrogenase activity, hydrogen production, and response to oxygen exposure. Appl Environ Microbiol 2018; 84: e01208-18.
[http://dx.doi.org/10.1128/AEM.01208-18] [PMID: 29915110]

[54]   Tec-Campos D, Zuñiga C, Passi A, et al. Modeling of nitrogen fixation and polymer production in the heterotrophic diazotroph *Azotobacter vinelandii* DJ. Metab Eng Commun 2020; 11: e00132.
[http://dx.doi.org/10.1016/j.mec.2020.e00132] [PMID: 32551229]

[55]   Hala Y, Ali A. Isolation and characterization of *Azotobacter* from neems rhizosphere. J Phys Conf Ser 2019; 1244: 012019.
[http://dx.doi.org/10.1088/1742-6596/1244/1/012019]

[56]   Cassán F, Coniglio A, López G, et al. Everything you must know about *Azospirillum* and its impact on agriculture and beyond. Biol Fertil Soils 2020; 56: 461-79.
[http://dx.doi.org/10.1007/s00374-020-01463-y]

[57]   Lindström K, Mousavi SA. Effectiveness of nitrogen fixation in rhizobia. Microb Biotechnol 2020; 13: 1314-35.
[http://dx.doi.org/10.1111/1751-7915.13517] [PMID: 31797528]

[58]   Soumare A, Diedhiou AG, Thuita M, et al. Exploiting biological nitrogen fixation: A route towards a sustainable agriculture. Plants 2020; 9: 1011.
[http://dx.doi.org/10.3390/plants9081011] [PMID: 32796519]

[59]   Ondieki DK, Nyaboga EN, Wagacha JM, et al. Morphological and genetic diversity of rhizobia nodulating cowpea (*Vigna unguiculata* L.) from agricultural soils of lower eastern Kenya. Int J

Microbiol 2017; 2017: 1-9.
[http://dx.doi.org/10.1155/2017/8684921] [PMID: 29463983]

[60] Herbert RA, Eng T, Martinez U, *et al.* Rhizobacteria mediate the phytotoxicity of a range of biorefinery-relevant compounds. Environ Toxicol Chem 2019; 38: 1911-22.
[http://dx.doi.org/10.1002/etc.4501] [PMID: 31107972]

[61] Gauri SS, Mandal SM, Pati BR. Impact of *Azotobacter* exopolysaccharides on sustainable agriculture. Appl Microbiol Biotechnol 2012; 95: 331-8.
[http://dx.doi.org/10.1007/s00253-012-4159-0] [PMID: 22615056]

[62] Berg G. Plant–microbe interactions promoting plant growth and health: Perspectives for controlled use of microorganisms in agriculture. Appl Microbiol Biotechnol 2009; 84: 11-8.
[http://dx.doi.org/10.1007/s00253-009-2092-7] [PMID: 19568745]

[63] Dobbelaere S, Vanderleyden J, Okon Y. Plant growth-promoting effects of diazotrophs in the rhizosphere. Crit Rev Plant Sci 2003; 22: 107-49.
[http://dx.doi.org/10.1080/713610853]

[64] Katiyar V, Goel R. Siderophore mediated plant growth promotion at low temperature by mutant of fluorescent pseudomonad. Plant Growth Regul 2004; 42: 239-44.
[http://dx.doi.org/10.1023/B:GROW.0000026477.10681.d2]

CHAPTER 3

Role of Soil Microbes in the Sustainable Development: Agriculture, Recovery of Metals and Biofuel Production

Anurag Singh[1], Priya Bhatia[1], Shreya Kapoor[1], Simran Preet Kaur[1], Sanjay Gupta[2], Nidhi S. Chandra[1] and Vandana Gupta[1,*]

[1] *Department of Microbiology, Ram Lal Anand College, University of Delhi, Benito Juarez Road, New Delhi - 110021, India*

[2] *Department of Biotechnology, Jaypee Institute of Information Technology, Sector 62, Noida, UP-201309, India*

Abstract: Indiscriminate use of agrochemicals to ramp up production capabilities has caused a considerable decline in soil health status. The growing awareness of their ill effects on the environment and human health has called for a reversion to old organic agricultural practices blended with modern-day science and technology. Soil microorganisms with an identified ability to support plant growth are now being deployed in the form of biofertilizers and microbial biocontrol agents. Other than augmenting nutrition supply, these bio-inoculums can synthesize phytohormones and can also enhance the micronutrient and organic content of the soil. They can further induce resistance in plants against phytopathogens and compete against them by secreting secondary metabolites to keep the pathogenic population in check. Soil microorganisms, due to their omnipresence and survivability on varied substrates and in different environmental conditions, also find their use in other applications such as in the mining and energy industries. Unlike conventional metallurgical practices that deplete high-grade mineral ore reserves and cause wide-scale destruction of habitats, bioleaching provides a safe and cheap prospect for the recovery of metals. Other than the extraction of precious metals from low-grade ores, they also find their use in metal recovery from e-waste and can even remove heavy metals from soil. Moreover, the rapidly developing mining and the agrochemical industry count upon fossil fuels to meet their energy needs. In the final section of this chapter, we discuss a yet fascinating aspect of how non-conventional sources of energy are produced by the action of soil microorganisms to minimize strains on fossil fuel reserves. These biofuels, produced by the transformation of organic biomass, have an edge over fossil fuels as they emit low levels of particulate matter, sulphur dioxide, and carbon monoxide.

---

[*] **Corresponding author Vandana Gupta:** Department of Microbiology, Ram Lal Anand College, University of Delhi, Benito Juarez Road, New Delhi - 110021, India; E-mail: vandanagupta72@rediffmail.com

Shampi Jain, Ashutosh Gupta and Neeraj Verma
All rights reserved-© 2023 Bentham Science Publishers

Keywords: Bioleaching, Biocontrol, Biopesticide, Bioethanol, Biogas, Biofertilizer, Nitrogen.

INTRODUCTION

While the world was witnessing a serious food shortage owing to an exploding population and limited arable land, the increased yield losses due to the pest attack added to the growing concern. It was around the 1950s when the green revolution started, which employed hybrid seeds that could diminish the shortage of food grains. These 'high yield' variety seeds were indeed instrumental in substantially increasing crop production. However, they demanded extensive use of expensive chemical fertilizers, pesticides, and better irrigation, which had an immediate financial impact [1]. To achieve self-sufficiency and food security, this new agricultural practice took place around the world, ignoring the long-term environmental and health issues that we are facing today.

Soil is described as "the most complicated biomaterial on the planet" [2]. This complexity is associated with the chemical and biological heterogeneity of the soils around the globe. Overuse of inorganic fertilizers has induced a microclimatic change in the soil and reduced its organic carbon content, affecting the efficient uptake of nutrients and depleting micronutrient content. Moreover, it causes soil erosion and adds to environmental pollution. These chemicals deplete soil health by disturbing indigenous microbial diversity, soil fertility, pH, salinity, and other physico-chemical properties [3]. Thus, there is a mandate for an alternative that is sustainable in the long term, is cost-effective, and still increases the crop yield. The use of bio-inoculants comprising plant growth-promoting bacteria, fungi, and cyanobacteria effectively makes nitrogen, phosphorus, and other nutrients available to plants. Moreover, microorganisms are capable of secreting various compounds with antimicrobial properties or releasing certain factors that induce resistance in plants, suggesting their role as an alternate biocontrol agent. Further, large amounts of agrochemicals end up in the soil unutilized. This accumulation pollutes surface and groundwater and poisons tens of thousands of people worldwide [1, 3]. The use of bioinoculum could also prove significant in nutrient management and could save up to 30% of conventional chemical fertilizers [4, 5].

Other than their use in agriculture, soil microbes are also used to resolve environmental woes. For instance, unsustainable mining practices have not only depleted high-grade mineral ore reserves but have also caused wide-scale destruction of habitats. Traditional metallurgical practices, such as pyrometallurgy (smelting), are costly and may degrade the quality of the environment further. Furthermore, the mining industry has extensively contributed to the emission of greenhouse gases, which is associated with severe degradation of the environment. Nevertheless, it is the inadequacy of the legislature and the biohydrometallurgical procedures that have amplified the carbon footprints. These processes employ a variety of autotrophic bacteria and archaea, which are dexterous enough to fix carbon dioxide. Moreover, biomining, which works under two heads of bioleaching and bio-oxidation at the commercial scale, not only reduces the build-up of toxic compounds but also operates at lower temperatures and pressures, and hence reduces the rate of energy consumption.

The rapidly developing mining and agrochemical industries depend upon fossil fuels for their energy requirements. In their Annual Energy Outlook 2020 report, the Energy Information Administration (EIA) projected a nearly 50% increase in global energy consumption between 2018 and 2050, led by developing African and South Asian countries [6]. The industrial revolution has also heightened energy requirements, which ultimately increased the pressure on these conventional sources of energy. These non-renewable resources are extracted from the depths of the earth in an unsustainable manner and at an extravagant cost. Eventually, the cost of extraction will further escalate due to the depletion of these reserves. Therefore, the need for large-scale production of renewable energy is imperative for addressing the global energy crisis and negating the ill effects of conventional fossil fuels [7]. In the final couple of sections, we shall discuss how the over-exploitation of resources has led to a decline in the quality of the environment and the biotechnological methods for establishing a sustainable bioeconomy.

SOIL MICROBES AS BIOFERTILIZERS

Plant roots release exudates that directly influence the microorganisms in a narrow range of soil around roots called the rhizosphere. The rhizosphere microbiome is also influenced by many different factors, including environmental conditions, interaction with other microbes, and agricultural practices [8]. These microorganisms have a positive effect on the survival and growth of plants and can be free-living, or can form symbiotic relationships with plants, or can even live inside them as endophytes. They can improve nutrient availability and have a direct impact on the acquisition process, produce various phytohormones, decompose organic matter, and compete with pathogens [5, 9]. Biofertilizers contain a consortium of such microorganisms that can colonize the plant's rhizosphere and help in its growth and development by various natural processes to mobilize nutrients, including nitrogen fixation, phosphate and potassium solubilization, among others [10]. These microbial inoculants produce several other growth promoters like enzymes, vitamins, and bioactive compounds, and induce stress and disease resistance in plants [8].

Nitrogen-Fixing Microorganisms

As already introduced, nitrogen-fixing microorganisms can be classified as either free-living, which fix nitrogen independently, or symbiotic, which are associated in a mutualistic relationship with plants [11]. *Rhizobia* are Gram-negative bacteria of the Rhizobiaceae family, which establish a symbiotic relationship with legumes and fix nitrogen [12]. These bacteria get attracted by the flavonoid signals and attach themselves to the root hairs, subsequently forming an infection thread. Later, they develop into irregularly shaped bacteroides and establish themselves within nodules. *Azorhizobium* fixes nitrogen by forming stem nodules and produces phytohormones like IAA (Indole Acetic Acid), while *Bradyrhizobium* has been shown to increase total organic carbon, nitrogen and phosphorus content when inoculated in mucuna seeds [13]. Other species like *Rhizobium trifolii, R. meliloti, R. phaseoli*, and *R. leguminosarum* are known to fix N symbiotically with leguminous plants [5].

In rice production, *Sinorhizobium meliloti* induces stress tolerance by enhancing the level of endogenous phytohormones and protects the plant from pathogenic microorganisms (non-leguminous association) [14]. Cyanobacteria too play an important role in enhancing soil fertility and agricultural productivity. *Anabaena, Nostoc, Calothrix, Anabaenopsis, Scytonema, Fischerella, Westiellopsis, Hapalosiphon*, and *Aulosira*, when isolated from a different soil sample and used as biofertilizers, can supplement up to 25-35% nitrogen in the rice field [15]. *Frankia* can form a symbiotic association with several angiosperms (known as actinorhizal symbiosis) and fix nitrogen in root nodules [11].

Further, free-living nitrogen fixers include both aerobic [*Azotobacter, Beijerinckia, Derxia, Azospirillum*, and *Herbaspirillum* (a microaerophile)] and anaerobic bacteria (*Clostridium* and *Desulfovibrio*). Apart from fixing nitrogen, *Azotobacter* (Gram-negative bacteria) also produces vitamins and phytohormones and inhibits pathogens around the roots. Different species of *Azotobacter*, viz., *A. chroococcum*, and *A. vinelandii*, are used as biofertilizers for wheat, barley, mustard, sunflower, and cotton [9, 11]. *Azotobacter* and *Azospirillum* are also efficiently used in rice fields. *A. lipoferum, A. brasilense, A. halopraeferens, A. irakense*, and *A. amazonense* have been reported to enhance root formation in rice by producing siderophores and thus facilitating better iron absorption [16]. Other than fixing nitrogen, these free-living bacteria help to tide over stress and to sequester iron and other important mineral nutrients [17]. When inoculated in maize, sugarcane, and sorghum fields, they produce growth promoters and improve nutrient uptake [12]. Micro-aerobic bacteria, *Herbaspirillum seropedi-*

acae, can fix around 31–54% of nitrogen in rice fields [18]. On the other hand, *Clostridium*, is an obligate anaerobe often found in rice fields with capabilities for fixing nitrogen [11].

Phosphate Solubilizing Microorganisms

Phosphorous (P) is the second most important nutrient for plants after nitrogen. It is abundantly present in the soil in both organic and inorganic forms. However, only about 0.1% of these are in soluble form and can be directly utilized by plants. This makes it a limiting nutrient in plant growth. Phosphorus plays an extremely important role in nearly all metabolic processes [19], including respiration, photosynthesis, and even nitrogen fixation in legumes. This limiting nutrient accounts for 0.2-0.8% of the plant's dry weight and has a direct impact on seed germination, root development, disease resistance, and early maturation of crops.

Around 75-90% of P supplemented by chemical fertilizers is lost unutilized by getting fixed in a process that readily removes available soluble P by sorption on soil solids or precipitation by cations like Al^{+3} and Fe^{+3} [20]. Apart from increasing costs and loss of soil fertility, this also leads to eutrophication of surface water. Phosphate solubilizing microorganisms (PSMs) (Table **1**) can thus provide a bio-friendly alternative and convert this unavailable P to a soluble, usable form. Many bacteria, fungi, several actinomycetes, and cyanobacteria can solubilize P [11, 20]. *Rhizobium*, which forms a symbiotic association to fix nitrogen in leguminous and non-leguminous plants, has also shown the capability to solubilize P with *R. leguminosarum* bv. *trifolii* [21]. Due to their ability to travel long distances and produce more acid, fungi show better P solubilizing capabilities than bacteria and can increase plant growth by up to 20%. Other than *Aspergillus* and *Penicillium* [22], certain strains of yeast like *Yarrowia lipolytica* can also solubilize phosphorus [23]. Moreover, arbuscular mycorrhizal fungi (AMF) like *Glomus manihotis* and *Entrophospora colombiana*, when used in synergy with phosphate solubilizing bacteria (PSB) like *Pseudomonas* sp., can act as a potential biofertilizer [24]. Several PSMs also promote plant growth by secreting phytohormones, antibiotics, and/or siderophores [20].

Plant Growth Promoter

Upon infection with *Rhizobium*, the ethylene level surges in legumes. This increased level of ethylene inhibits subsequent nodule formation. ACC (1-aminocyclopropane-1-carboxylate) is an immediate precursor of ethylene and is synthesized by ACC synthase in plants. To counter this, some rhizobial strains synthesize ACC deaminase to limit the ethylene level and promote nodulation in legumes [26]. Different plant hormones are responsible for different aspects of their growth and development. Rhizosphere microorganisms too can modulate the level of these hormones and hence can play a significant role in plant growth. Indole 3-acetic acid (IAA) produced by *Pseudomonas putida* and *P. fluorescens*, is involved in major plant processes like seed germination, cell division, root germination and elongation, and vascular bundle formation. *Azospirillum* and *Paenibacillus* also synthesize other auxins. Gibberellins (GA) are produced by a wide range of plant growth promoting rhizobacteria (PGPR) such as *Azospirillum brasilense*, *Acetobacter diazotrophicus*, and *Rhizobium phaseoli*, among others. Similarly, *Paenibacillus polymyxa* and *Rhodospirillum rubrum* produce cytokinins (CK) [11]. Several species of *Azotobacter* secrete vitamin B complex, plant hormones like naphthalene acetic acid (NAA) and GA, and other bioactive compounds to promote root growth and aid in better water and nutrient uptake. The activities of PGPRs are more prominent in the rhizosphere soil and are found to be positively correlated with its overall fertility. Turmeric plant growth is promoted by several mutualistic non-symbionts like *Bacillus subtilis*, *Trichoderma atroviride*, *Pseudomonas fluorescens*, and actinomycetes like *Streptomyces* sp [8]. Several other species like *Streptomyces setonii*, *S. champavatii*, and *S. roseosporus* also enhance root length, volume, and dry weight in rice fields [27].

Table 1. Phosphate-solubilizing microorganisms: a variety of bacteria, actinomycetes, fungi, and cyanobacteria have demonstrated the ability to solubilize phosphorus and make it available for plant acquisition [5, 11, 20, 25].

Bacteria		Actinomycetes	Fungi
Bacillus	*Rhizobium*	*Streptomyces*	*Aspergillus*
Pseudomonas	*Micrococcus*	*Micromonospora*	*Penicillium*
Rhodococcus erythropolis	*Sarcina*	*Nocardia*	*Mucor*
Serratia	*Escherichia*	*Streptoverticillium*	*Trichoderma*
Arthrobacter ureafaciens	*Azospirillum*	*Thermoactinomyces*	*Rhizoctonia solani*
Phyllobacterium	*Burkholderia*	-	*Arthrobotrys oligospora*
Delftia sp.	*Alcaligenes*	-	*Pichia fermentans*
Chryseobacterium	*Acinetobacter*	-	*Yarrowia lipolytica*
Azotobacter	*Flavobacterium*	-	*Schizosaccharomyces*
Xanthomonas	*Kurthia*	-	-
Xanthobacter agilis	*Erwinia*	Cyanobacteria	AMF
Enterobacter	-	*Anabaena* sp.	*Glomus manihotis*
Klebsiella	-	*Calothrix braunii*	*Entrophospora colombiana*
Vibrio proteolyticus	-	*Nostoc* sp.	-

Further, a vital nutrient, iron, is abundant in the earth's crust but remains unavailable for use due to its presence as Fe^{+3} ions in an aerobic environment, which tends to form an insoluble solid. Bacteria synthesize siderophores, a high-affinity iron chelator. These low molecular weight molecules scavenge iron and solubilize it under iron-limiting conditions, which are then actively taken up by bacteria and converted into an Fe^{+2} ion state. Plants can directly take up this siderophore-iron complex or can use the ligand exchange method to assimilate iron from bacterial siderophores. When *Arabidopsis thaliana* absorbed the Fe-pyoverdine complex synthesized by *Pseudomonas fluorescens*, its iron levels increased. Siderophores also form stable complexes with other heavy metals and radioactive particles. This increases soluble metal concentration and helps plants tolerate heavy metal stress [12]. Moreover, *P. fluorescens* augments drought tolerance in peanut crops by improving the concentration of phytohormones, plant nutrient uptake and alleviating oxidative stress [28].

However, it is important to note here that PGPRs can also produce antibiotics like HCN, butyrolactones, viscosinamide, cepaciamide A, phenazines, and several lytic enzymes to antagonise potential pathogens [11, 12]. *Azotobacter* spp. produce antifungals that can inhibit pathogenic fungi and protect seedlings from dying [29]. It has also been reported that *Bacillus subtilis*, *Trichoderma harzianum*, and *P. fluorescens* can protect plants from pathogens like *Fusarium*, *Pythium*, or *Rhizoctonia*. Some PGPRs prevent the proliferation of pathogens by competing for iron and other nutrients. *Bacillus lentus* and *Azospirillum brasilense* can synergistically express antioxidants under stress and also increase the chlorophyll content of leaves [12].

Other Potential Bioinoculum

Cyanobacteria are one of the major inhabitants of the upper soil crust. They synthesize phytohormones, enzymes, and enhance the micronutrient and organic content of the soil. They have been used efficiently as biofertilizers in the production of rice, maize, and wheat. When used along with chemical fertilizers, it can independently fix up to 20–30 kg of nitrogen/ha in submerged rice fields, reducing the latter's use by up to 30% [15]. Some synergistic approaches, including *Anabaena–Trichoderma*, *Anabaena–Serratia* and *Anabaena–Pseudomonas*, showed high nitrogen content and better yield [30].

Further, fungi can colonize plant roots either intracellularly (invade cortical cells) or intercellularly to form endomycorrhizae (also known as arbuscular mycorrhizae) and ectomycorrhizae, respectively. *Glomus versiforme* plays an important role in augmenting phosphorus acquisition by plants [12]. Arbuscular mycorrhizal fungi (AMF) (phylum Glomeromycota) can colonize the roots of a wide range of plant species. This association improves nutrient uptake and provides tolerance to various biotic (pathogen attack) and abiotic (drought and salinity) stresses. Apart from nitrogen and phosphorus, AMF also provides sulfate, potassium, and other vital micronutrients, and thus improves crop quality [31]. Moreover, AMF harbour metal transporters (Zn, Cu, and Fe) and can alleviate heavy metal toxicity. These factors make them a suitable candidate as biofertilizers. However, their obligate symbiont nature constrains the mass production of inoculum [31]. Bacteria like *Bacillus subtilis* and *Thiobacillus thiooxidans* are also known to solubilize zinc (a micronutrient) and can be exploited as biofertilizers [5].

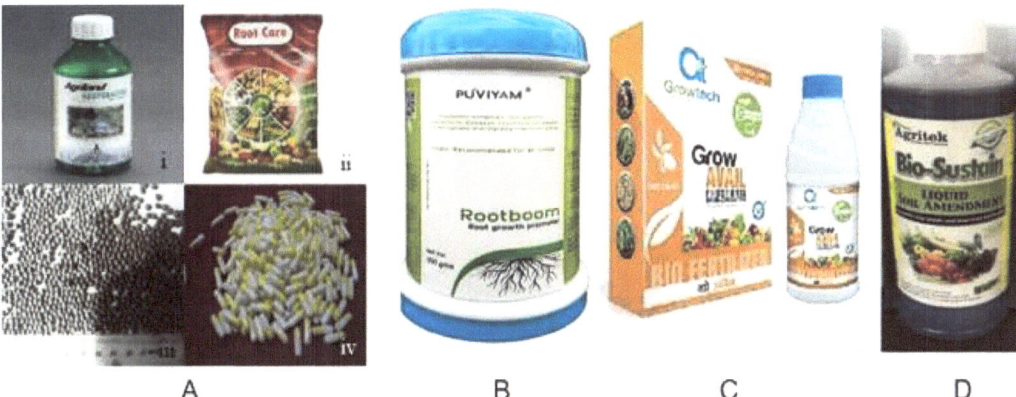

Fig. (1). Types and examples of commercially available biofertilizer preparations. **(A)** Types of biofertilizers formulations: (i) liquid; (ii) peat; (iii) granules and (iv) encapsulated freeze-dried powders (https://biofit.eu/q4/lo5-production-of-biofertilizers?showall=1); **(B)** Root Growth Promoter: Rootboom by i-Orgo (https://www.iorgo.in/product/rootboom-certified-organic-input-root-growth-promoter-pgpr-biofertilizer-bio-fungistatic-cultures-mycorrhizae-seaweed-derivatives-humic-and-fulvic-substances-and-vitamins/); **(C)** Phosphate solubilizing biofertilizer (https://www.biofertilizer.co.in/grow-avail-phosphate-solubilizng-bio-fertilizer.htm); **(D)** Liquid biofertizer by Agritek (http://www.agritek.co.nz/bio-sustain/).

Future Aspects

Despite the immense potential, biofertilizers are still not widely accepted over their chemical counterparts. However, with a developing conception of the mode of action and the diversity of its functional spectrum, it holds promise to become a reliable component in sustainable agricultural practices. Several commercial formulations are available now all over the world (Fig. **1**). Obtaining higher yields on limited arable land could be achieved by using better strains of bio-inoculants, improvised delivery systems, and finding stable and competitive rhizosphere colonizers. Genetic engineering tools could present an important prospect for this approach. Moreover, strict quality control norms, streamlined supply and reliable in-field performance are required to gain confidence in this new range of fertilizers among the end-users, *i.e.*, farmers. Thus, this organic way of augmenting crops bestows a wide scope for research to build on existing knowledge and prepare competitive multifunctional combinations of bio-inoculum for a variety of crops.

SOIL MICROBES AS BIOCONTROL AGENT

Biocontrol can be defined as the suppression of a living organism (like pests: pathogens, weeds, insects) by another [32], using the negative interactions *viz.*, parasitism, predation, amensalism, competition or antibiosis. Natural and modified cultivable microorganisms or their secondary metabolites are used for this purpose and are termed microbial biological control agents (MBCAs) [33], which keep a check on the pathogenic population by using biological mechanisms. The biocontrol agents have an edge over their chemical substitutes because of their target specificity, long-lasting benefits, ability to promote plant growth, biodegradability, and environmental friendly nature [34, 35].

An ideal biocontrol agent is expected to cause no or minimal harm to non-target organisms. It should be able to rapidly infect the target, have resistance to adverse environmental conditions, and should efficiently compete with its prey for available nutrients and space. Furthermore, they should have a short handling time and a life cycle similar to their targets in order to thrive even in low prey density [36, 37]. Several factors, like temperature, pH, moisture content, organic matter, and nutrient concentration, are known to affect the potency of biocontrol agents [34].

Microbial Biological Control Agents (MBCA)

The antagonistic activity of microorganisms against phytopathogens was first reported in 1926 by Sanford. He observed the suppression of *Streptomyces scabies* causing potato scab disease by the microorganisms present in green manure. Later in 1932, the inhibitory action of *Trichoderma lignorum* against *Rhizoctonia solani* was reported by others [38].

With the ability to replicate rapidly and survive in varied environmental conditions, soil microbes have been shown to counter phytopathogens by different mechanisms and have been exploited for their application in pest management. Table **2** enlists a wide array of microorganisms that have been identified for their role as biocontrol agents, and some of them have been further commercialized [39].

Table 2. Soil microorganisms as biocontrol agents [11, 34, 36, 37, 40 - 44].

S. No.	Organism	Comments
	Bacteria	
1	a. *Bacillus pumilus* GB34 b. *B. licheniformis* SB3086 c. *B. subtilis* d. *B. amyloliquefaciens* e. *B. thuringiensis* f. *B. lentimorbus* and *B. papillae*	a. Targets *Rhizoctonia* and *Fusarium* b. For control of leaf spot and blight disease caused by fungi (commercial name: Green-Releaf) c. Synthesizes bacillomycin D; inhibits *Aspergillus flavus* by destructing the cell walls and the cell membrane of the spores; also targets *Sclerotinia sclerotiorum* infecting tomatoes. d. Inhibits the growth of *Fusarium oxysporum* and other fungal pathogens; targets *Acidovorax avenae* infecting watermelons e. The most widely used MBCA; inhibits *Sclerotium rolfsii* (pathogen infecting soya bean), *Pectinophora gossypiella* (pink bollworm), and other fungal pathogens; effective against larvae of mosquitoes (*Anopheles* and *Culex*), butterflies, blackflies, and moths (commercial name: Dipel®). f. Cause 'milky disease' in scarabid Japanese beetles (commercial name: Doom®)
2	*Agrobacterium radiobacter* K15	Targets *A. tumefaciens* responsible for causing crown gall disease (commercial name: Galltrol)
3.	a. *Pseudomonas chlororaphis* K15 b. *P. fluorescens* A506 c. *P. putida*	a. Treats stem and root rots, and wilt diseases caused by *Verticillium* (commercial name: AtEze) b. Targets *Erwinia amylovora* (commercial name: Frostban A) c. Pathogenic to *Verticillium*, which causes wilt in strawberries
4.	*Arthrobacter* A6 and A16	Targets pitch canker causing *Fusarium moniliforme* and wilt causing *Penicillium*
5.	a. *Streptomyces kasugaensis* b. *S. philanthi* c. *S. vinaceusdrappus* d. *S. diastatochromogenes* e. *S. lydicus*	a. Synthesizes Kasugamycin; inhibits *E. amylovora* responsible for fire blight of pear b. Antagonistic to *Magnaporthe oryzae* causative agent of rice blast. c. Infects fungal pathogens like *Curvularia oryzae* and *F. oxysporum* d. Inhibits *Botrytis cinerea*, *Aspergillus niger* by secreting oligomycin A e. Antagonistic to root rot causing fungi *e.g.*, *Pythium*, *Aphanomyces* (commercial name: Micro108®)
	Fungi	
6.	*Beauveria bassiana*	Keeps a check on fire ants, beetles and a wide range of insects: Coleoptera, Lepidoptera, Hemiptera, and Thysanoptera (commercial name: Biopower, Bb plus, BotaniGard ES)
7.	*Chondrostereum purpureum*	Found in the loamy soils of Europe, West, and mid-Asia; targets *Prunus serotina* and other similar woody weeds; (commercial name: BIOCHON™)

(Table 2) cont.....

S. No.	Organism	Comments
8.	*Metarhizium anisopliae*	Antagonistic to Coleoptera, planthoppers, Hemiptera; (commercial name: Granment-P, BioMagic, Metadieca)
9.	*Trichoderma viride*	Targets *F. oxysporum* (infecting soya beans), *Nectria galligena* (inducing apple canker), and *Stereum purpureum* (cause silver leaf disease in plum and other fruits)
10.	*Fusarium lateritium*	Inhibits *Eutypa armeniaceae* responsible for causing gummosis disease of apricots
11.	*Conidiobolus thromboides*	Found in soil rich in organic matter; targets Hemiptera (commercial name: Vektor 25 SL)
12.	**Yeast** a. *Candida oleophila* b. *Saccharomyces cerevisiae* RC008 and RC009	a. Antagonistic to *B. cinerea* and *Penicillium expansum* there by controlling post-harvest diseases in strawberries, cherries, citrus fruits, and pome fruits (commercial name: Aspire®, Nexy®) b. target *Fusarium graminearum*, *Aspergillus ochraceus*, and other plant pathogens
13	*Arthrobotrys dactyloides*, *A. oligospora*, *Coprinus comatus* and *Dactylella* spp.	Nematode trapping/ nematophagous fungi parasitize and kill plant root damaging nematodes through the formation of constrictive rings around the nematode, and by immobilizing nematodes through adhesive knobs and nets or by producing toxins. Fungal hyphae penetrate the nematode and suck the nutrients ultimately kill it.
	Baculoviruses	Infect insects from Arthropods, Lepidoptera, Hymenoptera, Diptera, and Decapoda and kill them and thus are widely used as bioinsecticides

Mechanism of Action of Microbial Biocontrol Agents

Direct Interaction with the Microbes

MBCAs directly interact with pathogenic organisms and inhibit their growth by secreting secondary metabolites with antimicrobial properties. Bioactive secondary metabolites (BSM) like fengycins and iturins synthesized by *Bacilli* facilitate hyphal death and alter the cell membrane integrity of fungal cells, causing cytoplasm leakage and inhibiting spore germination. Rapamycin produced by *S. hygroscopicus* inhibits *Microsporum gypseum* [34, 45]. Also, enzymes like chitinases and glucanases secreted by certain yeast and *Streptomyces* spp. destroy the fungal cell wall [34, 43, 45]. Baculoviruses and nematophagous fungi parasitize their hosts directly and kill them.

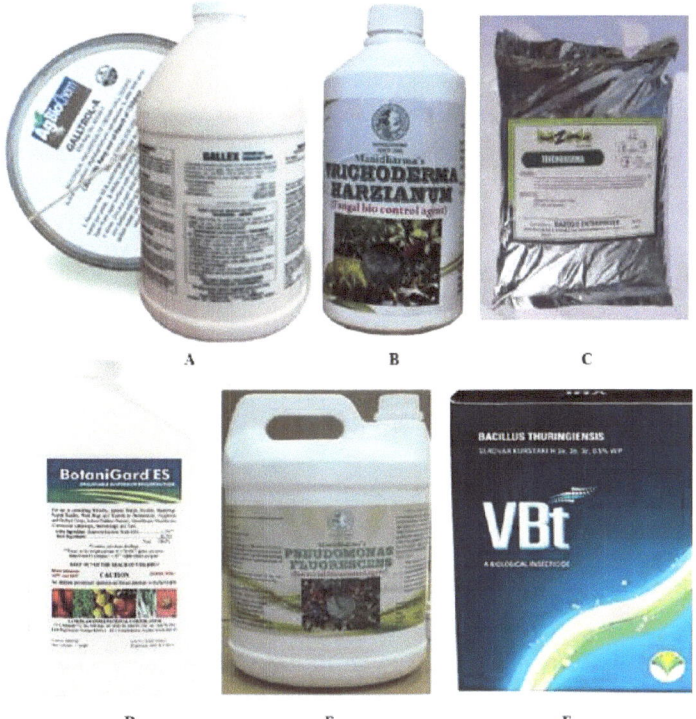

Fig. (2). Examples of commercially available biocontrol preparations. **(A)** Biocontrol agents: Galltroll-A® and Gallex® byAgBioChem (https://agbiochem.com/); **(B)** *Trichoderma harzianum* fungal biocontrol agent by Manidharma Biotech Private Limited (https://www.indiamart.com/proddetail/trichoderma-harzianu--fungal-biocontrol-agent-20498453773.html); **(C)** *Trichoderma viride* fungal biocontrol agent by Bazodo Enterprises; **(D)** Bioinsecticide: Botanigard® ES by Bioworks (https://www.bioworksinc.com/botanigard-es/); **(E)** *Pseudomonas fluorescens* bacterial biocontrol agents by Manidharma Biotech Private Limited; **(F)** *Bacillus thuringiensis* bacterial biocontrol agent by Varsha Bioscience and Technology India Private Limited.

Indirect Interaction with the Microbes

Here, MBCAs compete with the pathogen for available nutrients and space, thereby indirectly protecting the plant from pathogens [34, 37, 45]. For example, *B. pumilus* competes with *Fusarium oxysporum*, a pathogen of pea plants [34, 45].

Induced Resistance

MBCAs induce an immune response in the host by imparting systemic resistance referred to as ISR (induced systemic resistance). *Bacillus subtilis* and *B. amyloliquefaciens* trigger ISR in the host against pathogens causing root rot in tomatoes and *Panax ginseng*, respectively [44, 45].

Future Aspects

The use of biocontrol agents for pest management provides an eco-friendly alternative to chemical pesticides. However, this application of soil microorganisms is relatively unexplored (Fig. **2**). Moreover, various drawbacks like a narrow range of activity, a prerequisite of technical expertise and specific formulation techniques, high establishment costs, relatively low efficiency, and shorter shelf life make them less appealing to agricultural experts and farmers. Multidisciplinary tools can be used for designing model organisms to have a better insight into their functional and structural potential. Investigating their modes of actions, survival in varied environmental conditions, interactions with different organisms, and persistence over host organisms will allow the recognition and selection of potential strains of bioinoculums, ensuring the formulation of effective products with a longer shelf life that would lead to wide acceptance of MBCAs over their chemical substitutes.

SOIL MICROBES IN BIOLEACHING

Unsustainable mining practices have not only depleted high-grade mineral ore reserves but have also caused wide-scale destruction of habitats. Pyrometallurgy (smelting) and other conventional metallurgical practices are expensive as well as they may further degrade the quality of the ores and environment. Bioleaching presents a cost-effective technique for the extraction of low-grade ores and can be performed onsite with little technical knowledge [46]. Unlike ecologically irresponsible traditional mining, this method does not release any toxic byproducts. Bioleaching is the process of conversion of elements into more soluble forms with the help of microorganisms so that they can be recovered from the ore [36].

Mesophilic Bacteria

Acidithiobacillus

Previously known as *Thiobacillus*, they are aerobic, Gram-negative, non-spore-forming bacteria. Being chemolithoautotrophs, they can utilize an inorganic carbon source such as carbon dioxide and derive energy from the conversion of reduced sulfur compounds such as sulfide, thiosulfate, or elemental sulfur into sulfate (oxidation) [47, 48]. The most commonly used species, *Acidithiobacillus ferrooxidans* and *A. thiooxidans*, are acidophilic and can carry out bioleaching in the environment at pH 1.5 to 3.

Waksman and Joffe [49] isolated *A. thiooxidans*. This organism quickly oxidizes elemental sulfur along with additional partially reduced sulfur compounds. The acid production not only alters the pH but also allows acid-soluble metal compounds to leach into the solution as sulfates due to rock decomposition. It has a vital role in creating favourable acidic conditions for the growth of ferrous iron-oxidizing bacteria, such as *A. ferrooxidans* and *Leptospirillum ferrooxidans* [50]. *A. ferrooxidans* was isolated from acid coal mine drainage by Colmer and Hinkle [51]. It can utilize ferric iron as an electron donor, which allows it to carry out the oxidation of reduced inorganic sulfur compounds even in anaerobic conditions [52]. Other relatively newer microorganisms with potential applications include *Thiobacillus prosperus* (now known as *Acidihalobacter prosperus*), a halotolerant metal-mobilizing bacterium, and *T. cuprinus*, a facultative chemolithoautotrophic bacterium. The latter can only oxidise metal sulphides but not ferrous iron and extract copper selectively from its ore chalcopyrite [54, 55].

Leptospirillum

L. ferrooxidans was isolated from Armenian mines by Markosyan. It is an acidophilic obligate chemolithotrophic ferrous iron-oxidizing bacterium [56]. *L. ferrooxidans* can not act on mineral sulfides independently and is unable to oxidize sulfur or its compounds without the aid of *A. ferrooxidans* or *A. thiooxidans*. This organism can, however withstand lower pH values and greater concentrations of uranium, molybdenum and silver than *A. ferrooxidans*, but is sensitive to copper [57, 58].

Thermophilic Bacteria

Thiobacillus-like bacteria, also known as *Th*-bacteria, are moderately thermophilic iron-oxidizing microorganisms that can tolerate temperatures of around 50°C. They thrive on ores such as pyrite, pentlandite, and chalcopyrite, but only in the presence of yeast extract [59, 60]. There are many extremely thermophilic bacteria with the ability to grow on ferrous iron, elemental sulfur, and metal sulfides above 60°C. One such organism, *Acidianus brierleyi*, previously classified under the genus *Sulfolobus* [61], is a chemolithoautotrophic, facultative aerobic, acidophilic archaeon, which uses elemental sulfur as an electron acceptor and reduces it to hydrogen sulfide. Another genus with applications in bioleaching is *Sulfolobus*. They are aerobic, facultative chemolithotrophic bacteria that can oxidize ferrous iron, elemental sulfur, and sulfide minerals. Similarly, *Sulfobacillus thermosulfidooxidans*, a spore-forming facultatively autotrophic bacteria, can oxidize iron ores in the presence of yeast extract [62].

Moreover, microbial succession is observed as temperature shifts in the leach dump. *A. ferrooxidans* is a mesophile, and when heat is generated by microbial activities, it raises temperatures above about 30°C. This bacterium is outcompeted by mildly thermophilic chemolithotrophs such as *Leptospirillum ferrooxidans* and *Sulfobacillus*. Hyperthermophilic archaea such as *Sulfolobus* predominate in the leach dump at even higher temperatures (60–80°C) [63].

Heterotrophic Microorganisms

Heterotrophic microorganisms can also contribute to bioleaching, though they do not derive any benefits from it. Metal solubilization may be due to enzymatic reduction of highly oxidized metal compounds [64] or to the production of organic acids. It may also occur by displacement of metal ions by hydrophilic groups, which are excreted into the medium [62, 65]. Among the bacteria, members of the genus *Bacillus* are the most effective in metal solubilization, whereas *Aspergillus* and *Penicillium* are the most active among fungi [62].

Bioleaching Mechanisms

In the first step, low-grade ore is arranged in a large heap known as the leach dump, and a sulfuric acid solution with a pH of 2 is allowed to trickle down through the pile. A liquid with abundant dissolved metals gets collected at the bottom, which is transported to a precipitation plant where the desired metal is precipitated and purified. The liquid is then driven back to the top of the pile and the cycle gets repeated. This process takes place at an acidic pH, which is maintained by adding acid to the heap.

The key reactions in this process can be described as follows:

Bacterial or Chemical Oxidation of Metal Sulfides to Sulfate

In bacterial leaching, bacteria such as *A. ferrooxidans*, *A. thiooxidans*, and *L. ferrooxidans* oxidize soluble metal sulfides into sulfate, releasing the metal ions in solution. This process can occur in two ways: direct or indirect bioleaching.

Direct bioleaching is an oxygen-dependent process where there is direct contact between the surface of the metal sulfide and the bacterial cell. The metal sulfide is oxidized to metal sulfate, giving rise to ferric and sulfate ions in an aqueous solution.

For example, the conversion of pyrite to ferric sulfate involves the following steps:

$$4FeS_2 + 14O_2 + 4H_2O \rightarrow 4FeSO_4 + 4H_2SO_4$$

$$4FeSO_4 + O_2 + 2H_2SO_4 \rightarrow 2Fe_2(SO_4)_3 + 2H_2O$$

The overall reaction is as follows [66]:

$$4FeS_2 + 15O_2 + 2H_2O \rightarrow 2Fe_2(SO_4)_3 + 2H_2SO_4$$

In indirect bioleaching, the bacteria are not in direct contact with the mineral surface but release a liquid called the lixiviant, which is generally ferric iron at an acidic pH (<5). It chemically oxidizes the sulfide mineral. The following is the metal solubilization reaction:

$$MeS + Fe_2(SO_4)_3 \rightarrow MeSO_4 + 2FeSO_4 + S°$$

The sulfur released as a byproduct in the above reaction may be oxidized to sulfuric acid by *A. ferrooxidans*, but oxidation by *A. thiooxidans* is relatively much faster.

$$2S° + 3O_2 + 2H_2O \rightarrow 2H_2SO_4$$

The ferrous iron generated can be converted back to ferric iron by *A. ferrooxidans* or *L. ferrooxidans*. Bacteria only play a catalytic role as they speed up the re-oxidation of ferrous ions, which in the absence of bacteria takes a very long time.

Furthermore, chemical oxidation can occur spontaneously even in anoxic conditions, such as deep inside the leach dump. For example, in the case of copper ores, ferric ions carry out the conversion of copper sulfides to cupric (Cu^{2+}) ions and sulfate. The ferric ions are generated from bacterial oxidation of FeS_2 which is usually present in copper ores. The following is the reaction [63]:

$$CuS + 8\ Fe^{3+} + 4\ H_2O \rightarrow Cu^{2+} + 8\ Fe^{2+} + SO_4^{2-} + 8\ H^+$$

Nonetheless, bacterial oxidation of ferrous iron is about 105–106 times faster than chemical oxidation of ferrous iron in the pH range of pH 2-3 [67].

Precipitation/ Metal Recovery

The metal is recovered from the leaching solution at the precipitation plant. This process is based on the principle that a more reactive metal will displace a less reactive metal from its solution. For instance, shredded scrap iron (a source of Fe°) is added to the precipitation pond to recover copper from the leached liquid by the following reaction:

$$Fe° + Cu^{2+} \rightarrow Cu° + Fe^{2+} \text{ (Fe° from scrap steel)}$$

Fe^{2+} rich liquid is released from the precipitation pond and is pumped to a shallow oxidation pond where bacteria such as *A. ferrooxidans* and *L. ferrooxidans* oxidize it to Fe^{3+}. Now ferric iron-rich acidic liquid is again pumped back to the top of the pile and the Fe^{3+} is used to oxidize more mineral sulfides [63]:

$$Fe^{2+} + O_2 + 4H^+ \rightarrow Fe^{3+} + 2H_2O$$

Industrial Applications

Bioleaching is one of the most valuable techniques in extractive metallurgy and biohydrometallurgy. It is now used extensively in the copper and uranium industries, especially for metal recovery from low-grade ores. The mineral oxidation susceptibility must be taken under consideration to decide on the suitability of a mineral for microbial leaching. Iron and copper sulfide ores, such as pyrrhotite (FeS) and covellite (CuS), are readily oxidized, whereas lead and molybdenum ores are much harder to oxidize making them less ideal candidates for bioleaching. A quarter of the copper mined across the globe is recovered through microbial leaching as copper sulfate ($CuSO_4$) formed during the oxidation of copper sulfide ores is highly water-soluble.

Another application is uranium leaching, where *A. ferrooxidans* oxidizes U^{4+} to U^{6+} with oxygen as an electron acceptor. However, the major reaction in this process is the chemical oxidation of U^{4+} by Fe^{3+}. The reaction observed is as follows:

$$UO_2 + Fe_2(SO_4)_3 \rightarrow UO_2SO_4 + 2FeSO_4$$

The uranyl sulfate (UO_2SO_4) formed is highly soluble and is concentrated by other processes.

Gold is generally found along with minerals containing arsenic (As) and iron sulphide (FeS_2). Almost 95% of the gold trapped in arsenopyrite minerals can be leached with the help of *A. ferrooxidans* and other metal oxidizing bacteria. This process is carried out in small bioreactor tanks.

$$2\ FeAsS[Au] + 7\ O_2 + 2\ H_2O + H_2SO_4 \rightarrow Fe_2(SO_4)_3 + 2\ H_3AsO_4 + [Au]$$

Toxic arsenic is removed as a ferric precipitate, and CN-is removed by its bacterial oxidation to CO_2 and urea [63].

Future Aspects

The applications of bioleaching may also be extended to the recovery of zinc, nickel, cobalt and molybdenum. Investment and operating costs are much lower than for conventional metallurgical processes. The cost of transportation can be further reduced as the processing plant can be built near the ore deposits. The procedures are not complex and can be easily controlled even with very little technical knowledge. This process is especially suitable for developing countries that are rich in mineral ores but lack funds. Small-scale microbial-bioreactor leaching has become a popular alternative to the ecologically destructive gold-mining techniques that lead to the accumulation of potentially lethal contaminants such as As and CN at the extraction site. Aside from the metals recovered in the leachate, there are trace amounts of insoluble metals left in the residues, such as lead, that can be concentrated and purified. This can be achieved by leaching metals (*e.g.*, zinc, cadmium, and copper) that interfere with conventional processes for the recovery of lead. Similarly, techniques are being researched for the extraction of silver and other precious metals that are finely disseminated in iron, arsenic, copper, and zinc sulfides.

SOIL MICROORGANISMS IN BIOFUEL PRODUCTION

Biofuel, according to the United Nations Environment Program (UNEP), is an energy-dense compound that is directly or indirectly produced from biomass by the action of microorganisms. It is a budding source of sustainable energy. This renewable energy source, established upon the microbial transformation of organic material, has the potential to substitute fossil fuels, which are constantly utilized by the industrial and transportation sectors. Furthermore, biofuels have an edge over fossil fuels as they emit low levels of particulate matter, oxides of sulfur and carbon [68].

Biofuels are classified into four generations. The biofuels prepared by the action of microbial enzymes on sugars (from barley, wheat, corn, and potato), starch (from sugarcane, sugar beet), and vegetable oils are classified as first-generation biofuels. However, this led to increased pressure on arable lands, severe food shortages, and intensive use of fertilizers and water, which caused significant environmental problems [69]. The second-generation biofuels are produced from lignocellulosic biomass such as switchgrass, industrial waste like wood chips and pulp from food processing. The novel starch, oils, and sugar crops like *Jatropha curcas*, *Miscanthus*, or other non-food crops are a part of this category as they do not challenge food production. Nevertheless, the process of conversion of the lignocellulosic woody biomass to fermentable sugar requires pre-treatment with cellulases and ligninases, which also requires certain overpriced technologies. Second-generation biofuels, including cellulosic ethanol, biohydrogen, algae fuel, wood diesel, *etc.*, are currently under development [69]. The third-generation biofuels include biodiesel and bioethanol derived from microalgae and macroalgae, and hydrogen from green microalgae and other microbes [70]. The fourth-generation biofuels are mainly the laboratory-scale or theoretical fuels that can be obtained by the metabolic engineering of the biosynthetic pathways of microbes to generate the biofuel precursors or ready-to-use biofuels [71].

Microbial Communities Involved in Biofuel Production

Microbial communities survive, multiply, and coexist within the soil to form the soil microflora. These organisms are responsible for various biochemical reactions that have a huge impact on soil health and hence human health. These biochemical reactions of microbes are being extensively exploited by the industry to produce various commercial products [72].

Biomass is considered a renewable energy source that has the potential to resolve energy-related issues and fulfill the energy demands of the growing population. The exploitation of agronomic materials and microorganisms using existing technology can produce sustainable, environmentally friendly and profuse amounts of biofuel [73]. Table 3 enlists the organisms involved in the production of various biofuels.

Bacteria

Bacteria show a considerably high growth rate under modest cultivation conditions, so they exhibit great potential in biofuel production. Some bacterial species tend to store intracellular lipid molecules at varying concentrations, which indicates that these bacterial strains are appropriate feedstock for biodiesel production. Furthermore, various archaeal (*Methanosarcina barkeri* and *Methanococcus mazei*) and bacterial (*Clostridium aceticum*) communities are involved in biogas production, wherein they act as syntropic organisms and perform various stages of anaerobic biodegradation of the waste material to release varying amounts of methane and carbon dioxide [69]. While the anoxygenic phototrophic bacteria produce hydrogen at higher rates for a prolonged duration. The anoxygenic phototrophic bacterial species, which have been studied for photobiological hydrogen production, include *Rhodobacter capsulatus*, *R. sphaeroides*, *Rhodospirillum rubrum*, etc [75]. Nevertheless, thermotolerant β-1,4-exoglucanases are found in abundant quantities in *Clostridium thermocellum*. These cellulases are of great industrial importance wherein the raw material has to tolerate extremely high temperatures [76].

Fungi

In the past few years, yeasts and filamentous fungi have been extensively explored for biofuel production because of their ability to grow on varying sources of carbon and lignocellulosic substances. They tend to accumulate lipids and sugars, which form the feedstock for biodiesel and bioethanol, respectively [69]. Oleaginous filamentous fungi (*Mortierella alpine*, *Mucor rouxii*, *Trichosporon*, and *Mortierella isabellina*) and various yeasts (*Rhodosporidium toruloides*) can accumulate lipids under environmental stress. These groups can be readily exploited for the production of mycodiesel. Furthermore, cellulases obtained from filamentous fungi are extensively used in the preliminary steps of ethanol production [74].

Table 3. Organisms involved in the production of various biofuels [69, 74 - 76].

Organism	Comments	Biofuel
Bacteria		
Chromatium vinosum	Purple sulphur bacteria	Biohydrogen
Rhodospirillum rubrum	Purple non-sulphur bacteria	Biohydrogen
Nocardia coralline	It accumulates neutral lipids and other fatty acid molecules under nitrogen-limiting conditions.	Biodiesel
Zymomonas mobilis	These ethanologenic bacteria utilize wheat bran to form ethanol by fermenting various sugars.	Bioethanol
Fungi		
S. cerevisiae RL-11	It has been used for ages for the fermentation of sugars to release ethanol. While growing, yeasts form flocs in the fermentation vessel. These flocs settle easily and make the process of separation uncomplicated. Separated yeast cells can be used as a single-cell protein (SCP).	Bioethanol
Kluyveromyces marxianus K21	A thermotolerant yeast capable of co-fermenting hexoses and pentoses from taro-waste. It produces ethanol at a concentration of nearly 49g/L within 22 hours. It is capable of producing cellulolytic enzymes, and this allows the yeast to act directly upon the cellulosic biomass.	Bioethanol
Rhodosporidium toruloides	This oleagenic yeast under the stress of phosphorus limitation can accumulate lipids up to 80% of its dry weight by utilizing the laminaria residue hydrolysates.	Biodiesel
Mortierella isabellina	These filamentous fungi are known to accumulate lipids up to 86%, which is nearly 50% of their dry weight.	Biodiesel
Algae		
Arthrospira spp.	It requires low energy, requiring screen filtration for the process of biomass concentration.	Bioethanol
Cyanobacteria	The fragile structure aids in the physical disintegration of the cell wall, helping in reducing the costs involved in the process of hydrolysis, thus decreasing the overall costs.	Bioethanol
Nannochloropsis gaditana	This microalga can grow in wastewater and store a large amount of lipids.	Biodiesel

Algae

The phototrophic microalgae (*Chlorella*, *Scenedesmus*, *Chlorococcum*, and *Tetraselmis*) have attracted wide attention for bioethanol production and it has achieved significant advancements. Microalgae under stressful conditions can produce nearly 50-60% carbohydrate content, which has the potential to provide sufficient biomass for the process of fermentation. Moreover, the lack of lignified material in microalgae may aid in cutting costs for biofuel production. Furthermore, oleaginous algal genera (*Schizochytrium*, *Nannochloropsis*, and *Botryococcus*) respond to stressful conditions by storing intracellular triacylglycerols, which can be subsequently converted into usable fuels like biodiesel. However, the commercialization of this process remains elusive due to various inadequacies in microalgal cultivation [70].

Alcohol Production

According to the data provided in 2019 by the Renewable Energy Association, the United States of America has a colossal production of ethanol, with a yearly production of 15.7 billion gallons. The United States, along with Brazil, produces nearly 84% of the total ethanol produced worldwide. Ethanol, or ethyl alcohol, is used as a biofuel additive for gasoline in the United States of America and Brazil, with varying concentrations for running vehicles to limit the emission of exhaust gases. The blending of ethanol has improved the fuel properties, like higher heat of vaporization, higher octane number, wider flammability limits with less toxicity and higher biodegradability.

The production of cellulosic ethanol or bioethanol comprises various steps, including pre-treatment, hydrolysis, fermentation, and distillation (Fig. 3). The initial three steps are distinctively dependent upon the microbial activity.

Pre-Treatment and Hydrolysis

The second-generation biofuels depend upon low-maintenance non-crop plants producing large amounts of lignocellulosic material, which presumably grow on relatively unacceptable land. Lignocellulose is primarily composed of cellulose, hemicellulose, and lignin. Cellulose microfibrils are composed of a linear polymer of glucopyranose molecules linked in alternating orientations. These linear polymers form the inter-strand hydrogen bonds, which contribute to the crystalline structure of the microfibril along with the Van der Waals forces. This strong crystalline structure, along with the immediate association with cellulose, lignin, pectin, and hemicellulose, deters enzyme accessibility, thus limiting the efficiency of hydrolysis. Hemicellulose is the major source of pentoses. Therefore, there is an increasing focus on engineering the pentose utilization pathways into the ethanologenic microorganisms. This inability of the ethanologenic microbes to act directly upon the primary substrate (lignocellulose) has led to the incorporation of the steps of pre-treatment and hydrolysis before fermentation [76].

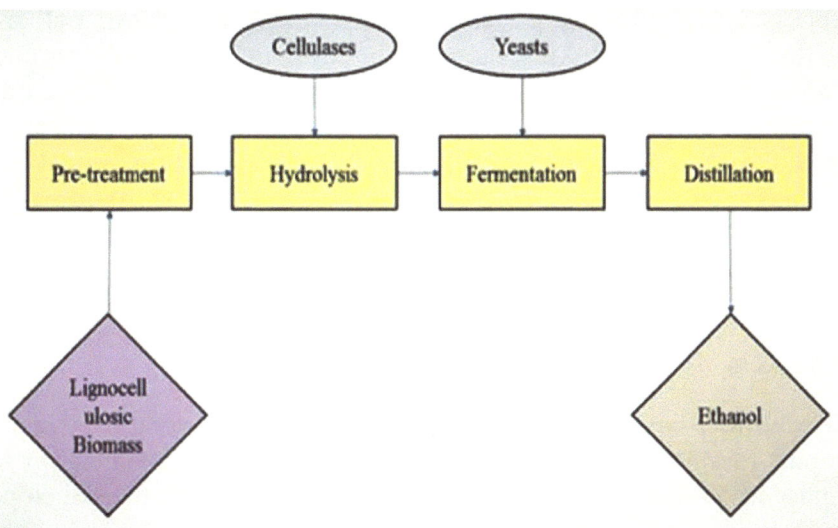

Fig. (3). Steps involved in the production of bioethanol from lignocellulosic biomass.

Various cellulolytic enzymes producing filamentous fungi like *Trichoderma reesei*, *Penicillium*, *Aspergillus*, *Humicola*, *Myceliophthora*, etc., have been utilized for the hydrolysis of the pretreated biomass. The use of cellulases from *Penicillium* is encouraged as it produces copious amounts of β-glucosidases, which is responsible for the higher efficiency in biomass saccharification. The higher efficiency of the enzyme brings about a reduction in the cellulase demand and hence a reduction in the overall cost of the production process [69].

Fermentation

The fermentation of the pretreated biomass is carried out by yeasts, which are responsible for the conversion of the monosaccharides or oligosaccharides into bioethanol. Mussatto and colleagues [77] demonstrated that yeast strains such as *S. cerevisiae* RL-11, *Pichia stipitis* NRRL-Y-7124, and *Kluyveromyces fragilis* Kf1 are powerful ethanol producers with an ethanol production greater than 90% and an ethanol tolerance greater than 40 g/L. These organisms typically grow on simple and economical media and have the ability to suppress the growth of the contaminating microflora. Baker's yeast, on the other hand, cannot withstand the stress induced by industrial processes like increased ethanol concentrations, elevated temperatures, osmotic stresses, the presence of inhibitory compounds, and the contaminating microflora. Furthermore, the wild-type *S. cerevisiae* is not capable of performing the co-fermentation of hexoses and pentoses [69].

Distillation

The process of distillation is employed to separate bioethanol from the fermentation broth. The mixture is transported to the multi-column distillation systems, which separates 93-95% pure ethanol, which is then dehydrated to 99% pure ethanol using the molecular sieving approach [78]. Nevertheless, ethanol concentrations are critical in determining energy consumption during the process of distillation and dehydration. However, distillation may require massive heating processes, so it can offset the carbon-neutral value of the biofuel. Therefore, energy-efficient alternatives like cyclic distillation, membrane techniques, gas stripping, *etc.*, are utilized [79].

Future Aspects

Microbial biofuels have the potential to overcome the challenges associated with conventional fossil fuels. Microbes are required to channelize the energy stored in the biomass into biofuels. Currently, research is focused on the development of bioethanol, biogas, biodiesel, and hydrogen. However, it has been suggested that ethanol is not a carbon-neutral product as the entire ethanol production process requires more energy than the product itself can yield. Therefore, the need of the hour is to formulate mechanisms and strategies to produce other biofuels with the aid of diverse microbial communities from various recalcitrant lignocellulosic but energy-rich feedstock. This can be established by an extensive understanding of the microbial ecology and biosynthetic mechanisms of different microbes present in the environment. Nevertheless, the production costs of both the upstream and downstream processes are still significantly higher than the production costs of conventional fossil fuels. Therefore, further research can be directed towards reducing production costs along with increasing production efficiency. Metabolic engineering of the biosynthetic routes of microbes in the genera *Pseudonocardia*, *Rhodococcus*, and *Streptomyces* can be used in the development of isoprenoids, which can then be used for biofuel production. Furthermore, future research can be focused on harnessing the gastrointestinal flora of various animals, which can convert the lignocellulosic material into biofuels. Thus, there is a need to work on integrated policies for energy, land use, and water management and lead to a future based on renewable, sustainable, and low-carbon economies.

CONCLUDING REMARKS

According to a United Nations report, about one in nine people across the world suffered from food shortages in 2017 [80]. This is after we have state-of-the-art machinery and a battery of different chemicals to enhance yield and streamline the supply chain. These agrochemicals are meant to supplement plants' growth requirements. To augment crop yields, most of these chemicals are applied in greater amounts than required and have caused a loss in soil health [5]. However, with growing awareness of environmental and health compromises that we are making, it is imperative to turn towards a more organic future.

In this chapter, we discussed how a wide array of microorganisms in the rhizosphere can fix nitrogen, solubilize phosphates, produce phytohormones, fight pathogens, and alleviate stress. They include plant growth-promoting bacteria, fungi, cyanobacteria, and actinomycetes, which could synergistically supplement the plant's nutrient requirements [8, 10]. Secondly, phytopathogens pose a serious threat to crop production, indicating a desperate need for new methods to ameliorate the severe losses caused by them. A diverse group of microorganisms referred to as MBCAs can antagonize these phytopathogens, by directly or indirectly interacting with the target organisms or by inducing resistance against them in the host plant itself. Different combinations of such microorganisms are slowly being commercialized (Dipel®, Galltrol, Frostban A, *etc.*) to provide better target specificity and lower toxicity than their chemical counterparts. However, these products are yet to gain widespread acceptance owing to their unreliable performance, lack of awareness and other shortcomings in in-field settings [34]. Nevertheless, biofertilizers and MBCAs bestow immense scope for further delineating better multifunctional bio-inoculums that can support the overall growth and development of plants. MBCAs can even be a key player in mitigating the rising risk of resistance in pests to chemical pesticides. Candidates for biofertilizers should be able to compete with indigenous microflora for available nutrients and space, as well as establish themselves in the rhizosphere, whereas MBCAs should be able to rapidly infect the target organism while causing no or little harm to non-target organisms [36, 37]. Biotechnological attempts to produce stable strains and innovative culturing techniques would allow for their large-scale production. Furthermore, these products should be easy to handle and have an improvised delivery system to further augment their usage, which could help in a smooth transition from agrochemicals to their biological substitute.

Furthermore, due to their omnipresence and ability to survive on varied substrates and in different environmental conditions, microbes also find their use in different other applications, *e.g.*, metal recovery and biofuel production. Bioleaching solves several issues associated with traditional metallurgical practices by extracting minerals from low-grade ores in a safe, environmentally friendly manner and at a lower cost. It is an ever-growing area of industrial research, with many studies on the discoveries of microbial species and the development of even more efficient techniques. It is currently being utilized not only for the recovery of precious metals from low-grade ores but also for the removal of heavy metals from soils. Moreover, it has also been considered for metal recovery from electronic waste, metal concentrates, and mine tailings. However, the biggest drawback of this technique is the time required for metal extraction and calls for the development of new catalysts that would hasten the interaction of the microbes with the metal surface and ensure a sufficient supply of nutrients for the microorganisms [81].

With new technological developments arising to ease human life, there is an equivalent rise in energy demands. Nevertheless, strategies like enhancing the efficiency of energy-consuming equipment can be adopted to manage the energy requirements, which can buy us more time to develop renewable energy sources. However, to assure the maintenance of economic activity and energy security, the scientific community has to seek alternatives in the form of biofuels. This non-conventional fuel can remarkably replace conventional fossil fuels and eventually combat climate change and compensate for higher energy demands [82]. Furthermore, research should be anchored for the optimization of technologies, processes, and feedstock variety to ameliorate the productivity of biofuels. Development of plant varieties like 'Energy-cane' with twice the sugar content and biomass production efficiency of 120%, when contrasted with conventional sugarcane, should be promoted [78]. Thus, the need of the hour is to establish a bioeconomy by furnishing the biological resources, processes, and services in the economic sector to promote sustainable development.

ACKNOWLEDGEMENTS

The authors would like to thank Dr. Rakesh Kumar Gupta, Principal, Ram Lal Anand College, for providing all the necessary infrastructure and support in compiling this chapter.

REFERENCES

[1] Pimentel D. Green revolution agriculture and chemical hazards. Sci Total Environ 1996; 188: S86-98.
[http://dx.doi.org/10.1016/0048-9697(96)05280-1] [PMID: 8966546]

[2] Young IM, Crawford JW. Interactions and self-organization in the soil-microbe complex. Science 2004; 304: 1634-7.
[http://dx.doi.org/10.1126/science.1097394] [PMID: 15192219]

[3] Savci S. An agricultural pollutant: Chemical fertilizer. Int J Environ Sci Dev 2012; 3: 77-80.
[http://www.ijesd.org/papers/191-X30004.pdf]

[4] Adesemoye AO, Kloepper JW. Plant–microbes interactions in enhanced fertilizer-use efficiency. Appl Microbiol Biotechnol 2009; 85: 1-12.
[http://dx.doi.org/10.1007/s00253-009-2196-0] [PMID: 19707753]

[5] Mazid M, Khan T. Future of bio-fertilizers in Indian agriculture: An overview. Ir J Agric Food Res 2014; 3: 10-23.

[6] US Energy Information Administration. Annual Energy Outlook 2020 with projections to 2050. www.eia.gov/aeo

[7] Louime C, Uckelmann H. Cellulosic ethanol: Securing the planet future energy needs. Int J Mol Sci 2008; 9: 838-41.
[http://dx.doi.org/10.3390/ijms9050838] [PMID: 19325787]

[8] Ponmurugan P, Babu RG, Mathivanan N, *et al.* Studies on Population Density of Different PGPRs in Turmeric Rhizosphere Soils for Biocontrol Activity. In: Reddy MS, Ilao RI, Faylon PS, Eds. Recent Advances in Biofertilizers and Biofungicides (PGPR) for Sustainable Agriculture. UK: Cambridge Scholars Publishing 2014; pp. 94-104.

[9] Bhardwaj D, Ansari MW, Sahoo RK, *et al.* Biofertilizers function as key player in sustainable agriculture by improving soil fertility, plant tolerance and crop productivity. Microb Cell Fact 2014; 13: 66.
[http://dx.doi.org/10.1186/1475-2859-13-66] [PMID: 24885352]

[10] Vessey JK. Plant growth promoting rhizobacteria as biofertilizers. Plant Soil 2003; 255: 571-86.
[http://dx.doi.org/10.1007/978-81-322-2259-0]

[11] Saxena S, Ed. Applied Microbiology. India: Springer 2015; p. 190.
[http://dx.doi.org/10.1007/s11356-016-8104-0]

[12] Mahanty T, Bhattacharjee S, Goswami M, *et al.* Biofertilizers: A potential approach for sustainable agriculture development. Environ Sci Pollut Res Int 2017; 24: 3315-35.
[http://dx.doi.org/10.1007/s11356-016-8104-0] [PMID: 27888482]

[13] Youssef MMA, Eissa MFM. Biofertilizers and their role in management of plant parasitic nematodes. A review. E3 J Biotechnol Pharm Res 2014; 5: 1-6. [https://www.e3journals.org/cms/articles/13904 91799_Youssef%20%20and%20%20Eissa.pdf]

[14] Chi F, Yang P, Han F, *et al.* Proteomic analysis of rice seedlings infected by *Sinorhizobium meliloti* 1021. Proteomics 2010; 10: 1861-74.
[http://dx.doi.org/10.1002/pmic.200900694] [PMID: 20213677]

[15] Hashem MA. Problems and prospects of cyanobacterial biofertilizer for rice cultivation. Aust J Plant Physiol 2001; 28: 881-8.
[http://dx.doi.org/10.1071/PP01052]

[16] Sahoo RK, Ansari MW, Pradhan M, *et al.* Phenotypic and molecular characterization of native *Azospirillum* strains from rice fields to improve crop productivity. Protoplasma 2014; 251: 943-53.
[http://dx.doi.org/10.1007/s00709-013-0607-7] [PMID: 24414168]

[17] Bashan Y, Holguin G, de-Bashan LE. *Azospirillum* - plant relationships: Physiological, molecular, agricultural, and environmental advances (1997-2003). Can J Microbiol 2004; 50: 521-77.
[http://dx.doi.org/10.1139/w04-035] [PMID: 15467782]

[18] Divan Baldani VL, Baldani JI, Döbereiner J. Inoculation of rice plants with the endophytic diazotrophs *Herbaspirillum seropedicae* and *Burkholderia* spp. Biol Fertil Soils 2000; 30: 485-91.
[http://dx.doi.org/10.1007/s003740050027]

[19] Khan MS, Zaidi A, Ahemad M, *et al.* Plant growth promotion by phosphate solubilizing fungi – current perspective. Arch Agron Soil Sci 2010; 56: 73-98.
[http://dx.doi.org/10.1080/03650340902806469]

[20] Sharma SB, Sayyed RZ, Trivedi MH, *et al.* Phosphate solubilizing microbes: Sustainable approach for managing phosphorus deficiency in agricultural soils. Springerplus 2013; 2: 587.
[http://dx.doi.org/10.1186/2193-1801-2-587] [PMID: 25674415]

[21] Abril A, Zurdo-Piñeiro jl, Peix A, *et al.* Solubilization of Phosphate by a Strain of *Rhizobium leguminosarum* bv. *trifolii* Isolated from *Phaseolus vulgaris* in El Chaco Arido Soil (Argentina). In: Velázquez E, Rodríguez-Barrueco C, Eds., First International Meeting on Microbial Phosphate Solubilization. Developments in Plant and Soil Sciences, vol 102. Springer, Dordrecht, 2007; pp. 135-138.
[http://dx.doi.org/10.1007/978-1-4020-5765-6_19]

[22] Khan MR, Khan SM. Effects of root-dip treatment with certain phosphate solubilizing microorganisms on the fusarial wilt of tomato. Bioresour Technol 2002; 85: 213-5.
[http://dx.doi.org/10.1016/S0960-8524(02)00077-9] [PMID: 12227549]

[23] Vassilev N, Vassileva M, Azcon R, *et al.* Preparation of gel-entrapped mycorrhizal inoculum in the presence or absence of *Yarrowia lipolytica.* Biotechnol Lett 2001; 23: 907-9.
[http://dx.doi.org/10.1023/A:1010599618627]

[24] Widada J, Damarjaya DI, Kabirun S. The Interactive Effects of Arbuscular Mycorrhizal Fungi and Rhizobacteria on the Growth and Nutrients Uptake of *Sorghum* in Acid Soil. In: Velázquez E, Rodríguez-Barrueco C, Eds., First International Meeting on Microbial Phosphate Solubilization. Developments in Plant and Soil Sciences, vol 102. Dordrecht: Springer 2007; pp. 173-177.
[http://dx.doi.org/10.1007/978-1-4020-5765-6_26]

[25] Kulkarni SW, Deshmukh AM. Phosphate Solubilizing Soil Actinomycetes as Biofertilizers. In: Deshmukh AM, Khobragade RM, Dixit PP, Eds. Handbook of Biofertilizers and Biopesticides Jaipur, India: Oxford Book Comp 2007; pp. 186-188.

[26] Zahir ZA, Zafar-ul-Hye M, Sajjad S, *et al.* Comparative effectiveness of *Pseudomonas* and *Serratia* sp. containing ACC-deaminase for coinoculation with *Rhizobium leguminosarum* to improve growth, nodulation, and yield of lentil. Biol Fertil Soils 2011; 47: 457-65.
[http://dx.doi.org/10.1007/s00374-011-0551-7]

[27] Gopalakrishnan S, Rao GVR, Kumari BR, *et al.* Development of Broad-Spectrum Actinomycetes for Biocontrol and Plant Growth Promotion of Food Crops. In: Reddy MS, Ilao RI, Faylon PS, Eds. Recent Advances in Biofertilizers and Biofungicides (PGPR) for Sustainable Agriculture. UK: Cambridge Scholars Publishing 2014; pp. 75-100.

[28] Sudhakar P, Kumar KVK, Latha P, *et al.* Efficacy of *Pseudomonas fluorescens* Strains in Enhancing Drought Tolerance and Yield in Peanut. In: Reddy MS, Ilao RI, Faylon PS, Eds. Recent Advances in Biofertilizers and Biofungicides (PGPR) for Sustainable Agriculture. UK: Cambridge Scholars Publishing 2014; pp. 250-6.

[29] Martin XM, Sumathi CS, Kannan VR. Influence of agrochemicals and *Azotobacter* sp. application on soil fertility in relation to maize growth under nursery conditions. Eurasian J Biosci 2011; 5: 19-28.
[http://dx.doi.org/10.5053/ejobios.2011.5.0.3]

[30] Abinandan S, Subashchandrabose SR, Venkateswarlu K, *et al.* Soil microalgae and cyanobacteria: The biotechnological potential in the maintenance of soil fertility and health. Crit Rev Biotechnol 2019; 39: 981-98.
[http://dx.doi.org/10.1080/07388551.2019.1654972] [PMID: 31455102]

[31] Berruti A, Lumini E, Balestrini R, *et al.* Arbuscular mycorrhizal fungi as natural biofertilizers: Let's benefit from past successes. Front Microbiol 2016; 6: 1559.
[http://dx.doi.org/10.3389/fmicb.2015.01559] [PMID: 26834714]

[32] Pal KK, Gardener BMS. Biological control of plant pathogens. 2006; p. 25.
[http://dx.doi.org/10.1094/PHI-A-2006-1117-02]

[33] Köhl J, Kolnaar R, Ravensberg WJ. Mode of action of microbial biological control agents against plant diseases: Relevance beyond efficacy. Front Plant Sci 2019; 10: 845.
[http://dx.doi.org/10.3389/fpls.2019.00845] [PMID: 31379891]

[34] Law JWF, Ser HL, Khan TM, *et al.* The potential of *Streptomyces* as biocontrol agents against the rice blast fungus, *Magnaporthe oryzae* (*Pyricularia oryzae*). Front Microbiol 2017; 8: 3.
[http://dx.doi.org/10.3389/fmicb.2017.00003] [PMID: 28144236]

[35] Yuliar, Nion YA, Toyota K. Recent trends in control methods for bacterial wilt diseases caused by *Ralstonia solanacearum.* Microbes Environ 2015; 30: 1-11.
[http://dx.doi.org/10.1264/jsme2.ME14144] [PMID: 25762345]

[36] Atlas RM, Bartha R, Eds. Microbial Ecology: Fundamentals & Applications. 4th ed. Pearson Education

Asia 2000; p. 694.

[37] Deketelaere S, Tyvaert L, França SC, *et al.* Desirable traits of a good biocontrol agent against *Verticillium* wilt. Front Microbiol 2017; 8: 1186.
[http://dx.doi.org/10.3389/fmicb.2017.01186] [PMID: 28729855]

[38] Babbal A. Khasa YP. Microbes as Biocontrol Agents. In: Kumar V, Kumar M, Sharma S, Eds. Probiotics and Plant Health. Singapore: Springer 2017; pp. 507-52.
[http://dx.doi.org/10.1007/978-981-10-3473-2_24]

[39] Saxena AK, Pal KK, Tilak KVBR. Bacterial Biocontrol Agents and their Role in Plant Disease Management. In: Upadhyay RK, Mukerji KG, Chamola BP, Eds. Biocontrol Potential and its Exploitation in Sustainable Agriculture. Boston, MA: Springer 2000; pp. 25-37.
[http://dx.doi.org/10.1007/978-1-4615-4209-4_3]

[40] Adaskaveg JE, Förster H, Wade ML. Effectiveness of kasugamycin against *Erwinia amylovora* and its potential use for managing fire blight of pear. Plant Dis 2011; 95: 448-54.
[http://dx.doi.org/10.1094/PDIS-09-10-0679] [PMID: 30743334]

[41] Bisen PS, Debnath M, Prasad GB, Eds. Microbes: Concepts and Applications. Wiley-Blackwell 2012; p. 716.
[http://dx.doi.org/10.1002/9781118311912]

[42] Barrows-Broaddus J, Dwinell LD, Kerr TJ. Evaluation of *Arthrobacter* sp. for biological control of the pitch canker fungus (*Fusarium moniliforme* var. *subglutinans*) on slash pines. Can J Microbiol 1985; 31: 888-92.
[http://dx.doi.org/10.1139/m85-166]

[43] Freimoser FM, Rueda-Mejia MP, Tilocca B, *et al.* Biocontrol yeasts: Mechanisms and applications. World J Microbiol Biotechnol 2019; 35: 154.
[http://dx.doi.org/10.1007/s11274-019-2728-4] [PMID: 31576429]

[44] Shafi J, Tian H, Ji M. *Bacillus* species as versatile weapons for plant pathogens: A review. Biotechnol Biotechnol Equip 2017; 31: 446-59.
[http://dx.doi.org/10.1080/13102818.2017.1286950]

[45] Andrić S, Meyer T, Ongena M. *Bacillus* responses to plant-associated fungal and bacterial communities. Front Microbiol 2020; 11: 1350.
[http://dx.doi.org/10.3389/fmicb.2020.01350] [PMID: 32655531]

[46] Acevedo F, Gentina JC, Bustos S. Bioleaching of minerals - a valid alternative for developing countries. J Biotechnol 1993; 31: 115-23.
[http://dx.doi.org/10.1016/0168-1656(93)90141-9]

[47] Vishniac W, Santer M. The thiobacilli. Bacteriol Rev 1957; 21: 195-213.
[http://dx.doi.org/10.1128/br.21.3.195-213.1957] [PMID: 13471458]

[48] Trudinger PA. The metabolism of inorganic sulphur compounds by thiobacilli. Rev Pure Appl Chem 1967; 17: 1-24.

[49] Waksman SA, Joffe JS. Microörganisms concerned with the oxidation of sulphur in soil. II. *Thiobacillus thiooxidans*, a new sulphur oxidising organism isolated from the soil. J Bacteriol 1922; 7: 239-56.
[http://dx.doi.org/10.1128/jb.7.2.239-256.1922] [PMID: 16558952]

[50] Torma AE. Complex Lead Sulfide Concentrate Leaching by Microorganisms. In: Murr LE, Torma AE, Brierley JA, Eds. Metallurgical Application of Bacterial Leaching and Related Microbiological Phenomena. New York, NY: Academic Press 1978; pp. 375-88.
[http://dx.doi.org/10.1016/B978-0-12-511150-8.50027-7]

[51] Colmer AR, Hinkle ME. The role of microorganisms in acid mine drainage; a preliminary report. Science 1947; 106: 253-6.
[http://dx.doi.org/10.1126/science.106.2751.253] [PMID: 17777068]

[52] Sugio T, Domatsu C, Munakata O, *et al.* Role of ferric reducing system in sulfur oxidation of *Thiobacillus ferrooxidans.* Appl Environ Microbiol 1985; 49: 1401-6.
[http://dx.doi.org/10.1128/aem.49.6.1401-1406.1985] [PMID: 16346806]

[53] Pablo Cárdenas J, Ortiz R, Norris PR, *et al.* Reclassification of '*Thiobacillus prosperus*' Huber and Stetter 1989 as *Acidihalobacter prosperus* gen. nov., sp. nov., a member of the family Ectothiorhodospiraceae. Int J Syst Evol Microbiol 2015; 65: 3641-4.
[http://dx.doi.org/10.1099/ijsem.0.000468] [PMID: 26198437]

[54] Huber H, Stetter KO. *Thiobacillus prosperus* sp. nov., represents a new group of halotolerant metal-mobilizing bacteria isolated from a marine geothermal field. Arch Microbiol 1989; 151: 479-85.
[http://dx.doi.org/10.1007/BF00454862]

[55] Huber H, Stetter KO. *Thiobacillus prosperus* sp. nov., a novel facultatively organotrophic metal-mobilizing bacterium. Appl Environ Microbiol 1990; 56: 315-22.
[http://dx.doi.org/10.1128/aem.56.2.315-322.1990] [PMID: 16348110]

[56] Markosyan GE. Ein neues eisenoxidierendes Bakterium *–Leptospirillum ferrooxidans* nov. gen. nov. spec. Translated from. Akad Nauk Armj SSR Biol Z Armen Bol 1972; XXV: 26-9.

[57] Sand W, Rohde K, Sobotke B, *et al.* Evaluation of *Leptospirillum ferrooxidans* for leaching. Appl Environ Microbiol 1992; 58: 85-92.
[http://dx.doi.org/10.1128/aem.58.1.85-92.1992] [PMID: 16348642]

[58] Norris PR, Parrott L, Marsh RM. Moderately thermophilic mineral-oxidizing bacteria. Biotechnol Bioeng Symp 1986; 16: 253-62.

[59] Brierley JA, Le Roux NW. A Facultative Thermophilic Thiobacillus-Like Bacterium: Oxidation of Iron and Pyrite. In: Schwartz W, Ed. Conference Bacterial Leaching. Verlag Chemie: Weinheim 1977; pp. 55-66.

[60] Brierley JA. Thermophilic iron-oxidizing bacteria found in copper leaching dumps. Appl Environ Microbiol 1978; 36: 523-5.
[http://dx.doi.org/10.1128/aem.36.3.523-525.1978] [PMID: 16345316]

[61] Segerer A, Neuner A, Kristjansson JK, *et al. Acidianus infernus* gen. nov., sp. nov., and *Acidianus brierleyi* Comb. nov.: Facultatively aerobic, extremely acidophilic thermophilic sulfur-metabolizing archaebacteria. Int J Syst Bacteriol 1986; 36: 559-64.
[http://dx.doi.org/10.1099/00207713-36-4-559]

[62] Bosecker K. Bioleaching: Metal solubilization by microorganisms. FEMS Microbiol Rev 1997; 20: 591-604.
[http://dx.doi.org/10.1111/j.1574-6976.1997.tb00340.x]

[63] Madigan MT, Martinko JM, Stahl DA, Eds. *et al.*Brock Biology of Microorganisms. Benjamin-Cummings Pub Co 2010; p. 1152.

[64] Ehrlich HL. Bacterial Leaching of Manganese Ores. In: Trudinger PA, Walter MR, Ralph BJ, Eds. Biogeochemistry of Ancient and Modern Environments. Berlin, Heidelberg: Springer 1980; pp. 609-14.
[http://dx.doi.org/10.1007/978-3-642-48739-2_67]

[65] Bosecker K. Bioleaching of Non-Sulfide Minerals with Heterotrophic Microorganisms. In: Durand G, Bobichin L, Florent J, Eds. 8th International Biotechnology Symposium, Paris 1988, Proceedings. Société Française de Microbiologie. 1988; 2: pp. 1106-18.

[66] Silverman MP. Mechanism of bacterial pyrite oxidation. J Bacteriol 1967; 94: 1046-51.
[http://dx.doi.org/10.1128/jb.94.4.1046-1051.1967] [PMID: 6051342]

[67] Lacey DT, Lawson F. Kinetics of the liquid-phase oxidation of acid ferrous sulfate by the bacterium *Thiobacillus ferrooxidens.* Biotechnol Bioeng 1970; 12: 29-50.
[http://dx.doi.org/10.1002/bit.260120104]

[68] Bringezu S, Schütz H, O'Brien M, *et al.* Towards sustainable production and use of resources: Assessing biofuels. Paris: UNEP, Kenya 2009; p. 40.

[69] Bardhan P, Gupta K, Mandal M. Microbes as Bio-Resource for Sustainable Production of Biofuels and Other Bioenergy Products. In: Singh JS, Singh DP, Eds. New and Future Developments in Microbial Biotechnology and Bioengineering: Microbial Biotechnology in Agro-environmental Sustainability. Elsevier 2019; pp. 205-22.
[http://dx.doi.org/10.1016/B978-0-444-64191-5.00015-8]

[70] Lakatos GE, Ranglová K, Manoel JC, *et al.* Bioethanol production from microalgae polysaccharides. Folia Microbiol (Praha) 2019; 64: 627-44.
[http://dx.doi.org/10.1007/s12223-019-00732-0] [PMID: 31352666]

[71] Aro EM. From first generation biofuels to advanced solar biofuels. Ambio 2016; 45: 24-31.
[http://dx.doi.org/10.1007/s13280-015-0730-0] [PMID: 26667057]

[72] Aislabie J, Deslippe JR. Soil Microbes and Their Contribution to Soil Services. In: Dymond JR, Ed. Ecosystem Services in New Zealand - Conditions and Trends. Lincoln, New Zealand: Manaaki Whenua Press 2013; pp. 143-61.

[73] Perea-Moreno MA, Samerón-Manzano E, Perea-Moreno AJ. Biomass as renewable energy: Worldwide research trends. Sustainability (Basel) 2019; 11: 863.
[http://dx.doi.org/10.3390/su11030863]

[74] Babu PR, Sarma VV. Fungi as Promising Biofuel Resource. In: Singh JS, Singh DP, Eds. New and Future Developments in Microbial Biotechnology and Bioengineering: Microbial Biotechnology in Agro-environmental Sustainability. Elsevier 2019; pp. 149-64.
[http://dx.doi.org/10.1016/B978-0-444-64191-5.00011-0]

[75] Sasikala C, Ramana CV. Biotechnological potentials of anoxygenic phototrophic bacteria. II. Biopolyesters, biopesticide, biofuel, and biofertilizer. Adv Appl Microbiol 1995; 41: 227-78.
[http://dx.doi.org/10.1016/S0065-2164(08)70311-3] [PMID: 7572334]

[76] Yeoman CJ, Han Y, Dodd D, *et al.* Thermostable enzymes as biocatalysts in the biofuel industry. Adv Appl Microbiol 2010; 70: 1-55.
[http://dx.doi.org/10.1016/S0065-2164(10)70001-0] [PMID: 20359453]

[77] Mussatto SI, Machado EMS, Carneiro LM, *et al.* Sugars metabolism and ethanol production by different yeast strains from coffee industry wastes hydrolysates. Appl Energy 2012; 92: 763-8.
[http://dx.doi.org/10.1016/j.apenergy.2011.08.020]

[78] Manochio C, Andrade BR, Rodriguez RP, *et al.* Ethanol from biomass: A comparative overview. Renew Sustain Energy Rev 2017; 80: 743-55.
[http://dx.doi.org/10.1016/j.rser.2017.05.063]

[79] Bušić A, Marđetko N, Kundas S, *et al.* Bioethanol production from renewable raw materials and its separation and purification: A review. Food Technol Biotechnol 2018; 56: 289-311.
[http://dx.doi.org/10.17113/ftb.56.03.18.5546] [PMID: 30510474]

[80] WHO, Global Hunger Continues to Rise, new UN report says 2018. [https://www.who.int/news/item/11-09-2018-global-hunger-continues-to-rise---new-un-report-says]

[81] Borja D, Nguyen K, Silva R, *et al.* Experiences and future challenges of bioleaching research in South Korea. Minerals (Basel) 2016; 6: 128.
[http://dx.doi.org/10.3390/min6040128]

[82] Dale E, Holtzapple M. The need for biofuels. Chem Eng Prog 2015; 111: 36-40.

CHAPTER 4

Industrial Applications of Soil Microbes: Production of Enzymes, Organic Acids and Biopigments

Simran Preet Kaur[1], Tanya Srivastava[1], Anushka Sharma[1], Sanjay Gupta[2], Nidhi S. Chandra[1] and Vandana Gupta[1,*]

[1] *Department of Microbiology, Ram Lal Anand College, University of Delhi, Benito Juarez Road, Delhi - 110021, India*

[2] *Department of Biotechnology, Jaypee Institute of Information Technology, Sector 62, Noida, UP-201309, India*

> **Abstract:** Commodity chemicals are the intermediates that are generally involved in the synthesis of other high-end products. The increasing demand for various industrial products has upscaled the requirement for commodity chemicals. Originally, the industrial sector was dependent upon conventional and toxic chemicals to sustain its processes. However, the advent of biotechnology led to the development of numerous microbial processes producing enzymes, extremozymes, organic acids, organic solvents, *etc.*, Moreover, the soil environment has diverse forms of microbial communities performing assorted functions. As a result, a thorough understanding of the soil microbiota involved in providing regulatory ecosystem services can aid in the development of exceptional microbial strains capable of meeting the high demand for these commodity chemicals. In addition, the exploitation of these excellent manipulative microbial systems can improve and customize the synthesis of commodity chemicals and thereby reduce the reliance on synthetic and petroleum-based products. This chapter will inform the readers about the applications of soil microbes in industry and their involvement in enzymes, extremozymes, organic acids, and biopigments production.

Keywords: Amylase, Microbial enzymes, Microbial pigments, Phytase, Thermophilic enzymes.

INTRODUCTION

Soil is the primary physical covering of the earth's crust, which is a distinctive product of geological parent material, geomorphological history, natural disturbances, and the presence of natural biota. Soil provides the habitat for a vast

---

[*] **Corresponding author Vandana Gupta:** Department of Microbiology, Ram Lal Anand College, University of Delhi, Benito Juarez Road, New Delhi - 110021, India; E-mail: vandanagupta72@rediffmail.com

assortment of bacteria, fungi, algae, and many higher organisms. It is estimated that one gram of undisturbed soil may contain 10 billion microorganisms, representing an array of genomes. Despite the large numbers, these soil microbes comprise less than 1% of the total soil mass, but they play an indispensable role in supporting life on earth and influence countless ecosystems and commercial activities [1].

These microbial commodity chemicals find applications in diverse industries like medical, food and beverages, and agriculture and related fields. The enzymes like lipases, proteases, cellulases, amylases, *etc.*, obtained from these soil-dwelling microbial creatures represent the largest share in industrial applications. To list a few, (a) The textile industry utilizes a variety of proteases to eliminate undesirable fibres from products; (b) Enzymes from *Aspergillus* spp. are typically used in the processing of textiles, and (c) Lignin-degrading fungi like *Phanerochaete chrysosporium* have decolorizing properties which are also utilized by the textile industry for the treatment of wastewater [2]. Furthermore, microbial cultures play an important role in the production of a variety of food and dairy products. A wide range of organic acids are being procured by various fermentation processes. These acids are being utilized as acidulants, preservatives, emulsifiers, flavoring, sequestrants, and buffering agents.

Furthermore, metabolites obtained from the soil microbiota are being extensively manipulated for the development of various products of medical importance. Enzymes like asparaginase, which have a potential application against tumour cells, are produced by *Escherichia coli* and *Erwinia chrysanthemi* at a considerably lower cost [3]. Biofilms tend to confer antibiotic resistance and hence cause severe bacterial infections. Various biofilm inhibitory agents like cahuitamycins, obtained from soil microorganisms, find wide applications. Genetically engineered bioreporters, which have the potential to detect toxic compounds and generate quantifiable signals, are an emerging technology that has the potential to be utilized in various industries [4].

Techniques like microbiologically induced calcium carbonate precipitation are being employed in the cement and brick manufacturing industries. These alternative biobricks and biocement can possibly reduce environmental pollution by limiting the emission of carbon produced during the conventional manufacturing processes. Thus, microorganisms not only foster the products of commercial importance but also help in the sustainable development of economies [5]. Furthermore, growing public awareness has shifted research into microbiological alternatives to various products, such as various biological pigment molecules being used in the fabrication of bio-cosmetics [6].

The soil microbiota can be further explored for their potential to produce various powerful enzymes and drugs of medicinal value. Nevertheless, microengineering and synthetic biology tools can be employed to improve these soil microorganisms for the synthesis of the desired products. This chapter aims at providing an overview of various applications of soil microorganisms in industry and why there is a constant requirement for exploring soil microflora for novel products.

SOIL MICROORGANISMS IN ENZYME PRODUCTION

Modern enzyme technology began in 1874 when the Danish chemist Christian Hansen extracted rennet from calves' stomachs with the help of saline. This significant event was followed by a lengthy evolution, which brought about the parallel development of a multitude of industrial processes (Fig. 1), which ultimately led to the need for further expansion of the enzyme industry.

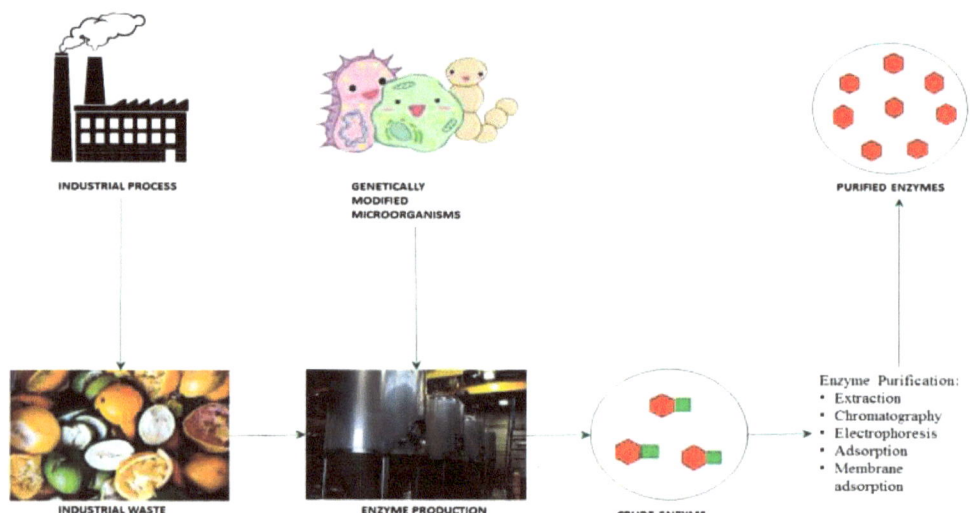

Fig. (1). Steps involved in the production of microbial enzymes from industrial waste.

Enzymes can be defined as organic compounds that catalyze various metabolic reactions occurring in a living organism. Proteins usually form the infrastructural backbone of enzymes. However, several enzymes contain non-protein components as well, which may be present in the form of a covalently attached carbohydrate group or a metal ion. Microbial enzymes have numerous industrial applications and hence have been applied to make several products. They are frequently used in the production of detergents, dyes, chemicals, wine and beer, laundry, food and beverages, paper and textiles, clothing, *etc.*

Fig. (2). Uses of microbial enzymes in various industries.

SOME IMPORTANT INDUSTRIAL ENZYMES

Microorganisms are extensively used for the production of enzymes on an industrial scale (Fig. **2**). The ability of microbes to grow rapidly, produce a variety of enzymes, and be susceptible to genetic manipulation allows the enzymes to be produced at the desired amount. Owing to their numerous benefits, microbial enzymes are increasingly replacing the enzymes produced by plants and animals, as well as conventional chemical catalysts. Nevertheless, industries are still looking for new microbial strains to meet the growing demand for enzymes. Soil, being a favourable environment for microbial communities, provides a wide array of microbial enzymes that are necessary for accelerating numerous chemical interconversions. Table **1** enlists a few enzymes that find applications in the industrial sector.

Table 1. Important industrial enzymes and their applications [7].

Amylases
• Mashing for beer making. • Sugar recovery from scrap candy in the candy industry. • Starch modification for paper coating in the paper industry. • Cold swelling laundry starch in the starch and syrup industry. • Wallpaper removal. • De-sizing of fabrics in textiles. • Degradation of protein, causing stains in the detergent industry. • Bread baking in the baking and milling industry. • Digestive aids in clinics and pharmaceuticals.
Protease
• Chill-proofing in the beer industry. • For condiments in the food industry. • Milk protein hydrolysate making in the dairy industry. • Dehairing and bating in the leather industry. • Recovery of silver from films in photography. • Degradation of fat stains in the detergent industry. • Bread baking in the baking and milling industry. • Spot removal in dry cleaning, the laundry industry. • Digestive aids in clinics and pharmaceuticals.
Glucose oxidase
• Oxygen removal in beer, beverage and dairy industries. • Paper test strips for diabetic glucose in pharmaceuticals. • In toothpaste to convert glucose into gluconic acid and hydrogen peroxide as both act as a disinfectant.
Pectinases
• Coffee bean fermentation, coffee concentrates in the coffee industry. • Clarification, filtration, concentration in fruits, and fruit juices. • Pressing, clarification, and filtration in the wine industry.
Lactase
• Whole milk concentrates. • Ice cream and frozen desserts. • Whey concentrates. • Lactose hydrolysis in the dairy industry.
Cellulase
• De-inking of papers for recycling in the paper and pulp industry. • Bio stonewashed denim in the textile industry. • Hydrolyzing cellulosic biomass to generate glucose for ethanol production in the biofuel industry • Loosening of cellulose fibers to easily remove dirt and color in the detergent industry.
Xylanase
• Bio-bleaching in the paper and pulp industry. • Fiber solubility in the animal feed industry.
Lipase and proteinase

• Contact lens.
• Brightening in the detergent industry.
• Ripening of cheese in the dairy industry.
Invertase
• Soft center candies and fondants.
• High test molasses.
Laccases and peroxidase
• Polymerize materials with wood-based fibers in the paper and pulp industry.
Hemicellulase
• Coffee concentrates.
• Hydrolyzing hemicellulosic biomass to generate glucose for ethanol production in the biofuel industry.
Catalase
• Bio polishing and bleach clean-up in the textile industry.

ENZYMES PRODUCED BY COMMON SOIL MICROBES

Various species of bacteria such as *Bacillus*, *Pseudomonas*, *Alcaligenes*, *Staphylococcus*, *Streptomyces*, and fungi such as *Aspergillus*, *Penicillium*, and *Mucor* are involved in enzyme production on an industrial scale.

Bacteria

Bacillus

Bacillus is a Gram-positive, aerobic, heterotrophic bacterium that constitutes 2-10% of the culturable soil population [8]. Most of the extracellular enzymes and several thermostable enzymes are derived from various *Bacillus* spp.

Proteases and amylases obtained from *Bacillus* are used quite widely in the baking, brewing, detergent, and textile industries. *B. amyloliquefaciens* majorly produces zinc-metalloproteases, while *B. licheniformis* is utilized for manufacturing alkaline protease subtilisin, which finds applications in the development of biological detergents and amylases, and *B. polymyxa* yields β-amylases along with glucoamylase. Other enzymes, such as hemicellulase and glucanase, as well as penicillin acylase, have been isolated from various *Bacillus* spp [7, 9].

Actinomycetes

Actinomycetes form an important component of the soil microbial community, especially under conditions of high temperature or water stress. They have delivered substantial advantages to the enzyme industry *via* their involvement in the production of proteases. *Streptomyces* is employed in the industrial production of β-amylases and glucose isomerase. Since glucose isomerase is used to manufacture high fructose syrups, *Streptomyces* has proven to confer considerable benefits to the sweetening industry. This enzyme can also be obtained from *Actinomyces missouriensis* [9].

Pseudomonas

Pseudomonas is a Gram-negative heterotrophic, aerobic, or facultative anaerobic bacterium. It is present abundantly in soil and forms 10-20% of culturable soil bacteria [8]. Apart from playing a critical role in nutrient cycling and biodegradation, *Pseudomonas* has high industrial significance as well and is used in the production of multiple enzymes. Phytase (used for the release of soluble phosphate in the cattle feed industry), proteinase and lipases, cephalosporin acylase, *etc.*, are obtained from *Pseudomonas* spp. Cephalosporin acylase is used to remove the side groups of penicillins or cephalosporins to form 6-aminopenicillanic acid or 7-aminocephalosporanic acid, respectively, the base molecules for the synthesis of semisynthetic penicillins and cephalosporins [7]. *P. aeruginosa* releases a protease called PA02, which does not get inhibited in the presence of specific inhibitors. *P. fluorescens* produces cholesterol esterase, which monitors serum cholesterol levels. Creatininase, an enzyme used for the determination of serum creatinine levels, is produced *by P. putida* and is also produced on a large scale using recombinant *E. coli* [9].

Staphylococcus

Staphylococcus is a facultative anaerobic Gram-positive bacterium. Staphylococci are known to play a vital role in the cheese, fruit processing, and detergent industries since they produce lipases and pectinases [7].

Arthrobacter

Arthrobacter is a heterotrophic, aerobic, Gram-variable bacterium, constituting up to 40% of culturable soil bacteria in soil [8]. *Arthrobacter* plays a pronounced role in the sweetening industry since it serves as one of the chief sources for the production of the enzyme glucose isomerase, an enzyme used to convert glucose to fructose. *Arthrobacter* is used for the industrial production of the amino acid L-glutamate as well. Moreover, *Arthrobacter* spp. produce enzymes such as inulinase II (which converts inulin into the medically-relevant nutrient difructose anhydride), and Alu (which cleaves Alu sequences, which are frequently repeated in human DNA). Uricase is obtained from *Arthrobacter globiformis* (used in the treatment and diagnosis of gout) [9].

Xanthomonas

Xanthomonas is a genus of Proteobacteria and is a prominent member of the soil microbial community. Many *Xanthomonas* species cause plant diseases. *Xanthomonas oryzae* is a chief producer of phytase that finds exclusive value in the cattle feed industry. *Xanthomonas* species produce an edible polysaccharide called xanthan gum that has a wide range of industrial uses, including foods, petroleum products, and cosmetics [7].

Corynebacterium

Corynebacterium, a notable soil bacteria, is used for very important industrial applications. It is Gram-positive and aerobic. The non-pathogenic species of *Corynebacterium* are particularly used in the production of enzymes. One of the most studied species is *C. glutamicum*, which produces glutamic acid under aerobic conditions and is thus exclusively utilized in the production of sauces. This species is also known to produce catalase [9].

Fungi

Aspergillus

Aspergillus is abundantly found in the soil and is chiefly involved in nutrient cycling, the development of soil texture and the degradation of complex plant polymers [8]. Several species of *Aspergillus* are largely exploited for manufacturing commercially valuable enzymes. Glucoamylase and α-amylase are obtained from *A. oryzae* and *A. niger* alongside neutral protease. *A. niger* is also employed in the production of glucose oxidase, inulinase, pectinase, hemicellulase, and amyloglucosidase (used in immobilized form for glucose production from starch). On the other hand, *A. oryzae* is used for obtaining xylanases, lipases, proteinases, and aminoacylase. Humicola lipase, an enzyme extensively used in the detergent manufacturing industry, is explicitly obtained from this species. Another commercially exploitable species, *A. nidulans*, facilitates us with laccases and peroxidases. Several other species of *Aspergillus* serve as the source of catalase and cellulase. *A. ficuum* is known to produce phytase. L-amino acid acylases derived from *Aspergillus* spp., are used for the resolution of L-amino acids from acylated racemic mixtures of D, L-amino acid [9].

Penicillium

Penicillium, yet another common genus of soil fungi, plays a crucial role in nutrient cycling, developing soil texture and degrading polymers such as lignin and cellulose [8]. Besides having such major importance in the natural environment, this ascomycete is widely used for producing several enzymes of industrial value. *P. chrysosporium* is the source of protease and glucose, while oxidase is produced by *P. notatum* besides the production of penicillin by these species. Furthermore, cellulase, hemicellulase, and pectinase are also manufactured using *Penicillium* spp. The fruit juice, wine, and beer industries are greatly dependent on *P. emersonii* since this species is solely used for the production of certain specific β-glucanases. *P. funiculosum* is mainly utilized in the cattle feed industry for producing phytase [7].

Mucor

Mucor is a genus comprising approximately 40 species commonly found in soil and is responsible for the production of several enzymes such as amylases, lipases, pectinases, and proteases. It has found a special significance in the lens cleaning, detergent, and cheese industries for its lipases and proteinases [7].

Yeasts

Yeasts are found at cfu of up to 103 per gram of soil. It is, however, quite obvious for yeasts to find a name in this section, concerning the wide array of extracellular, commercially exploitable enzymes produced by them, and it is quite veracious to state that yeasts as the cardinal group have lots of significance in industries. *Kluyveromyces* produces lactases and is hence largely used in the processing of ice cream, whole milk, whey, and other dairy products. For *Saccharomyces*, the most prominent yeast is used for industrial production of enzyme invertase and alcohol and has thus dominated the wine and beer industries for ages [7]. Hemicellulases obtained from *Cryptococcus* and *Trichosporon* are used in baking (pentosanases are used to alter dough consistency), brewing, animal feeds, nutraceuticals, and wood pulp processing [9]. *Candida albicans* is used to obtain lipase, proteinase, and β-galactosidase (lactase). Uricase is obtained from *Candida utilis*. *Kluyveromyces fragilis* produces invertase and β-galactosidase (lactase). *Candida* spp. and *Kluyveromyces* spp. also produce inulinase, an enzyme that causes hydrolysis of polyfructans and levans. *Rhodotorula* also produces lipase [9].

Others

Several other bacteria and fungi found in soil are also utilized for the industrial production of various commercial enzymes. *Alcaligenes*, a notable soil bacterium, is employed to obtain lipase and proteinase. *Trichoderma* is a prominent fungus that produces xylanase. *Trichoderma reesei* and *T. viride* produce cellulase. *T. reesei* also provides hemicellulase [7]. Rhodanase, derived from *Trichoderma* species, treats cyanide poisoning [9]. The fungus *Chrysosporium lucknowense* is known to produce cellulase. *Rhizopus* spp. are known to produce α-amylase and glucoamylase, whereas *R. oligosporus* produces proteases. Lipase produced by Rhizopus arrhizus is used as cocoa butter substitutes [7]. Hydantoinase is an enzyme produced by *Flavobacterium ammoniagenes* and is used in the production of d-and l-amino acids [9].

Moving ahead, we shall be learning about enzymes obtained from soil microbes that grow optimally in extreme conditions. Microbial communities that thrive in extreme environments are referred to as extremophiles, and they have established a diversity of molecular strategies to survive in extreme conditions. Since enzymes are used as biocatalysts in food and feed, pharmaceutical, textile, detergent, beverage, and several other industries, the industrial processes necessitate catalysis to be performed under harsh conditions. In this context, the best option is to use enzymes derived from extremophilic microorganisms called extremozymes. They have increased stability at high temperatures, extreme pH, in the presence of organic solvents and heavy metals, and against proteolytic attack, and can withstand harsh conditions during industrial processes. Owing to their great solidity, they pose new opportunities for biocatalysis and biotransformations, as well as for the development of the economy and a new line of research. Through improved modern technologies such as protein engineering, extremozymes can be modified for use in a diversity of biotechnological as well as commercial applications [10].

ENZYMES FROM ACIDOPHILES

Enzymes from microorganisms that can survive at an extremely low pH could be particularly useful for applications under highly acidic reaction conditions. Extremozymes obtained from acidophiles chiefly include amylases, glucoamylases, xylanases, cellulases, and proteases. These enzymes have been widely put to use in various industrial processes such as biofuel production, the food industry, mining, starch processing, and desulfurization of coal [10].

An acidophile that is most commonly employed in the production of a variety of useful extremozymes is *Sulfolobus* sp. It is generally found in regions near volcanic activity and hot springs. A few species, however, may be encountered in the highly acidic, sulfur-rich soil found near these regions, such as *Sulfolobus acidocaldarius* and *S. solfataricus*. *Sulfolobus* is a thermoacidophile that requires a high-temperature (>55°C) for it's survival [11].

Yet another group of industrially significant acidophiles is Acidobacteria. This acidophile, which is particularly abundant in soil habitats, represents up to 52% of the total bacterial community. They are, however, under-represented in culture

[12]. Alicyclobacilli form another crucial group of acidophilic, soil-dwelling organisms and serve as the source of several commercially valuable enzymes [13]. Although fungi, on the whole, are more acidotolerant than bacteria and actinomycetes, a few fungal genera, however, qualify well to be classified as true acidophiles. *Mucor* is the most notable one and is used to produce certain industrially significant enzymes. Furthermore, enzymes derived from *Aspergillus, Trichoderma, Fusarium, Penicillium,* and *Bispora antennata* have found broad-scale utilization in industries [14]. Table 2 enlists the microorganisms utilized for extracting the acidophilic enzymes.

Table 2. Acidophilic enzymes, their microbial sources and applications [10, 14 - 18].

cidophilic Enzyme	Bacterial Source	Fungal Source	Applications
Proteases	*Sulfolobus solfataricus* and *Sulfolobus acidocaldarius*	*Mucor racemosus*	Used in the food and feed industry and are used to degrade complex proteins, pre-digested baby foods or soft meat
Collagenase	*Alicyclobacillus sendaiensis*	–	Medicinal uses and therapeutics
Amylase	*Alicyclobacillus acidocaldarius*	*Aspergillus niger* strain RBP7	Sugar syrup production
Lipase	*Sulfolobus* sp.	*Rhizomucor miehei*	Applications in the food industry and is used for the biocatalysis of stereoselective transformations
β–mannanases	*Bacillus* sp. and *Streptomyces*	*Penicillium oxalicum, P. pinophilum, P. occitanis, Bispora antennata,* and *Aspergillus sulphurous*	Animal feed, pulp and paper, textile, and food industries as well as in agriculture
Polygalacturonases	*Bacillus*	*Penicillium, Bispora, Trichoderma, Fusarium,* and *Aspergillus*	Clarification of fruit, vegetable juices and wines, animal feed, fermentation of coffee and tea, and production of baby foods
Cellulose-degrading enzymes (endoglucanase and β-glucosidase)	*S. solfataricus*	*Thielavia* and *Bispora*	Pulp and paper industry
Xylanase	*S. solfataricus*	*Thielavia, Penicillium, Bispora,* and *Aspergillus*	Paper and textile industries, treatment of pectic wastewater

(Table 2) cont.....

cidophilic Enzyme	Bacterial Source	Fungal Source	Applications
Enzyme SuaI	*Sulfolobus acidocaldarius*	–	Used as a thermostable restriction enzyme
Thermostable phosphotriesterase-like lactonase (PLL)	*Sulfolobus acidocaldarius* and *Sulfolobus solfataricus*	–	Used for hydrolyzing the organophosphates (Ops)(highly toxic compounds that irreversibly inhibit the acetylcholinesterase (AChE) and compromise the functionality of the nervous system of the target organisms)
Peroxidases	Acidobacteria	–	Dye coloration
Esterase EST2	*Alicyclobacillus acidocaldarius*	–	Milk and cheese production industry in the pasteurization process

ENZYMES FROM ALKALIPHILES

Extremozymes derived from alkaliphiles primarily consist of proteases, cellulases, pectinases, lipases, and amylases that are resistant to and active at high pH and chelator concentrations found in modern detergents [19]. This has prompted the screening of alkaliphilic bacteria for their ability to produce such enzymes [20]. The main industrial applications of alkaliphilic enzymes are in the detergent industry. Detergent enzymes account for approximately 30% of total worldwide enzyme production [21]. Several alkaline proteases have been produced by alkaliphilic *Bacillus* strains and are commercially available. Alkaline enzymes are also used to decompose the gelatinous coating of X-ray films, from which silver can be recovered [21].

The predominant soil alkaliphiles include species of bacteria like *Micrococcus*, *Pseudomonas*, *Streptomyces*, and several *Bacillus* spp. such as *B. alcalophilus*, *B. agaradherens*, *B. clarkii*, *B. clausii*, *B. gibsonii*, *B. haemophilus*, *B. halodurans*, *B. horikoshii*, *B. pseudoalcaliphilus*, and *B. pseudofirmus* [21]. The fungal genera *Acremonium* and *Stilbella* are the dominant sources of alkaline enzymes. *Aspergillus* sp., yeast, and actinomycetes also act as prolific sources for alkaline proteases, mostly either serine proteases or metalloproteases, and show optimal pH around 8–12 and optimal temperature around 40–65 °C [22]. An alkalophilic strain of *Streptomyces albidoflavus* produces extracellular proteases that are capable of hydrolyzing keratin [23]. Table 3 enlists the important alkaliphilic enzymes and their microbial sources.

ENZYMES FROM THERMOPHILES

The ability of thermophiles to survive under harsh conditions has attracted major attention from researchers around the globe to derive and exploit enzymes that are stable and active at elevated temperatures. At higher temperatures, the solubility of many reaction components, in particular polymeric substrates, is significantly improved. Moreover, the risk of contamination, leading to undesired complications, is also significantly reduced. Thermostable enzymes are perhaps the most extensively utilized extremozymes, which possess enumerable properties suitable for countless biotechnological and commercial applications [20, 25]. Notable soil-dwelling thermophilic bacterial species and the enzymes produced by them are mentioned in Table **4**.

Table 3. Alkaliphilic enzymes, their microbial sources, and applications [10, 21, 22, 24].

Alkaliphilic Enzyme	Microbial Source	Applications
Protease	*Bacillus* str. 221, *Bacillus* strains AB42 and PB12 (pH 9.0 to 12.0), *Bacillus* str. AH-101(active toward casein at pH 12 to 13 and stable 60°C and pH 5 to 13), *Bacillus* str. NKS21, and *Bacillus* str. B21-e2	Used as detergent additives
Cyclodextrin glucanotransferase (CGTase)	*Bacillus* sp. No. HA3-3-2 and *Mycobacterium terrae* KNR 9	Used for the production of cyclodextrin from starch
Cellulase	*Bacillus* sp. KSM-635	Used as laundry detergent additive
Xylanases	*Bacillus* str. 41M-l, *Bacillus* TAR-I, and *Cephalosporium* str. NCL 87.11.9	Used in the paper industry on a large scale
Polygalacturonate lyase	*Bacillus* sp. str. RK9 and *Bacillus subtilis* 7--3	Textile industry
Alkalphilic tributyrin esterase	*Bacillus* str. A30-1	Food, beverage, textile, paper industry
Amylases	*Bacillus halodurans* A-59 (ATCC 21591)	High potential in starch degradation
Pullanases	*Bacillus* sp. S-1 and *Micrococcus* sp. Y-1	Production of ethanol and sweetener
Lipases	*Pseudomonas nitroreducens*, *P. thermotolerans*, and *P. fragi*	Detergent manufacture

Table 4. Thermophilic enzymes, their bacterial sources and application [26, 27].

Enzyme	Bacterial Source	Applications
Cellulases	*Clostridium owensensis* and other species	Paper, textile, and bioethanol industry
Collagenase	*Clostridium histolyticum*	Medical uses and the debridement of infected wounds
Proteases	*Bacillus, Brevibacillus*, and *Streptomyces rectus* var. *proteolyticus*	Pharmaceutical, leather, food and waste processing, and detergent industry
Thermostable group II intron reverse transcriptase (TGIRT)	*Geobacillus stearothermophilus*	Useful for reverse transcribing long and highly structured RNA molecules. A method for determining RNA secondary structure DMS-MaPseq uses this enzyme
DNA polymerase	*Geobacillus stearothermophilus*	Loop-mediated isothermal amplification (LAMP) and polymerase chain reaction (PCR)
Thermostable lipases	*Thermoactinomyces vulgaris, Pseudomonas stutzeri,* and *Bacillus sporothermodurans*	Textile, brewing, food, pharmaceutical, and paper industry, and detergent production
Amylase and 1,4--xylanase	*Bacillus amyloliquefaciens, Bacillus licheniformis, Bacillus stearothermophilus, Bacillus subtilis, Bacillus megaterium, Bacillus circulans,* and *Thermoactinomyces vulgaris*	The textile industry, high potential in starch saccharification
Thermostable esterase estUT1	*Ureibacillus thermosphaericus*	Production of flavor esters, emulsifiers, and pharmaceutical intermediates
Pullulanase	*Ureibacillus thermosphaericus*	Starch saccharification (in the production of high glucose and high maltose syrup)
Chitinases	*Brevibacillus*	Chitin modification for food and health products

Thermophilic fungi develop in heaped masses of plant material and piles of agricultural and forestry products and require a warm, humid, and aerobic environment. Such fungi are only moderately thermophilic and have, therefore, not received much attention. Secretory enzymes from some of these thermophilic fungi, however, have been used in bioprocesses [28] (Table 5).

Table 5. Thermophilic enzymes, their fungal sources, and applications [28].

Thermophilic Enzyme	Fungal Source	Applications
Rennin enzyme	*Mucor pusillus* and *Mucor miehei*	Cheese Production
Lipases	*Mucor miehei* and *Thermomyces lanuginosus*	Detergent Production
Glucoamylase	*Humicola grisea* var. *thermoidea* and *T. lanuginosus*	High potential in starch saccharification
α-amylase	*Thermomyces lanuginosus*	Used for hydrolyzing starch, amylose and amylopectin, and can withstand lethal temperatures and then return to the native state with full enzymatic activity
Thermostable protease	*Penicillium dupontii*	Hydrolysis in food and feed, brewing, baking industry, and in detergent production
Cellulases and Xylanases	*Chaetomium thermophilum* and *Humicola grisea* var. *thermoidea*	Starch, cellulose, chitin, pectin processing, textiles
Laccase	*Chaetomium thermophilum*	Bleaching and paper industry
β-galactosidase	*Chaetomium thermophilum*	Lactose-free milk production
Trehalase	*Humicola grisea* var. *Thermoidea* and *T. lanuginosus*	Possesses the unique property of stabilizing membranes and enzymes against drying and thermal denaturation
Protein Disulfide Isomerase	*Sporotrichum thermophile*	Production of cellulase, β-glycosidase, and cellobiose dehydrogenase (CBD), an oxidative enzyme
Phytase	*Thermomyces lanuginosus*	Processing and reduction of phytate in the food industry, production of myo-inositol phosphate derivative for human health and medicine
Invertase	*Thermomyces lanuginosus*	Hydrolysis of raffinose and stachyose and to obtain fructo-oligosaccharide

FUTURE PERSPECTIVES

As research goes on, microorganisms have been characterized, purposely manipulated, and genetically optimized to produce a high-quality enzyme with the desired special characteristics on large scales for industrial applications. It is expected that in the future, alkaline enzymes will find additional uses in various industries, such as chiral-molecule synthesis, biological wood pulping, and the production of more sophisticated enzymatic detergents. Moreover, alkaliphiles can prove to be genetic resources for the production of signal peptides for secretion and promoters for the hyperproduction of enzymes. As global demand increases, more and more emphasis is being laid on the use of recombinant enzymes, which are produced on a large scale and can be easily purified. The latest obligations demand the implication of modern technologies such as site-directed mutagenesis, directed evolution, rational design, and metabolic engineering to generate a new generation of industrial biocatalysts that are more efficient and well-structured. There is a need to generate better molecular tools for the overexpression of some of these novel enzymes, which are poorly expressed in the currently available molecular tools. The construction of gene expression libraries with fast and accurate detection technologies will lead to the discovery of several advanced extremozymes which may be used in novel biocatalytic processes that are faster, more accurate, specific, and environmentally friendly.

INDUSTRIAL PRODUCTION OF ORGANIC ACIDS

Organic acids are weak acids with pKa values ranging from 3 (for carboxylic acids) to 9 (for phenolic acids). They find a wide distribution in animal, plant, and microbial sources. The foundation of the organic acid industry is the acids of microbial origin. The biological production of organic acids from bacterial and fungal sources has an economic edge over chemical synthetic processes. In this section, we shall discuss commercially important acidic compounds produced industrially by microorganisms.

Organic acids represent a remarkable section of the fermentation market in the world. Microbial production is an important economic substitute for chemical synthesis for many of these acids. Owing to the rising demand for organic acids over time, it needs to evolve new strategies and discoveries for novel microbial strains to carry out high-level production of commercially important organic acids such as malate, gluconate, and citrate [29].

Soil organic acids are a noteworthy part of the water-soluble fractions of organic molecules released in the rhizosphere. They are normal components of most agricultural soils and play crucial roles in soil nutrient availability, ecology, and productivity. They are greatly linked to the carbon cycle with lichens and algae, playing major roles as primary organic acid producers, and are generally the first colonizers of the rock surface [30].

Organic acids have attracted appreciable attention for their role in natural ecology and their potential industrial applications as food additives, pharmaceutical and cosmetic excipients [29]. They are completely degradable molecules and might be used as chemical intermediates for the production of biodegradable polymers, potentially replacing petroleum-based or synthetic chemicals. Some soil fungi are well known to produce high amounts of several useful organic acids [31]. These acids act as valuable building blocks for future bio-refining applications. Such applications involve the conversion of inexpensive renewable resources to platform sugars, which are then converted to platform chemicals *via* fermentation. They are further derivatized into large-volume chemicals through typical catalytic routes [32].

The use of biotechnological applications to enhance existing strains and develop new ones for these fermentations is an active research area globally [33].

WHY ARE SOIL MICROORGANISMS USED FOR ACID PRODUCTION?

Microorganisms are the preferred source of organic acids for the following reasons:

- They help overcome the toxic effects associated with chemical acid synthesis, for example, in unpredictable side reactions, while using of cyanides, in the presence of a copper-containing catalyst, in the production of toxic waste, and for a reduction in product synthesis due to different by-products [29]. It avoids the use of heavy metals, organic solvents, strong acids and bases, thus allowing the synthetic process to take place in an environmentally benign way [34].
- They are truly a sustainable alternative.
- They economize the bio-process, an advantage over chemical synthesis.
- Fungal natural production of organic acids is believed to have numerous key roles in nature depending on the species of fungi producing them [29].
- Production of organic acids using soil microbes is a promising way to obtain building-block chemicals from renewable carbon sources [35].
- Filamentous fungi can degrade vegetable biomass, which is a favourable feedstock for the biotechnological production of organic acids. *Postia stiptica* (brown-rot fungus) and *Ganoderma weberianum* (white-rot fungus) produce oxalic acid [31].
- Aspergillus forms a distinct cluster among Ascomycota due to their various organic acid productions. This exemplifies the relevance of organic acids in the chemotaxonomy of *Aspergillus*. Some of these strains also possess the ability to produce malic, oxalic, gluconic, and citric acids [31].
- Proper knowledge and exploitation of these soil microbes can lead to novel, efficient, economically feasible, and environmentally responsible fermentations for organic acid production [29].
- Better success rates are observed using microbial fermentation, and with time, this technique has become the method of ultimate choice for commercial production [36].
- Some compounds with complex structures already have natural synthetic

pathways, while establishing chemical synthetic routes for these complex compounds is not easy [34].
- However, with the use of metabolic engineering, ways to further improve the yield and productivity of a target organic acid have been found. Additionally, combinatorial biosynthesis allows the creation of novel derivatives [34].
- In certain cases, for instance, traditional lactic acid bacteria (LAB) of Africa increase food palatability and improve the quality of food by increasing the availability of proteins and vitamins. Moreover, LAB confers preservative and detoxifying effects on food. Its regular usage boosts the immune system and strengthens the body [37].

Important Organic Acids Derived from Soil Microbes

Alpha Ketoglutaric Acid

Initially, alpha-ketoglutaric acid (α-KGA) was synthesized chemically through various steps using oxalic acid and succinic acid diethyl esters with cyanohydrins with a yield of 75% or by hydrolysis of acyl cyanides. However, microbial synthesis has now taken over. In particular, the yeast *Yarrowia lipolytica* has been examined and metabolically engineered for the production of α-KGA since the 1960s [38].

It is used as a preliminary material for the advent of novel pharmaceutical products possessing activities against tumours. It has a role in the detoxification of ammonia in the brain, owing to which Reye's syndrome, cirrhosis, and urea cycle disorders are prevented. Furthermore, according to recent studies, α-KGA also plays a vital role in significantly increasing the lifespan of nematode worms [29].

Citric Acid

The global production of citric acid in 2007 was 1.6 million tons, with annual growth in demand and consumption of 3.5–4.0%. Owing to new genetic engineering production units mostly located in China, the worldwide supply of citric acid in twenty years has increased from less than 0.5 to more than 2 million tons. Citric acid is employed extensively across many industries, especially in food, beverage, pharmaceutical, nutraceutical, and cosmetic products. It acts as an acidulant, preservative, emulsifier, flavoring, sequestrant, and buffering agent [39].

It also has universal recognition of safety, nice acid taste, high water solubility, buffering and chelating properties. Many complex molecules and commercially valuable products are also formed by their carboxyl and hydroxyl groups [29].

Fumaric Acid

Fumaric acid was discovered by Felix Ehrlich in 1911 in *Rhizopus nigricans*, *Circinella*, *Mucor*, *Cunninghamella*, and *Rhizopus* species. It is presently employed in rye bread, refrigerated biscuit dough, corn tortillas, wheat, sourdough, fruit juices, nutraceutical drinks, gelatin desserts, pie fillings, gelling aids, wine, and as a supplement in animal feed [29].

Gluconic Acid

It has applications in dairy and meat products, mainly in baked items. It also found some applications in dropping fat absorption in cones and doughnuts [29].

Malic Acid

Owing to its better taste retention and a greater acid taste than citric acid, malic acid is used as a flavour enhancer and an acidulant, mainly consumed in beverages, food, and candy. Also, malic acid is used as one component of antimicrobial agents [29].

Itaconic Acid

Itaconic acid has wide applications in industries such as improved adhesion to cellophane, a detergent, industrial cleaner, dental fillers, dental adhesives, and thickeners in lubricating grease. Moreover, it is also used as an antifungal drug in ophthalmic drug delivery [29].

Lactic Acid

Lactic acid plays a crucial role due to its versatile application in the food, medical, and cosmetics industries, as well as a potential raw material for the manufacture of biodegradable plastics. The production of optically pure lactic acid using microbial fermentation has increased due to the prospects of environmental friendliness and cost-effectiveness [40].

Butyric Acid

Butyric acid has many uses in various industries. At present, its use as a precursor to biofuels is of great interest [41].

Propionic Acid

Propionic acid is chiefly used in the food industry, but it also finds applications in the cosmetic, plastics, and pharmaceutical industries [42]. Furthermore, algae that colonize rock surfaces produce carbonic acid, acetic acid, and oxalic acid, which enable the mineralization of rocks. The ability of mosses and lichens to produce organic acids (OAs) has also been reported. According to Fujii and co-workers, up to 57% of the organic acids produced by microbes in the soil are consumed in the respiration of lichenous fungi, hence contributing to the "C" flux in the soil [30]. Table **6** summarizes soil microorganisms involved in the production of different organic acids.

Table 6. List of the soil microorganisms used to produce diverse organic acids [29, 41, 43 - 52].

Acid	Bacteria	Yeast	Fungi
Alpha keto glutaric acid	*Arthrobacter paraffineus, Bacillus subtilis* var. *natto, B. megaterium, B. mesentericus, Pseudomonas fluorescens,* and *Corynebacterium glutamicum*	*Candida rugosa, Torulopsis glabrata, Pichia* spp., and *Yarrowia lipolytica*	-
Citric acid	*B. licheniformis, Arthrobacter paraffinens, Corynebacterium* sp., *Pseudomonas fluorescens, P. poae, P. trivialis,* and phosphate-solubilizing fluorescent *Pseudomonas* strains	*Candida oleophila, C. guilliermondii, C. parapsilosis, C. tropicalis, Hansenula anomala,* and *Yarrowia lipolytica*	*Penicillium janthinellum* and the species of *Aspergillus* (*A. foetidus, A. awamori, A. wentii, A. fonsecaeus,* and *A. phoenicis*)
Fumaric acid	-	*Saccharomyces cerevisiae*	*Rhizopus nigricans, Circinella, Mucor, Aspergillus niger,* and *A. flavus*
Gluconic acid	*Gluconobacter oxydans, Acetobacter, Morexella, Tetracoccus, Enterobacter* spp., *Pseudomonas fluorescens, P. poae, P. trivialis* (phosphate-solubilizing fluorescent *Pseudomonas* strains), and *Burkholderia multivorans* WS-FJ9	*Aureobasidium,* and *Pullulans*	*Aspergillus* spp., *Gliocladium* spp., *Penicillium* spp., and *Scopulariopsis* spp.
Malic acid	Engineered strains, *E. coli* WGS-10, *E. coli* XZ658, *E. coli* XZ658, *B. subtilis, Pseudomonas fluorescens, P. poae,* and *P. trivialis*	*Zygosaccharomyces rouxii, A. pullulans* ZD-3, *A. pullulans* ZX-10, and *S. cerevisiae*	*A. niger, A. oryzae, A. flavus, S. commune,* and *Penicillium rubens*
Itaconic acid	-	*Rhodotorula* spp.	*Candida* spp., *Ustilago zeae, Aspergillus terreus* TN-484-M1 82, and *A. terreus* SKR10 20

(Table 6) cont.....

Acid	Bacteria	Yeast	Fungi
Glyceric acid	*Gluconobacter oxydans*, *G. frateurii* NBRC3262, *Acetobacter* strains, and *Gluconacetobacter* strains	-	-
Acetic acid	*Gluconobacter suboxydans*, *Komagataeibacter europaeus*, *K. intermedius*, *Acetobacter aceti*, *A. pasteurianus*, and *A. pomorum*	-	-
Hyaluronic acid	*B. subtilis*, Streptococci and *Pasteurella multocida*	-	-
Lactic acid	*B. coagulans*, *Leuconostoc*, *Weissella*, *Lactobacillus plantarum*, *Pediococcus acidilactici*, *Pseudomonas fluorescens*, *P. poae*, and *P. trivialis*	*Candida utilis*	*Aspergillus niger* and *Rhizopus* spp.
Butyric acid	*Clostridium butyricum*	-	-
Ascorbic acid	*Gluconobacter* sp.	-	-
Trans-cinnamic acid	*Cellulomonas galba*	-	-
Cinnamic acid	*Nocardia*, *Pseudomonas pseudomallei*, *Alcaligenes* spp., and *Cellulomonas galba*	*Rhodotorula* spp.	-
Dipicolinic acid	*Bacillus natto*, *B. subtilis*, and *B. megaterium*	-	*Penicillium* spp.
3-hydroxy propionic acid	*Rhodococcus erythropolis*	-	-
Oxalic acid	*Pseudomonas fluorescens*, *P. poae*, and *P. trivialis*	-	-
Succinic acid	*Pseudomonas fluorescens*, *P. poae*, and *P. trivialis*	-	-
Indole acetic acid	*Rhizobium* spp. and *Bacillus* spp.	-	-
Propionic acid	*Propionibacterium* spp., *P. acidipropionici*, *P. shermanii*, and *P. jensenii*	-	-

Future Aspects

Microbial production is a profitable alternative to chemical synthesis for many of the acids. The toxic effects associated with chemical synthesis make the use of microbes all the more necessary and useful. Although the in-depth knowledge of microbial production of some acids has been known and further microbial production seems convenient, large-scale production has not yet been achievable [35]. The information from microbial genome sequences can help accelerate the recognition of metabolic pathways and the collection of catabolic enzymes accessible to an organism. Such valuable information can also simplify the process. These developments in information and tools can speed up the development of economically feasible and environmentally benign fermentations for organic acid production. The fastest way to initiate an industrial process for microbial organic acid production is through the exploitation of natural producers with rigorous bioprocess engineering. Furthermore, the choice of the organism also influences by-product formation and consequently influences costs.

The discovery that soil microbes may translate into advantages for genetic engineering technology, maximized the need for their isolation and characterization by scientists. Activities such as the production of acids, antibiotics, and enzymes from microorganisms of soil constitute the prime objective of industries to cope with the growing world's population. It is therefore quite evident that the soil is extremely rich in useful microorganisms, the full potential of which has not been realised yet. However, with all the new techniques, many new microbial species and even families have been discovered. New insights are being obtained into the composition of microbial communities, the interactions between microorganisms and the factors influencing microbial communities in soil.

Taking such aspects into consideration, the route leading to a sustainable society becomes more perceptible. Therefore, organic acids produced using soil microbes constitute a class of molecules with a great future ahead of them.

BIOPIGMENTS

Biopigments or biochromes are the dye molecules manufactured by living organisms that absorb a specific wavelength of light. Microorganisms are typically known to produce biopigments as a part of their innate defense systems against UV radiation, which enhances their survivability and adaptability over non-pigmented ones [53]. These microbial pigments, in addition to their function as colorants, possess various biological properties like anti-inflammatory, anti-bacterial, anti-fungal, antineoplastic, immunosuppressive, anti-viral, anti-aging, anti-fouling, herbicidal, insecticidal, and anti-diabetic activities. Therefore, microbial pigments are utilized according to their color and biological functions [6].

Synthetic dyes have been used for various industrial processes. However, these compounds are found to possess carcinogenic properties, which has shifted our focus to eco-friendly microbial pigments. Biochromes are therefore utilized in various industries, including textiles, cosmetics, food, pharmaceuticals, plastic, paint, *etc* (Fig. **3**). Furthermore, the microbial sources of pigments are favoured over the plant pigments due to their diligent productivity along with the optimizable culture conditions. Various industrially significant pigment molecules (flavins, monascins, melanins, phenazine, violacein, carotenes, *etc.*) are released as a byproduct by a variety of microbial species [6]. Table **7** enlists a few industrially important biopigments along with their microbial sources.

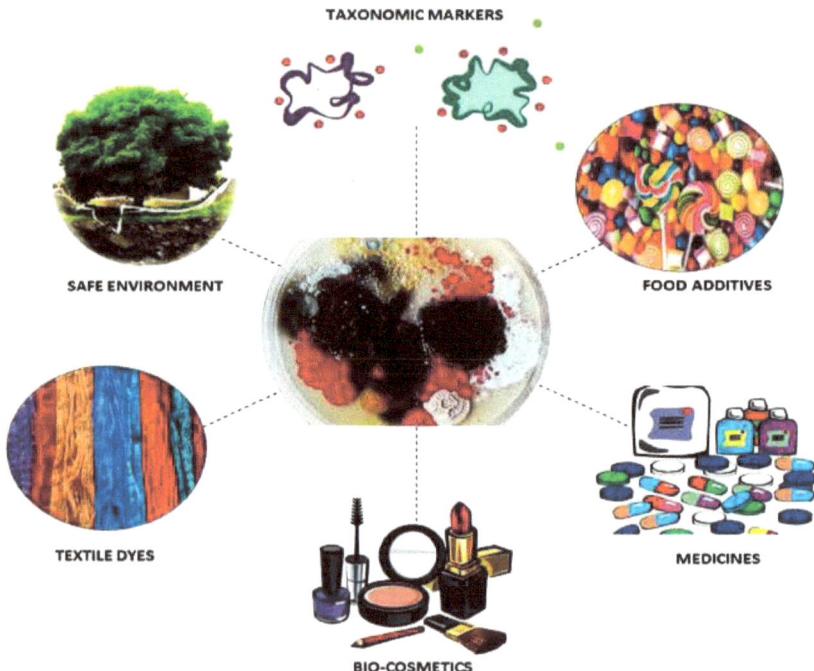

Fig. (3). Uses of biopigments in various industries.

Table 7. Biopigments used in various industries [6].

Pigment	Microorganism	Application
Carotenoids	*Haematococcus pluvialis* and *Phaffia rhodozyma*	Used in pharmaceutical, nutraceutical, food and beverage and aquaculture industries.
Astaxanthins	*H. pluvialis*	Utilized in the aquaculture feeds and play a key role in developing memory and improving anti-aging properties.
Phycocyanin	*Spirulina* spp.	The cosmetic industries manipulate phycocyanins for the development of bio-makeup. It is also used in research facilities in fluorescent microscopy.
Arpink red™	*Penicillium oxalicum*	It is widely used as a food colorant

(Table 7) cont.....

Pigment	Microorganism	Application
Prodiginines	*Pseudoalteromonas tunicata*	Prodiginines and related compounds are fungicidal and anti-bacterial agents which can be used in combatting antimicrobial resistance. They also exhibit anti-metastatic activity and have the potential to be used in cancer treatment.
Violacein	*Janthinobacterium lividum*	The pigment depicts anti-viral properties against the Herpes simplex virus and poliovirus. It shows an exceptional affinity towards silk fibers therefore, it is a potential dye for the textile industries.
Anthraquinones	*Phoma exigua* var. *exigua*	It depicts herbicidal activities against a variety of weeds.
Flexirubin	*Chryseobacterium* spp.	It is used in the treatment of various dermatological disorders.

Future Aspects

A plethora of pigment synthesizing microbes are known. However, there is a huge population that is still unexplored. Terrestrial ecosystems are a storehouse of microorganisms that can exhibit a wide range of biosynthetic pathways that are influenced by the biological, physical, and chemical environments of the organism. Therefore, the generation of biological data of uninvestigated microorganisms is essential for the development of novel colourants and compounds of medicinal value. Biotechnological advancements should be directed towards scaling up these commodity molecules by optimizing the culture conditions, which can essentially enhance the visual appeal of the industrial products.

CONCLUDING REMARKS

Traditionally, microbial utilization for various industrial processes was restricted to a few microorganisms. However, the development of culturing procedures, molecular biology tools, metagenomics, and various biotechnological methods has led to the isolation and development of a lot of competent strains. These biological systems are specifically employed in various industrial processes wherein we harness microbial expertise to synthesize compounds of industrial importance.

Customarily, industrial processes are predominantly dependent upon chemical compounds for catalysis, which fosters toxic compounds. Biocatalysts, which introduce minor changes in the substrate molecules for bioconversion, are emerging as alternatives to chemical catalysts. However, mesophilic enzymes tend to have low stability at high temperatures or extremes of pH. Nonetheless, extremozymes that can withstand high temperatures, unusual conditions, and high salt concentrations can be widely used in biotech industries. The properties of enantio-selectivity and stereo-specificity exhibited by these biocatalysts ensure that the product formed is not a concoction of isomers but is in the biologically active form. Consequently, the extremozymes can ameliorate the fabrication of cellulose, chitin, xylan, and starch along with the production of enantiomerically pure molecules. These biocatalysts are extensively used for the synthesis of organic acids like itaconic acid, succinic acid, fumaric acid, malic acid, and citric acid. The organic acid biosynthesis by the immobilized enzyme system or the whole-cell catalysts provides high yields at low cost and lower energy in contrast to chemical catalysis. Furthermore, organic acids are the most versatile ingredients in processed foods and are obtained as the end product or the intermediate components of a biogeochemical cycle. Exhaustive research for the development of microbial strains and enzymes has contributed to the current production processes and their optimization.

Notwithstanding the increasing demands, the scientific community needs to generate better molecular tools for the overexpression of novel enzymes. Therefore, the need of the hour is to explore more microbial populations and augment their systems through modern molecular technologies like site-directed mutagenesis, directed evolution, metabolic engineering, *etc.*, These recombinant systems can give rise to industrial biocatalysts, which can pave the way for the development of greener industrial processes by reducing the reliance on conventional chemical methods, thereby saving energy and the environment.

ACKNOWLEDGEMENT

Authors would like to thank Dr. Rakesh Kumar Gupta, Principal, Ram Lal Anand College, for providing all the necessary infrastructure and support in compiling this chapter.

REFERENCES

[1] Raynaud X, Nunan N. Spatial ecology of bacteria at the microscale in soil. PLoS One 2014; 9: e87217.
[http://dx.doi.org/10.1371/journal.pone.0087217] [PMID: 24489873]

[2] Karnwal A, Singh S, Kumar V, *et al.* Fungal Enzymes for the Textile Industry. In: Yadav A, Mishra S, Singh S, *et al.*, Eds. Recent Advancement in White Biotechnology Through Fungi Fungal Biology Springer, Cham 2019; pp. 459-82.
[http://dx.doi.org/10.1007/978-3-030-10480-1_14]

[3] Kwok KK, Vincent EC, Gibson JN. Antineoplastic Drugs. In: Dowd FJ, Johnson BS, Mariotti AJ, Eds. Pharmacology and Therapeutics for Dentistry Elsevier 2017; pp. 530-562.
[http://dx.doi.org/10.1016/B978-0-323-39307-2.00036-9]

[4] Pham JV, Yilma MA, Feliz A, *et al.* A review of the microbial production of bioactive natural products and biologics. Front Microbiol 2019; 10: 1404.
[http://dx.doi.org/10.3389/fmicb.2019.01404] [PMID: 31281299]

[5] Dhami NK, Alsubhi WR, Watkin E, *et al.* Bacterial community dynamics and biocement formation

during stimulation and augmentation: Implications for soil consolidation. Front Microbiol 2017; 8: 1267.
[http://dx.doi.org/10.3389/fmicb.2017.01267] [PMID: 28744265]

[6]   Ramesh C, Vinithkumar NV, Kirubagaran R, *et al.* Multifaceted applications of microbial pigments: Current knowledge, challenges and future directions for public health implications. Microorganisms 2019; 7: 186.
[http://dx.doi.org/10.3390/microorganisms7070186] [PMID: 31261756]

[7]   Brahmachari G, Demain A, Adrio J, Eds. Biotechnology of Microbial Enzymes: Production, Biocatalysis and Industrial Applications. Elsevier, Academic Press 2017; p. 608.

[8]   Pepper IL, Gerba CP, Gentry TJBT-EM, Eds. Environmental Microbiology. 3rd ed. Elsevier, Academic Press 2015; p. 705.

[9]   Waites MJ, Morgan NL, Rockey JS, *et al.* Eds. Industrial Microbiology: An Introduction. Oxford and Malden (Massachusetts): Blackwell Science Ltd 2001; p. 288.

[10] Cabrera MÁ, Blamey JM. Biotechnological applications of archaeal enzymes from extreme environments. Biol Res 2018; 51: 37.
[http://dx.doi.org/10.1186/s40659-018-0186-3] [PMID: 30290805]

[11] Brock TD, Brock KM, Belly RT, *et al. Sulfolobus*: A new genus of sulfur-oxidizing bacteria living at low pH and high temperature. Arch Microbiol 1972; 84: 54-68.
[http://dx.doi.org/10.1007/BF00408082] [PMID: 4559703]

[12] Dunbar J, Barns SM, Ticknor LO, *et al.* Empirical and theoretical bacterial diversity in four Arizona soils. Appl Environ Microbiol 2002; 68: 3035-45.
[http://dx.doi.org/10.1128/AEM.68.6.3035-3045.2002] [PMID: 12039765]

[13] Chang SS, Kang DH. *Alicyclobacillus* spp. in the fruit juice industry: History, characteristics, and current isolation/detection procedures. Crit Rev Microbiol 2004; 30: 55-74.
[http://dx.doi.org/10.1080/10408410490435089] [PMID: 15239380]

[14] Thanh VN, Thuy NT, Huong HTT, *et al.* Surveying of acid-tolerant thermophilic lignocellulolytic fungi in Vietnam reveals surprisingly high genetic diversity. Sci Rep 2019; 9: 3674.
[http://dx.doi.org/10.1038/s41598-019-40213-5] [PMID: 30842513]

[15] Mandrich L, Manco G, Rossi M, *et al. Alicyclobacillus acidocaldarius* thermophilic esterase EST2's activity in milk and cheese models. Appl Environ Microbiol 2006; 72: 3191-7.
[http://dx.doi.org/10.1128/AEM.72.5.3191-3197.2006] [PMID: 16672457]

[16] Tsuruoka N, Isono Y, Shida O, *et al. Alicyclobacillus sendaiensis* sp. nov., a novel acidophilic, slightly thermophilic species isolated from soil in Sendai, Japan. Int J Syst Evol Microbiol 2003; 53: 1081-4.
[http://dx.doi.org/10.1099/ijs.0.02409-0] [PMID: 12892130]

[17] Restaino OF, Borzacchiello MG, Scognamiglio I, *et al.* High yield production and purification of two recombinant thermostable phosphotriesterase-like lactonases from *Sulfolobus acidocaldarius* and *Sulfolobus solfataricus* useful as bioremediation tools and bioscavengers. BMC Biotechnol 2018; 18: 18.
[http://dx.doi.org/10.1186/s12896-018-0427-0] [PMID: 29558934]

[18] Suzuki S, Kurosawa N. Disruption of the gene encoding restriction endonuclease *Sua*I and development of a host–vector system for the thermoacidophilic archaeon *Sulfolobus acidocaldarius*. Extremophiles 2016; 20: 139-48.
[http://dx.doi.org/10.1007/s00792-016-0807-0] [PMID: 26791382]

[19] Sarethy IP, Saxena Y, Kapoor A, *et al.* Alkaliphilic bacteria: Applications in industrial biotechnology. J Ind Microbiol Biotechnol 2011; 38: 769-90.
[http://dx.doi.org/10.1007/s10295-011-0968-x] [PMID: 21479938]

[20] van den Burg B. Extremophiles as a source for novel enzymes. Curr Opin Microbiol 2003; 6: 213-8.
[http://dx.doi.org/10.1016/S1369-5274(03)00060-2] [PMID: 12831896]

[21] Horikoshi K. Alkaliphiles: Some applications of their products for biotechnology. Microbiol Mol Biol Rev 1999; 63: 735-50.
[http://dx.doi.org/10.1128/MMBR.63.4.735-750.1999] [PMID: 10585964]

[22] Kladwang W, Bhumirattana A, Hywel-Jones N. Alkaline-tolerant fungi from Thailand. Fungal Divers 2003; 13: 69-83.

[23] Singh CR, Mishra RM, Kango N, *et al.* Microbial enzymes and their applications in biotechnological processes. Int J Sci Res Publ 2017; 7: 860-82.

[24] Fentahun M, Kumari PV. Isolation and screening of amylase producing thermophilic spore forming *Bacilli* from starch rich soil and characterization of their amylase activity. Afr J Microbiol Res 2017; 11: 851-9.
[http://dx.doi.org/10.5897/AJMR2017.8543]

[25] Santana MM, Gonzalez JM. High temperature microbial activity in upper soil layers. FEMS Microbiol Lett 2015; 362: fnv182.
[http://dx.doi.org/10.1093/femsle/fnv182] [PMID: 26424766]

[26] Mehta R, Singhal P, Singh H, *et al.* Insight into thermophiles and their wide-spectrum applications.3 Biotech 2016; 6: 81.
[http://dx.doi.org/10.1007/s13205-016-0368-z]

[27] Novik G, Savich V, Meerovskaya O. *Geobacillus* Bacteria: Potential Commercial Applications in Industry, Bioremediation, and Bioenergy Production. In: Mishra M, Ed. Growing and Handling of Bacterial Cultures. IntechOpen 2019; 76053.
[http://dx.doi.org/10.5772/intechopen.76053]

[28] Maheshwari R, Bharadwaj G, Bhat MK. Thermophilic fungi: Their physiology and enzymes. Microbiol Mol Biol Rev 2000; 64: 461-88.
[http://dx.doi.org/10.1128/MMBR.64.3.461-488.2000] [PMID: 10974122]

[29] Khan I, Qayumm S, Maqbool F, *et al.* Microbial organic acids production, biosynthetic mechanism and applications - Mini review. IJMS 2017; 46: 2165-74. [http://nopr.niscpr.res.in/handle/123456789/42982]

[30] Adeleke R, Nwangburuka C, Oboirien B. Origins, roles and fate of organic acids in soils: A review. S Afr J Bot 2017; 108: 393-406.
[http://dx.doi.org/10.1016/j.sajb.2016.09.002]

[31] Liaud N, Giniés C, Navarro D, *et al.* Exploring fungal biodiversity: Organic acid production by 66 strains of filamentous fungi. Fungal Biol Biotechnol 2014; 1: 1.
[http://dx.doi.org/10.1186/s40694-014-0001-z]

[32] Warnecke T, Gill RT. Organic acid toxicity, tolerance, and production in *Escherichia coli* biorefining applications. Microb Cell Fact 2005; 4: 25.
[http://dx.doi.org/10.1186/1475-2859-4-25] [PMID: 16122392]

[33] McKay LL, Baldwin KA. Applications for biotechnology: Present and future improvements in lactic acid bacteria. FEMS Microbiol Lett 1990; 87: 3-14.
[http://dx.doi.org/10.1111/j.1574-6968.1990.tb04876.x] [PMID: 2271224]

[34] Du J, Shao Z, Zhao H. Engineering microbial factories for synthesis of value-added products. J Ind Microbiol Biotechnol 2011; 38: 873-90.
[http://dx.doi.org/10.1007/s10295-011-0970-3] [PMID: 21526386]

[35] Sauer M, Porro D, Mattanovich D, *et al.* Microbial production of organic acids: Expanding the markets. Trends Biotechnol 2008; 26: 100-8.
[http://dx.doi.org/10.1016/j.tibtech.2007.11.006] [PMID: 18191255]

[36] Vandenberghe LPS, Soccol CR, Pandey A, *et al.* Microbial production of citric acid. Braz Arch Biol Technol 1999; 42: 263-76.

[http://dx.doi.org/10.1590/S1516-89131999000300001]

[37] Chelule PK, Mokoena MP, Gqaleni N. Advantages of Traditional Lactic Acid Bacteria Fermentation of Food in Africa. In: Méndez-Vilas A, Ed. Current Research, Technology and Education Topics in Applied Microbiology and Microbial Biotechnology. Badajoz, Spain: Formatex 2010; pp. 1160-7.

[38] Beer B, Pick A, Sieber V. *In vitro* metabolic engineering for the production of α-ketoglutarate. Metab Eng 2017; 40: 5-13.
[http://dx.doi.org/10.1016/j.ymben.2017.02.011] [PMID: 28238759]

[39] Ciriminna R, Meneguzzo F, Delisi R, *et al.* Citric acid: Emerging applications of key biotechnology industrial product. Chem Cent J 2017; 11: 22.
[http://dx.doi.org/10.1186/s13065-017-0251-y] [PMID: 28326128]

[40] Wang Y, Tashiro Y, Sonomoto K. Fermentative production of lactic acid from renewable materials: Recent achievements, prospects, and limits. J Biosci Bioeng 2015; 119: 10-8.
[http://dx.doi.org/10.1016/j.jbiosc.2014.06.003] [PMID: 25077706]

[41] Dwidar M, Park JY, Mitchell RJ, *et al.* The future of butyric acid in industry. Sci World J 2012; 2012: 1-10.
[http://dx.doi.org/10.1100/2012/471417] [PMID: 22593687]

[42] Gonzalez-Garcia R, McCubbin T, Navone L, *et al.* Microbial propionic acid production. Fermentation (Basel) 2017; 3: 21.
[http://dx.doi.org/10.3390/fermentation3020021]

[43] Vyas P, Gulati A. Organic acid production *in vitro* and plant growth promotion in maize under controlled environment by phosphate-solubilizing fluorescent *Pseudomonas*. BMC Microbiol 2009; 9: 174.
[http://dx.doi.org/10.1186/1471-2180-9-174] [PMID: 19698133]

[44] Li GX, Wu XQ, Ye JR, *et al.* Characteristics of organic acid secretion associated with the interaction between *Burkholderia multivorans* WS-FJ9 and poplar root system. BioMed Res Int 2018; 2018: 1-12.
[http://dx.doi.org/10.1155/2018/9619724] [PMID: 30687759]

[45] Habe H, Shimada Y, Yakushi T, *et al.* Microbial production of glyceric acid, an organic acid that can be mass produced from glycerol. Appl Environ Microbiol 2009; 75: 7760-6.
[http://dx.doi.org/10.1128/AEM.01535-09] [PMID: 19837846]

[46] Štornik A, Skok B, Trček J. Comparison of cultivable acetic acid bacterial microbiota in organic and conventional apple cider vinegar. Food Technol Biotechnol 2016; 54: 113-9.
[http://dx.doi.org/10.17113/ftb.54.01.16.4082] [PMID: 27904401]

[47] Widner B, Behr R, Von Dollen S, *et al.* Hyaluronic acid production in *Bacillus subtilis*. Appl Environ Microbiol 2005; 71: 3747-52.
[http://dx.doi.org/10.1128/AEM.71.7.3747-3752.2005] [PMID: 16000785]

[48] Taskila S, Ojamo H. The Current Status and Future Expectations in Industrial Production of Lactic Acid by Lactic Acid Bacteria. In: Kongo JM, Ed. Lactic Acid Bacteria - R & D for Food, Health and Livestock Purposes. IntechOpen 2013; pp. 615-32.
[http://dx.doi.org/10.5772/51282]

[49] Gomes RJ, Borges MF, Rosa MF, *et al.* Acetic acid bacteria in the food industry: Systematics, characteristics and applications. Food Technol Biotechnol 2018; 56: 139-51.
[http://dx.doi.org/10.17113/ftb.56.02.18.5593] [PMID: 30228790]

[50] Douros JD Jr, Frankenfeld JW. Effects of culture conditions on production of *Trans*-cinnamic acid from alkylbenzenes by soil microorganisms. Appl Microbiol 1968; 16: 320-5.
[http://dx.doi.org/10.1128/am.16.2.320-325.1968] [PMID: 16349793]

[51] Arima K, Kobayashi Y. Bacterial oxidation of dipicolinic acid I. Isolation of microorganisms, their culture conditions, and end products. J Bacteriol 1962; 84: 759-64.
[http://dx.doi.org/10.1128/jb.84.4.759-764.1962] [PMID: 16561964]

[52] Lee SH, Park SJ, Park O-J, *et al.* Production of 3-hydroxypropionic acid from acrylic acid by newly isolated *rhodococcus erythropolis* LG12. J Microbiol Biotechnol 2009; 19: 474-81.
[http://dx.doi.org/10.4014/jmb.0808.473] [PMID: 19494695]

[53] Agogué H, Joux F, Obernosterer I, *et al.* Resistance of marine bacterioneuston to solar radiation. Appl Environ Microbiol 2005; 71: 5282-9.
[http://dx.doi.org/10.1128/AEM.71.9.5282-5289.2005] [PMID: 16151115]

CHAPTER 5

Applications of Microbial Biopesticides

Poonam Meena[1], Neelam Poonar[1], Sampat Nehra[2,*] and P.C. Trivedi[3]

[1] Department of Botany, University of Rajasthan, Jaipur, India

[2] Department of Biotechnology, Birla Institute of Scientific Research, Statue Circle, Jaipur, Rajasthan, India

[3] Deptartment of Botany, University of Rajasthan, Jaipur, Rajasthan, India

Abstract: Microbial biopesticides involve various microorganisms such as bacteria, fungi, viruses, nematode-associated bacteria, protozoans, and endophytes working against invertebrate pathogens in agro-ecosystems. Such novel biopesticidal products, after extensive research work, have been explored in the global market to combat synthetic pesticide application adverse problems. Recent academic and industrial efforts are involved in the discovery of toxins and virulence factors from microbial species for the synthesis of commercial formulations. The current review is the expansion of the application of various bacteria, fungi, viruses, nematodes, protozoans, and endophytes for biopesticide formulations and their role in pest management.

Keywords: Entomopathogenic bacteria, Entomopathogenic nematodes, Endophytes, Fungi, Integrated pest management (IPM), Virus.

INTRODUCTION

Biopesticides are natural biochemicals, produced by plants, animals, fungi, bacteria, and some minerals such as baking soda and canola oil, that exert pesticide properties *via* nontoxic pathways. Approximately 195 registered biopesticidal active ingredients and 780 products were recorded at the end of 2001 [1]. Biopesticides play a crucial role in plant protection, either individually or commonly in combination with chemical pesticides, also known as integrated pest management (IPM). Biopesticides are expected to grow 10-15% faster than conventional chemical pesticides over the next decade [2], while Dar and colleagues [3] reported a 2.5 percent estimated annual growth rate for biopesticides in India. These biopesticides are eco-friendly, target pest-effective, and minimally threatening to human or environmental health [4 - 7]. Gupta and

---

*Corresponding author Sampat Nehra: Department of Biotechnology, Birla Institute of Scientific Research, Statue Circle, Jaipur, Rajasthan, India; E-mail: nehrasampat@gmail.com

Shampi Jain, Ashutosh Gupta and Neeraj Verma
All rights reserved-© 2023 Bentham Science Publishers

Dixit [8] elaborated on the potential applications of biopesticides for public and human health. Some plants like *Annona squamosa* (anonaine in the stem, leaf and unripe fruit), *Anacardium occidentale* or cashew (phenolic compounds in shell oil), and *Albizia lebbeck* (caffeic acid and quercetin in bark, stem and leaf) have been reported with insecticidal properties. The floral extract of *Butea monosperma* has chalcones and aurones (termiticidal) and the seed oil of *Madhuca longifolia* var. *latifolia* contains saponins, which might also serve as an excellent insecticidal and repellent [9].

Biopesticides involved in disease management are important since they are (a) Eco-friendly- causing minimum harm to the ecosystem, (b) Have environmental utility- small doses are enough to control a pest with easy decomposition, (c) Pest-specific- highly specific interaction with the pest, (d) High suitability- biopesticides contribute maximum during integrated pest management. The most common biopesticides are living genera like bioinsecticides (*Bacillus thuringiensis*, Bt), biofungicides (*Trichoderma* spp.) and bioherbicides (*Phytophthora* spp.). Such biopesticides involve plant-incorporated protectants (plants inserted with genetic material), biochemical pesticides, microbial pesticides, *etc.* Biopesticides are being used in agriculture against insects, weeds, and nematodes to improve the productivity of plants [10]. Recently, microbial biopesticides have been popularized amongst farmers worldwide [11]. Biopesticides include the following types:

Plant-Incorporated Protectants (PIPs)

The pesticidal compounds, or plant-incorporated protectants (PIPs), are produced by the plants upon insertion of genetic material, like the incorporation of Bt pesticidal protein into a plant genome. Consequently, such genetically modified plants kill the pest directly rather than the Bt bacterium.

Biochemical Pesticides

Biochemical pesticides (BP) involve naturally existing substances like fatty acids and plant extracts, which regulate pests *via* different nontoxic reactions, while synthetic pesticides commonly kill or inactivate the pest. Such BPs contain compounds like pheromones, which either attract or repel pests from mating, and also contain plant growth regulators (PGRs) that stimulate plant growth itself.

Microbial Pesticides

Microbial pesticides (MPs) are derived from naturally existing or genetically modified fungi, bacteria, viruses, algae, and protozoans. Microbial pesticides were found to be effective and safe substitutes for chemical insecticide formulations.

Microbial toxins are biological poisons derived from microorganisms, and their effects are significantly pest-specific. The uptake of such microbial toxins by entomopathogens occurs *via* (i) Invasion of the gut or integument of insects and (ii) With the multiplication of the microbes, leading to the death of insects due to the accumulation of insecticidal toxins. Such MPs were found to have efficiency, safety for non-target organisms like humans and animals, residual incorporation in food produced from crops, eco-friendly and potential targeted pest-specificity, and thus facilitate the survival of useful insects in treated crops. Therefore, such microbial insecticides have been exploited as biological control agents for three decades. These microbes contain active components that can control a wide range of pathogens either alone or in groups. For example, the *Bacillus sphaericus* strain has been shown to be effective against various types of mosquitoes such as *Culex quinquefasciatus*, whereas *B. thuringiensis* is more effective on *Aedes aegypti* alone [12].

Besides *B. thuringiensis*, novel bacterial species such as *Xenorhabdus* spp., *Photorhabdus* spp. (entomopathogenic nematode), *Yersinia entomophaga*, *Serratia* spp., *Pseudomonas entomophila*, *Burkholderia* spp., *Chromobacterium* spp., *Saccharopolyspora* spp., and *Streptomyces* spp. have received attention from an industrial point of view due to the presence of a diversity of metabolites or toxins exerting insecticidal properties [13, 14]. Gouli and coworkers [7] reported extensive utilization of microbial pesticides in pest management. Similar to synthetic pesticides, microbial pesticides have been mass cultivated or formulated from bacteria, fungi, and protozoans (single-celled organisms), and such biopesticidal formulations are regulated by the environmental protection agency and their application is controlled by the federal insecticide, fungicide, and rodenticide acts [3, 10, 11, 15].

BACTERIAL BIOPESTICIDES

Bacterial biopesticides are relatively cheaper and more common in microbial pesticides. Bacterial insecticides are usually specific to individual species of beetles, flies, butterflies, mosquitoes, and moths. An effective bacterial insecticide must be ingested by the gut of the target pest, followed by contact. The *Bacillus* genus is a rod-shaped, soilborne, and spore-forming bacterial insecticide that has been used against insects. This bacterium contains important traits that make it a potential microbial biopesticide [16]. It has crystalline inclusion bodies or endotoxins or Cry proteins and cytolytic (Cyt) toxins or δ-endotoxins or insecticidal crystal proteins. *Bacillus subtilis* has been recognized as a safe, non-pathogenic bacteria with the approval of the USFDA (US Food and Drug Administration). It is applied as a biopesticide and its spores are highly resistant to heat, dormancy, nutritional scarcity, and unfavourable pH [17]. Bacterial insecticides may be effective against only one species or the entire group of insects (Table **1**). Various types of bacteria exist that are as follows:

Sporulating Bacteria

Bacillus thuringiensis (Bt)

Bacillus thuringiensis (Bt) is a Gram-positive, aerobic, soilborne, spore-forming insecticidal bacterium, exploited commercially worldwide against various plant pests like Lepidoptera (butterflies and moths), caterpillars, Simuliidae (black flies) and mosquito larvae. Bacterial cells along with spores also produce crystalline protein or endotoxin, which is later ingested by a target insect with or without spores, followed by activation by the alkaline condition of the gut and attaches to gut wall receptors or enters inside insect hemolymph, consequently breaking down the gut lining and the insect dies. Commercially, Bt products are a blend of toxic crystals and dried spores applied to the insect larvae feeding site. Since the discovery of microbial insecticide in 1901, Bt has been extensively used to manage the insect population in forestry, agriculture, and medicine. Over one hundred bioinsecticides have been developed, which are being effectively used against lepidopteran, coleopteran, and dipteran insect larvae [18]. Cry protein toxins are safe, specific and efficient, and substitute for chemical pesticides. Thus, nowadays, cry genes coded for insecticidal crystal proteins have been transferred to various crops *via* transgenic technology to control the pest.

Cry genes involved in the formation of ion channels or pores *via* the adenylyl cyclase/PKA signalling pathway cause insect death [19]. Broderick and coworkers [20] reported that *B. thuringiensis* is completely pathogenic when it is in commensalism with other bacteria inside the gut of the insect. As a pesticide, Bt accounts for more than 90% of the current bioinsecticide market for several decades [21]. Worldwide, different formulations are available for the control of caterpillars (*B. thuringiensis* var. *kurstaki*, *aizawai*, *galleriae*, and *entomocidus*), beetle larvae (*B. thuringiensis* var. *tenebrionis*), black fly and mosquito larvae (*B. thuringiensis* var. *israelensis*). The Bt formulations are stable over years, even in the presence of UV rays without any disintegration. During the spray of bacterial insecticides upon cabbage plants infested with white butterfly larvae, some bacteria will remain for a prolonged period and be found in the top 2 cm of soil for 120 days to a year in the dormant state till the production of the toxin, and thus toxicity will persist there for the next crop to some extent. Such spores germinate well in dead insect bodies and produce up to 100 million new spores per insect larvae, which can be spread by the fly movement [1, 22].

Bacillus sphaericus (Bs)

Entomopathogenic bacteria, namely *Bacillus sphaericus* (Bs), a toxic larvicidal strain, are used for the control of mosquitoes, *Psorophora*, and *Aedes* spp. Pasteurized soil is a selective medium for *B. sphaericus*. It has spherical spores, which are located terminally, and a pro-toxin produced by parasporal inclusions or crystalline bodies of 51 KDa and 42 KDa (binary toxin/bin toxin). These, upon ingestion, solubilize in the mid gut after 15 minutes, leading to the failure of neural and skeletal muscles and ultimately killing mosquitoes. Additionally, *B. sphaericus* contains another mosquitocidal toxin known as Mtx1.

Paenibacillus Species

The genus *Paenibacillus popilliae* (Dutky), spore-forming with parasporal inclusions (80 KDa), shows insecticidal property against insects causing American foulbrood (AFB) disease in honeybees and phytophagous coleopteran larvae [23].

Table 1. List of some bacterial insecticides from *Bacillus* spp [1].

S. No.	Pathogen	Product Name	Host Range	Comments
1.	*Bacillus thuringiensis* var. *kurstaki* (Btk)	Bactur, Bactospeine®, Bioworm®, Caterpillar Killer®, Dipel®, Futura®, Javelin®, SOK-Bt®, Thuricide®, Topside®, Tribactur®, and Worthy Attack®	Caterpillars, larvae of moths, and butterflies	Effective for foliage-feeding caterpillars and Indian meal moths in stored grain. Deactivated rapidly in sunlight, therefore applies in the evening or on overcast days and direct some spray to lower surfaces of leaves. It does not cycle extensively in the environment.
2.	*B. thuringiensis* var. *israelensis* (Bti)	Aquabee®, Bactimos®, Gnatrol®, LarvX®, Skeetal®, Teknar®, and VectoBac®	Larvae of *Aedes* and *Psorophora* mosquitoes, black flies, and fungus gnats	Effective against larvae only. Active only if ingested. *Culex* and *Anopheles* mosquitoes are not controlled at normal application rates and it does not cycle extensively in the environment.
3.	*B. thuringiensis* var. *tenebrionis* (Btt)	Foil®, M-One®, M-Track®, Novardo®, and Trident®	Larvae of colorado potato beetle and elm leaf beetle adults	Effective against Colorado potato beetle larvae and the elm leaf beetle. Like other Bt, it must be ingested. It is subjected to a breakdown in ultraviolet light and does not cycle extensively in the environment.

(Table 1) cont.....

S. No.	Pathogen	Product Name	Host Range	Comments
4.	*B. thuringiensis* var. *aizawai* (Bta)	B401/B402 (Certan®)	Wax moth caterpillars	Used only for the control of wax moth infestations in honeybee hives
5.	*B. popilliae* and *B. lentimorbus*	Doom, Japidemic®, Milky Spore Disease®, and Grub Attack®	Japanese beetle larvae (grubs)	The main Illinois lawn grub (the annual white grubs, *Cyclocephala* sp.) is not susceptible to milky spore disease.
6.	*B. sphaericus* (*Lysinibacillus sphaericus*)	VectoLex CG® and VectoLex WDG®	Larvae of *Culex*, *Psorophora*, and *Culiseta* mosquitoes, and larvae of some *Aedes* spp.	Active only if ingested and for use against *Culex*, *Psorophora*, *Culiseta* spp. and *Aedes vexans*. Remains effective in stagnant or turbid water

Non Sporulating Bacteria

Pseudomonas aeruginosa (Schroeter) Migula

Bucher and Stephenes [24] classified *P. aeruginosa* as a potent pathogenic bacteria that works against *Camnula pellucida* and *Melanoplus bivittatus* (grasshoppers) *via* multiplying inside hemocoele of the insect gut with an invasive mechanism. Scarpa [25] reported phospholipase production from *P. aeruginosa*, which might be toxic against a few strains of *Bacillus cereus*.

Serratia marcescens

S. marcescens is a facultative anaerobe, proteolytic, phospholipase-producing bacteria, isolated from dead and diseased insects like grasshoppers and locusts. *Pseudomonas* spp. and *Serratia* spp. have a limitation in that both bacteria exhibit some degree of pathogenicity in mammals.

Clostridium Species

Two obligate anaerobes have been isolated from diseased larvae of the western tent caterpillar (*Malacosoma pluvial*) [24], namely *Clostridium brevifaciens* and *C. malacosomae*. After ingestion of such protectants, larvae die within 72 hours. *C. bifermentans* exhibits toxicity against mosquitoes and black flies *via* the mosquitocidal Cbm71 protein, similar to Bt.

Gammaproteobacteria

Photorhabdus and Xenorhabdus

Additionally, other groups of entomopathogenic bacteria like *Photorhabdus* and *Xenorhabdus* synergistically worked with insecticidal nematodes. After nematode invasion in susceptible insects, symbiotic bacteria are released inside the gut and weaken the insect immune system with an insecticidal toxin complex (Tc).

Serratia Species

Serratia entomophila, associated with entomopathogenic nematodes or insects, produces insecticidal toxin Sep proteins (Sep A, B, and C), analogous to Bt against *Costelytra zealandica* (grass grub) [26].

Yersinia entomophaga

Y. entomophaga, a nonsporulating bacterium isolated from the New Zealand grass grub, *C. zealandica*, produces a Yen-Tc (insecticidal toxin complex) showing chitinase activity [27] against insects.

Pseudomonas entomophila

After ingestion, *Pseudomonas entomophila*, a unique bacterium exhibiting insecticidal activities against *Drosophila melanogaster*, leads to the death of the insect. After genome sequencing, Vodovar and coworkers [28] reported that it has specific multiple toxins that contribute to entomopathogenic properties.

Betaproteobacteria

Burkholderia Species

Burkholderia species are associated with insects in mutualism, specifically in the gut region and are reported to influence hemipteran bean bug (*Riptortus pedestris*) oviposition and fecundity. Cordova-Kreylos and coworkers [29] isolated a new strain of *Burkholderia rinojensis* from Japan, which exhibited oral toxicity in beet armyworms and mites *via* their strong insecticidal toxins.

Chromobacterium Species

Martin and coworkers [30] isolated a strain of *Chromobacterium subtsugae* from a soil sample and named it PRAA4-1T in Maryland (USA) with strong insecticidal activity against insects. *C. subtsugae* produces violacein (a tryptophan derivative) that confers a violet colour to all colonies; besides, it produces other molecules associated with the insecticidal action [31].

Actinobacteria

Streptomyces Species

Various *Streptomyces* spp. have cellulolytic properties, which are used against herbivorous insects and produce different insecticidal substances like antimycin A, prasinons, flavensomycin, piericidins, and macrotetralides as potent toxins against either phytopathogenic insects or pests [32]. Lepidopteran insects were inhibited by the toxins emamectin and milbemectin produced by *Streptomyces* species.

Saccharopolyspora spinosa

Mertz and Yao [33] discovered *Saccharopolyspora spinosa* with the insecticidal role of its isolate A83543. It has insecticidal toxicity due to the compounds spinosyn A and spinosyn D, which interact with the insect gut receptors nicotinic acetylcholine g- and g-aminobutyric acid receptors, resulting in neural activity damage, paralysis, and death. They also have a minimum risk to non-target species like aquatic organisms and mammals. Lewer and coworkers [34] isolated spinosyn-related compounds from *S. pogona*.

FUNGAL BIOPESTICIDES

More than 90 genera and 700 species of fungi have been recorded as insecticidal or entomopathogenic fungi like *Beauveria*, *Metarhizium*, *Paecilomyces*, *Lecanicillium*, *Isaria*, *Tolypocladium*, *etc.* against foliar insect pests in greenhouses or areas under high humidity (Table **2**). The entomopathogenic fungal group is an important biocontrol agent of insects with potential mycoinsecticide activity in agriculture. *Beauveria bassiana* can target a diversity of arthropods, while another fungus, *Metarhizium anisopliae* (insect-pathogenic), is effective against adult *Aedes albopictus* and *Aedes aegypti* mosquitoes. Sharaf [35] reported that crude chitinase obtained from *Alternaria alternata* exhibited 82% mortality in the fruit fly. These fungi (i) Invade their hosts' cuticle *via* mechanical pressure (appressorium) or enzymatic degradation (chitinases, proteases, lipoxygenases, upases, and quitobiases) (ii) Oral ingestion, gaining access inside hemocoele with the dissemination of hyphal bodies into the muscle tissues, Malpighian tubes, fatty bodies, haemocytes and mitochondria, causing the death of the host insect within 3-14 days of infestation. Under suitable environmental conditions, dead insects serve as the best option for fungi. They can penetrate and emerge from the surface under suitable environmental conditions.

Indeed, various fungi produce a wide range of secondary metabolites and toxins that possess insecticidal properties that lead to the death of pathogens ultimately. For instance, host degrading enzymes like acid phosphatases and phosphatase isozymes were obtained from *Metarhizium anisopliae* under fermentation conditions [36]. Shakeri and Foster [37] reported that *Trichoderma* produces chitinase (44 kDa) and protease (31 kDa), while Hoell and coworkers [38] mentioned various metabolites like trichodermol, harzianum A, trichodermin, harzianolide, and paptaibols, and found them effective against the insect *Tenebrio molitor*. Similarly, boverin (commercial mycoinsecticide), beauvericin, bassianolide, bassianin, bassiacridin, and oosporein derived from *B. bassiana* mimic further outbreaks of *Cydia pomonella*, *Culex pipiens*, and *Aedes aegypti* [39]. *Metarhizium anisopliae* showed insecticidal potential against *Crocidolomia pavonana*, *Plutella xylostella*, and *Spodoptera litura* insect populations infecting *Brassica oleracea* (Cruciferae) *via* various toxic metabolites like cytochalasin C, metacytofilin, myroridins, aurovertins, fungerins, destruxins (dtxs), swainsonine, hydroxyl-ovalicin, helvolic acid, serinocyclins A and B, and viridoxins [40]. Pecilomicine-B obtained from *Paecilomyces fumosoroseus* is highly effective against *Trialeurodes vaporariorum* [41]. Mites are effectively controlled by hirsutellin A and phomalatone, which are derived from the fungus *Hirsutella thompsonii* [42]. Whiteflies are significantly controlled by the toxins destruxins, dustanin, and homodestruxins produced by *Aschersonia aleyrodis* and *A. tubulata* [43]. Besides this, some volatile compounds were also produced from fungi like *Muscodor albus* and *Gliocladium* species used as biofumigants @ 15 or 30g against potato tuber moth, which revealed 72% control [44], while Daisy and coworkers [45] reported an insect repellent (naphthalene) produced from *Muscodor vitigenus* to control the stored pest.

Table 2. List of some fungal insecticides.

S. No.	Pathogen	Product Name	Host Range	References
1.	*Beauveria bassiana*	Boverin, Beauvericin, Bassianolide, Bassianin, Bassiacridin, and Oosporein	Caterpillars of *Cydia pomonella*, *Culex pipiens*, and *Aedes aegypti*	[39]
2.	*Aschersonia aleyrodis* and *A. tubulata*	Destruxins, Dustanin, and Homodestruxins	Larvae of *Aedes* and whiteflies	[43]
3.	*Hirsutella thompsonii*	Hirsutellin A and Phomalatone	Mites	[42]

(Table 2) cont.....

S. No.	Pathogen	Product Name	Host Range	References
4.	*Paecilomyces fumosoroseus*	Pecilomicine-B	*Trialeurodes vaporariorum*	[41]
5.	*Trichoderma*	Trichodermol, Harzianum A, Trichodermin, Harzianolide, and Paptaibols	*Tenebrio molitor*	[38]
6.	*Metarhizium anisopliae*	Destruxins (dtxs), Cytochalasin C, Helvolic acid, Myroridins, Swainsonine, Tyrosine Betaine, Serinocyclins A and B, Aurovertins, Hydroxy-ovalicin, Viridoxins, Fungerins, Metacytofilin, Hydroxyfungerins A and B, and 12-hydroxyovalicin	Termites, thrips, *Crocidolomia pavonana*, *Plutella xylostella*, and *Spodoptera litura*	[40]

VIRAL PESTICIDES

Insect-specific viruses serve as efficient biocontrol agents against caterpillar pests. An insect-feeding virus is allowed to spread during mating and egg-laying from one insect to another insect (Table 3). More than 1600 viruses have been reported, which target more than 1100 species of mites and insects [1]. Baculoviruses are DNA-containing rod-shaped particles with virus inclusion bodies formed by protein coats. After ingestion, alkaline midgut hydrolyzes inclusion bodies and releases viral proteins. Following this, viral proteins after multiplication fuse with epithelial cells and kill the host. Baculoviruses alone account for the control of more than 100 insect species. They are significantly effective on vegetables, rice, and cotton lepidopterous pests [15]. Many caterpillar pests were controlled with available nuclear polyhedrosis viruses and granulosis viruses [46]. Sandoz Inc. in 1975 introduced the first viral insecticide, Elcar™ prepared from *Heliothis zea* NPV or HzSNPV (range of baculovirus), which was effective not only against pests infecting maize, beans, soybeans, tomatoes, and sorghum but also cotton bollworm, while GVs (granulosis viruses) are effective against alfalfa looper and leaf roller [47]. Currently, more than fifty viral formulations (including baculovirus) are distributed under different trade names worldwide [48]. The velvet bean caterpillar in soybean was successfully controlled with *Anticarsia gemmatalis* nuclear polyhedrosis virus (AgNPV) as a biopesticide [1]. Mettenmeyer [49] reported that baculovirus pesticides have ecological, economic, and social benefits, making them superior to conventional chemical pesticides. However, baculovirus pesticide application is limited due to host specificity, high cost, instability, lack of exposure to ultraviolet rays, and slow action. Such constraints can be solved with genetic engineering or recombinant virus synthesis [10].

Table 3. List of some insecticidal virus [50].

S. No.	Nature of Virus	Host
1.	Nuclear polyhedrosis virus	*Helicoverpa armigera, Spodoptera litura, Amsacta albistriga, Spilosoma obliqua, Pericallia ricini, Pseudaletia separata, Spodoptera mauritia, Corcyra cephalonica, Plusia chalcites, Antheraea mylitta, Dasychira mendosa*, and *Plusia peponis*
2.	Granulosis virus	*Cnaphalocrocis medinalis, Pericallia ricini, Achaea janata, Phthorimaea operculella*, and *Chilo infuscatellus*
3.	Cytoplasmic polyhedrosis virus	*Helicoverpa armigera*
4.	Pox virus	*Amsacta moorei*

PROTOZOANS

The protozoa, Onidospora, and Sporozoa subphyla contain multiple entomophilic protozoans, which serve as the most promising examples of natural insect pest reduction episodes. The protozoan pathogen *Nosema locustae* naturally infects a wide range of insect hosts, particularly grasshoppers, killing or debilitating them. The protozoan infection in insects consequently reduces the number of offspring. Thus, they are suitable for insecticide formulations, but their applicability is confined to a small scale due to their very slow mechanism of action (days or weeks to harm insects) and only reducing pest reproduction or feeding rather than killing them [1]. However, their infections remain for a prolonged period as they target vitality reduction, life span, and fecundity of insects. Natural protozoan epizootics cause infection in corn borer, flies, Lepidoptera, and Diptera mosquitoes [15]. The *Mettasia grandis* is a promising pathogen infecting cotton boll weevil, while *Mettasia frogodermae* has been used against the khapra beetle pest. Their infective empty spores, producing motile sporozoites inside the gut followed by penetration into the hemocoel by infecting cells within 2 days, lead to the early death of diseased larvae and badly shrivel, and are thus utilized for their own spore production. Weinzieri and Henn [51] applied effective spores of *Nosema locustae* in bran bait to grasshopper or locust species.

NEMATODES

Nematodes are simple, colourless, un-segmented, roundworms that are free-living, parasitic, and lack appendages, which cause important diseases in humans, animals, and plants. Normally, nematodes are thought to be phytopathogens or plant parasites that cause significant crop losses, but nematodes are not of microbial origin. However, some of them, such as *Heterorhabditis* and *Steinernema* genus, which contain *Photorhabdus* and *Xenorhabdus* (symbiotic bacteria), play an important role in nematode-insect interaction [52]. Such nematodes do not kill insects directly but get inside insects from natural sites like mouths, spiracles, or sometimes cuticles, where symbiotic bacteria get active and release toxic compounds inside hemocoele. Infectious juveniles mature into adults at the J3 stage and leave dead insects to infect new hosts [1, 15]. Eilenberg and coworkers [53] synthesized common insecticidal preparations with entomogenous nematodes like *Heterorhabditis heliothidis*, *Steinernema feltiae*, and *S. scapteriscae*, *S. carpocapsae*, and *S. riobravis*. Such nematodes were effective against a variety of insects, including caterpillars, beetles, and fly larvae (over 400 pest species). The most infectious nematode stage is J3, or the "dauer" larvae, which requires moisture film to stay active or mobile to reach their target for prolonged periods, as either dry soil makes J3 stage dormant or wet soil reduces the grip of J3 stage.

ENDOPHYTES

Endophytic microbes serve as the best biopesticide as they are easily delivered inside pests and are cheap to introduce inside tissue culture plantlets, seeds, and other propagules, protecting microbes against abiotic or biotic environmental adversities. Endophytes significantly contribute to plant growth promotion and stress tolerance [44]. Lewis and coworkers [54] showed that endophytic *Beauveria bassiana* can be used to inhibit *Ostrinia nubilalis*, or European corn borer, and is equally compatible with carbofuran and *B. thuringiensis*. Similarly, endophytic *B. bassiana* mimics the mortality of *Coleomegilla maculata* (spotted lady beetle), a predator of eggs and larvae of the European corn borer [55] *via* fungal metabolites, which are detrimental upon feeding or antibiosis. A few fungal endophytes like *Neotyphodium*, infecting insects like *Rhopalosiphum padi* and *Metopopophium dirhodum* [56], *Acremonium strictum* infecting *Hypothenemus hampei* [57], and *Clonostachys rosea* and *B. bassiana* against *H. hampei* [58], have been reported.

ROLE OF MICROBIAL INSECTICIDES (MIS)

Microbial biopesticides are natural, easily available, biodegradable, less expensive, and least toxic to living organisms, and have variable mechanisms of action against pests. MIs always exhibit the following beneficial characteristics.

A. Microbes, utilized for microbial insecticide preparation, should be safe and nonpathogenic to humans, wildlife, and other unrelated organisms.
B. Microbial insecticides yield toxins that must have specificity for a single genus or species of insects only and be unaffected by the beneficial insects under-treated crops.

C. Microbial insecticides may be conjugated with synthetic (chemical-based) insecticides for better management; such products are comparatively susceptible to degradation and pose fewer hazards to humans, animals, and the environment. Therefore, their application is safe even at the time of crop harvest.
D. They not only control the pest problem but also encourage the soil microflora in treated areas, thus enhancing crop yields.

LIMITATION OF MICROBIAL INSECTICIDES

There are some naturally occurring limitations associated with the microbial insecticides:

i. Pest specificity: usually a single MI can control only a single or few species of insects in the field but is unable to affect other types of insects, while chemical insecticides may also have the same problem as they are not only killing predators but may be harmful to the beneficial population of insects in threatened crop areas.
ii. MIs are sensitive to desiccation, heat, and ultraviolet radiation exposure.
iii. MIs are pest-specific. So potential marketing of these products aides costs and are expensive.

CONCLUDING REMARKS

Currently, the population bomb pressurizes the farmers to fulfill the need for food crops, which is accompanied by chemical/synthetic pesticides and fertilizers, which are hazardous to human beings, animals, and the environment. Microbial pesticides/biopesticides are the best substitutes for haphazardly explored toxic chemical pesticides, which are not only detrimental to our health and the environment but also responsible for rising insect resistance to pesticides. Microbial pesticide demand is constantly increasing worldwide. Their use in integrated pest management is far more effective than individual applications. Therefore, the use of such integrated biopesticide applications, which are safe and effective with a wide host range and minimum application restrictions, will be fruitful in horticulture and agriculture in the future.

REFERENCES

[1] Usta C. Microorganisms in Biological Pest Control-A Review (Bacterial Toxin Application and Effect of Environmental Factors). In: Silva-Opps M, Ed. Current Progress in Biological Research. IntechOpen 2013; pp. 288-317.
[http://dx.doi.org/10.5772/55786]

[2] Menn JJ. Biopesticides: Has their time come? J Environ Sci Health B 1996; 31: 383-9.
[http://dx.doi.org/10.1080/03601239609372998]

[3] Dar SA, Khan ZH, Khan AA, *et al.* Biopesticides–Its Prospects and Limitations: An Overview. Perspective in Animal Ecology and Reproduction. In: Gupta VK, Ed. Perspective in Animal Ecology and Reproduction. New Delhi, India: Daya Publishing House 2011; vol 8: pp. 296-314.

[4] Bennett AE, Orrell P, Malacrinò A, *et al.* Fungal-Mediated Above–Below ground Interactions: The Community Approach, Stability, Evolution, Mechanisms, and Applications In: Ohgushi T, Wurst S, Johnson S, Eds. Aboveground–Belowground Community Ecology Ecological Studies (Analysis and Synthesis). Cham: Springer 2018; vol. 234: pp. 85-116.
[http://dx.doi.org/10.1007/978-3-319-91614-9_5]

[5] Pathak D, Yadav R, Kumar M. Microbial Pesticides: Development, Prospects and Popularization in India. In: Singh D, Singh H, Prabha R, Eds. Plant-Microbe Interactions in Agro-Ecological Perspectives. Singapore: Springer 2017; pp. 455-71.
[http://dx.doi.org/10.1007/978-981-10-6593-4_18]

[6] Matyjaszczyk E. Products containing microorganisms as a tool in integrated pest management and the rules of their market placement in the European Union. Pest Manag Sci 2015; 71: 1201-6.
[http://dx.doi.org/10.1002/ps.3986] [PMID: 25652108]

[7] Gouli VV, Marcdino JAP, Gouli SY, Eds. Microbial Pesticides-Biological Resources, Production and Application. 1st ed. Academic Press 2020; p. 348.

[8] Gupta A, Dikshit AK. Biopesticides: An ecofriendly approach for pest control. J Biopesticides 2010; 3: 186-8. [www.jbiopest.com/users/lw8/efiles/suman_gupta_v31.pdf]

[9] Shanker C, Solanki KR. Botanical insecticides: A historical perspective. Asian Agrihist 2000; 4: 221-32.

[10] Kachhawa D. Microorganisms as a biopesticides. J Entomol Zool Stud 2017; 5: 468-73. [www.entomoljournal.com/archives/2017/vol5issue3/PartG/5-3-16-781.pdf]

[11] Thakur N, Kaur S, Tomar P, *et al.* Microbial Biopesticides: Current Status and Advancement for Sustainable Agriculture and Environment. In: Rastegari AA, Yadav AN, Yadav N, Eds. New and Future Developments in Microbial Biotechnology and Bioengineering. Elsevier 2020; pp. 243-82.
[http://dx.doi.org/10.1016/B978-0-12-820526-6.00016-6]

[12] Lacey LA, Frutos R, Kaya HK, *et al.* Insect pathogens as biological control agents: Do they have a future? Biol Control 2001; 21: 230-48.
[http://dx.doi.org/10.1006/bcon.2001.0938]

[13] Ruiu L. Insect pathogenic bacteria in integrated pest management. Insects 2015; 6: 352-67.
[http://dx.doi.org/10.3390/insects6020352] [PMID: 26463190]

[14] Ruiu L. Microbial biopesticides in agroecosystems. Agronomy (Basel) 2018; 8: 235.
[http://dx.doi.org/10.3390/agronomy8110235]

[15] Sarwar M. Microbial insecticides- an ecofriendly effective line of attack for insect pests management. IJEART 2015; 1: 4-9.

[16] Ongena M, Jacques P. *Bacillus* lipopeptides: Versatile weapons for plant disease biocontrol. Trends Microbiol 2008; 16: 115-25.
[http://dx.doi.org/10.1016/j.tim.2007.12.009] [PMID: 18289856]

[17] Piggot PJ, Hilbert DW. Sporulation of *Bacillus subtilis*. Curr Opin Microbiol 2004; 7: 579-86.
[http://dx.doi.org/10.1016/j.mib.2004.10.001] [PMID: 15556029]

[18] Roh JY, Choi JY, Li MS, *et al. Bacillus thuringiensis* as a specific, safe, and effective tool for insect pest control. J Microbiol Biotechnol 2007; 17: 547-59. [www.jmb.or.kr/submission/Journal/017/JMB017-04-01.pdf]
[PMID: 18051264]

[19] Zhang X, Candas M, Griko NB, *et al.* A mechanism of cell death involving an adenylyl cyclase/PKA signaling pathway is induced by the Cry1Ab toxin of *Bacillus thuringiensis*. Proc Natl Acad Sci USA

2006; 103: 9897-902.
[http://dx.doi.org/10.1073/pnas.0604017103] [PMID: 16788061]

[20] Broderick NA, Robinson CJ, McMahon MD, *et al.* Contributions of gut bacteria to *Bacillus thuringiensis*-induced mortality vary across a range of Lepidoptera. BMC Biol 2009; 7: 11.
[http://dx.doi.org/10.1186/1741-7007-7-11] [PMID: 19261175]

[21] Chattopadhyay A, Bhatnagar NB, Bhatnagar R. Bacterial insecticidal toxins. Crit Rev Microbiol 2004; 30: 33-54.
[http://dx.doi.org/10.1080/10408410490270712] [PMID: 15116762]

[22] Marche MG, Mura ME, Falchi G, *et al.* Spore surface proteins of *Brevibacillus laterosporus* are involved in insect pathogenesis. Sci Rep 2017; 7: 43805.
[http://dx.doi.org/10.1038/srep43805] [PMID: 28256631]

[23] Zhang J, Hodgman TC, Krieger L, *et al.* Cloning and analysis of the first cry gene from *Bacillus popilliae*. J Bacteriol 1997; 179: 4336-41.
[http://dx.doi.org/10.1128/jb.179.13.4336-4341.1997] [PMID: 9209052]

[24] Bucher GE, Stephens JM. A disease of grasshoppers caused by the bacterium *Pseudomonas aeruginosa* (Schroeter) Migula. Can J Microbiol 1957; 3: 611-25.
[http://dx.doi.org/10.1139/m57-067] [PMID: 13437224]

[25] Scarpa B. Simplified method for the simultaneous performance of microbial phosphatase and lecithinase tests. Ann Sclavo 1959; 1: 655-8.

[26] Jackson TA, Pearson JF, O'Callaghan M, *et al.* Pathogen to product development of *Serratia entomophila* Enterobacteriaceae as a commercial biological control agent for the New Zealand grass grub *Costelytra zealandica*. In: Jackson TA, Glare TR, Eds. Use of Pathogens in Scarab Pest Management. Andover, UK: Hampshire, Intercept Ltd 1992; pp. 191-8.

[27] Landsberg MJ, Jones SA, Rothnagel R, *et al.* 3D structure of the *Yersinia entomophaga* toxin complex and implications for insecticidal activity. Proc Natl Acad Sci USA 2011; 108: 20544-9.
[http://dx.doi.org/10.1073/pnas.1111155108] [PMID: 22158901]

[28] Vodovar N, Vinals M, Liehl P, *et al. Drosophila* host defense after oral infection by an entomopathogenic *Pseudomonas* species. Proc Natl Acad Sci USA 2005; 102: 11414-9.
[http://dx.doi.org/10.1073/pnas.0502240102] [PMID: 16061818]

[29] Cordova-Kreylos AL, Fernandez LE, Koivunen M, *et al.* Isolation and characterization of *Burkholderia rinojensis* sp. nov., a non-*Burkholderia cepacia* complex soil bacterium with insecticidal and miticidal activities. Appl Environ Microbiol 2013; 79: 7669-78.
[http://dx.doi.org/10.1128/AEM.02365-13] [PMID: 24096416]

[30] Martin PAW, Hirose E, Aldrich JR. Toxicity of *Chromobacterium* subtsugae to southern green stink bug (Heteroptera: Pentatomidae) and corn rootworm (Coleoptera: Chrysomelidae). J Econ Entomol 2007; 100: 680-4.
[http://dx.doi.org/10.1603/0022-0493(2007)100[680:TOCSTS]2.0.CO;2] [PMID: 17598525]

[31] Asolkar R, Huang H, Koivunen M, *et al. Chromobacterium* bioactive compositions and metabolites. United States: Patent Application Publication 2016 US 2016/0095323 A1.

[32] Book AJ, Lewin GR, McDonald BR, *et al.* Cellulolytic *Streptomyces* strains associated with herbivorous insects share a phylogenetically linked capacity to degrade lignocellulose. Appl Environ Microbiol 2014; 80: 4692-701.
[http://dx.doi.org/10.1128/AEM.01133-14] [PMID: 24837391]

[33] Mertz FP, Yao RC. *Saccharopolyspora spinosa* sp. nov. isolated from soil collected in a sugar mill rum still. Int J Syst Bacteriol 1990; 40: 34-9.
[http://dx.doi.org/10.1099/00207713-40-1-34]

[34] Lewer P, Hahn DR, Karr LL, *et al.* Discovery of the butenyl-spinosyn insecticides: Novel macrolides from the new bacterial strain *Saccharopolyspora pogona*. Bioorg Med Chem 2009; 17: 4185-96.

[http://dx.doi.org/10.1016/j.bmc.2009.02.035] [PMID: 19324553]

[35] Sharaf EF. A potent chitinolytic activity of *Alternaria alternata* isolated from Egyptian black sand. Pol J Microbiol 2005; 54: 145-51.
[PMID: 16209108]

[36] Li Z, Wang Z, Peng G, *et al.* Regulation of extracellular acid phosphatase biosynthesis by culture conditions in entomopathogenic fungus *Metarhizium anisopliae* strain CQMa102. Ann Microbiol 2007; 57: 565-70.
[http://dx.doi.org/10.1007/BF03175356]

[37] Shakeri J, Foster HA. Proteolytic activity and antibiotic production by *Trichoderma harzianum* in relation to pathogenicity to insects. Enzyme Microb Technol 2007; 40: 961-8.
[http://dx.doi.org/10.1016/j.enzmictec.2006.07.041]

[38] Hoell IA, Klemsdal SS, Vaaje-Kolstad G, *et al.* Overexpression and characterization of a novel chitinase from *Trichoderma atroviride* strain P1. Biochim Biophys Acta Proteins Proteomics 2005; 1748: 180-90.
[http://dx.doi.org/10.1016/j.bbapap.2005.01.002] [PMID: 15769595]

[39] Quesada-Moraga E, Vey A. Bassiacridin, a protein toxic for locusts secreted by the entomopathogenic fungus *Beauveria bassiana*. Mycol Res 2004; 108: 441-52.
[http://dx.doi.org/10.1017/S0953756204009724] [PMID: 15209284]

[40] Melanie MM, Miranti M, Kasmara H, *et al.* Insecticidal activities of crude extact of *Metarhizium anisopliae* and conidia suspension against *Crocidolomia pavonana* fabricius. IOP Conf Ser Earth Environ Sci 2018; 166: 012017.
[http://dx.doi.org/10.1088/1755-1315/166/1/012017]

[41] Yankouskaya A. Application of biological insecticides pecilomicine-B for greenhouse pest control. Sodinink Darzinink 2009; 28: 249-58.

[42] Mazet I, Vey A, Hirsutellin A. A toxic protein produced *in vitro* by *Hirsutella thompsonii*. Microbiology (Reading) 1995; 141: 1343-8.
[http://dx.doi.org/10.1099/13500872-141-6-1343] [PMID: 7670635]

[43] Boonphong S, Kittakoop P, Isaka M, *et al.* A new antimycobacterial, 3 β-acetoxy-15 α,22-dihydroxyhopane, from the insect pathogenic fungus *Aschersonia tubulata*. Planta Med 2001; 67: 279-81.
[http://dx.doi.org/10.1055/s-2001-11997] [PMID: 11345704]

[44] Kumari BR, Vijayabharathi R, Srinivas S, *et al.* Microbes as interesting source of novel insecticides: A review. Afr J Biotechnol 2014; 13: 2582-92.
[http://dx.doi.org/10.5897/AJB2013.13003]

[45] Daisy BH, Strobel GA, Castillo U, *et al.* Naphthalene, an insect repellent, is produced by *Muscodor vitigenus*, a novel endophytic fungus. Microbiology (Reading) 2002; 148: 3737-41.
[http://dx.doi.org/10.1099/00221287-148-11-3737] [PMID: 12427963]

[46] Consigli RA, Tweeten KA, Anderson DK, *et al.* Granulosis viruses, with emphasis on the GV of the Indian meal moth, *Plodia interpunctella*. Adv Virus Res 1983; 28: 141-73.
[http://dx.doi.org/10.1016/S0065-3527(08)60723-X] [PMID: 6362364]

[47] Arthurs SP, Lacey LA, Fritts R Jr. Optimizing use of codling moth granulovirus: Effects of application rate and spraying frequency on control of codling moth larvae in Pacific Northwest apple orchards. J Econ Entomol 2005; 98: 1459-68.
[http://dx.doi.org/10.1093/jee/98.5.1459] [PMID: 16334311]

[48] Ignoffo CM, Couch TL. The Nucleopolyhedrosis Virus of *Heliothis* Species as a Microbial Pesticide. In: Burges HD, Ed. Microbial Control of Pests and Plant Diseases. London: Academic Press 1981; pp. 329-62.

[49] Mettenmeyer A. Viral insecticides hold promise for bio-control. Farming Ahead 2002; 124: 50-1.

[50] Ramakrishnan N, Kumar SK. Biological control of insects by pathogen and nematodes. Pesticides 1976; pp. 32-47.

[51] Weinzieri R, Henn T. Microbial insecticides, Circular 1295. Urbana-Champaign: Cooperative Extension Service, University of Illinois 1989; p. 24.

[52] Lewis EE, Clarke DJ. Nematode Parasites and Entomopathogens. In: Vega FE, Kaya HK, Eds. Insect Pathology. 2nd ed. Amsterdam, the Netherlands: Elsevier 2012; pp. 395-424.
[http://dx.doi.org/10.1016/B978-0-12-384984-7.00011-7]

[53] Eilenberg J, Vlak JM, Nielsen-LeRoux C, *et al.* Diseases in insects produced for food and feed. J Insects Food Feed 2015; 1: 87-102.
[http://dx.doi.org/10.3920/JIFF2014.0022]

[54] Lewis L, Berry EC, Obrycki JJ, *et al.* Aptness of insecticides (*Bacillus thuringiensis* and carbofuran) with endophytic *Beauveria bassiana*, in suppressing larval populations of the European corn borer. Agric Ecosyst Environ 1996; 57: 27-34.
[http://dx.doi.org/10.1016/0167-8809(95)01011-4]

[55] Pingel RL, Lewis LC. The fungus *Beauveria bassiana* (Balsamo) Vuillemin in a corn ecosystem: Its effect on the insect predator *Coleomegilla maculata* De Geer. Biol Control 1996; 6: 137-41.
[http://dx.doi.org/10.1006/bcon.1996.0017]

[56] Clement SL, Elberson LR, Bosque-Pérez NA, *et al.* Detrimental and neutral effects of wild barley-*Neotyphodium* fungal endophyte associations on insect survival. Entomol Exp Appl 2005; 114: 119-25.
[http://dx.doi.org/10.1111/j.1570-7458.2005.00236.x]

[57] Jallow MFA, Dugassa-Gobena D, Vidal S. Influence of an endophytic fungus on host plant selection by a polyphagous moth *via* volatile spectrum changes. Arthropod-Plant Interact 2008; 2: 53-62.
[http://dx.doi.org/10.1007/s11829-008-9033-8]

[58] Vega FE, Posada F, Catherine Aime M, *et al.* Entomopathogenic fungal endophytes. Biol Control 2008; 46: 72-82.
[http://dx.doi.org/10.1016/j.biocontrol.2008.01.008]

CHAPTER 6

Soil Microflora - A Potential Source of Antibiotics

S. Bharathi[1,*] and **Prajeesh P. Nair**[1]

[1] *Department of Microbiology, The Oxford College of Science, Bangalore, India*

Abstract: Soil is one of the principal ingredients of the universe, which supports all forms of life directly or indirectly. Soil consists of a mixture of various organic and inorganic matter, fluids, gases, and micro and macro-living systems and acts as a major living medium for a wide group of living organisms. One of the important soil inhabitants are microorganisms. Soil microorganisms can be categorized as bacteria, fungi, actinomycetes, protozoa, algae, and viruses. These microbes have varied features and functions. Most importantly, these microorganisms do not exist in isolation but interact with each other and contribute significantly to overall soil fertility. Many of these organisms have the capacity to produce antimicrobial substances as a defense mechanism to compete with other organisms for their survival and existence. Most of these antimicrobial substances, which are released as metabolites produced during trophophase as well as in idiophase, are medically significant in the treatment of many life-threatening infections in plants and animals. This chapter describes various soil microflora and their roles in the production of different kinds of antimicrobials. The primary goal is to familiarise readers with the various microflora found in soil and their ability to produce anti-microbial components.

Keywords: Actinomycetes, Antibiotics, Bacteria, Fungi, Isolation, Soil microflora.

INTRODUCTION

Certain groups of secondary metabolites released by specific groups of organisms possess antimicrobial properties against various microorganisms and are called antibiotics. These antimicrobial agents serve as a major compound in fighting against various infections caused by other microorganisms by either killing or inhibiting their growth and multiplication.

A variety of soil microorganisms are known to produce different types of antibiotics that are screened, isolated, purified and used against many life-threatening infections, mostly in human beings, animals and agriculture.

---

*Corresponding author **S. Bharathi**: Department of Microbiology, The Oxford College of Science, Bangalore, India; E-mail: bartiravi@gmail.com

The major groups of microbes capable of producing antibiotics include bacteria, fungi, and actinomycetes. These antibiotics serve as a primary defense mechanism against other groups of organisms in their vicinity for better survival [1].

Soil acts as a major habitat for a huge and diverse population of microorganisms because of its heterogeneous nature. It has been estimated that one gram of soil contains one to one hundred million fungi, fifty thousand to a million actinomycetes, one to ten million algae, ten thousand to fifty thousand protozoans, and many millions of bacteria [2]. The number and type of organisms vary depending on the type of the soil, the depth of the soil, climatic conditions, and various human, animal, and agricultural practices. The rhizosphere represents the thin layer of soil surrounding plant roots and the soil occupied by the roots, and it supports the growth of any number and variety of microorganisms [3]. The plant roots release a variety of biomolecules, including amino acids and sugars, that serve as food sources for microorganisms, greatly increasing microbial populations and activity in and around the rhizosphere [4]. Variation in the biotic and abiotic nature of soil makes its microbial inhabitants adapt and develop strategies for survival. The production of antimicrobial substances is one of the most powerful strategies employed in adaptation. The rhizosphere consists of three different zones based on their relative proximity to the root. The endorhizosphere consists of the endodermal along with the plant cortical region, in which the microbes interact significantly. It is followed by a region of rhizoplane, which is adjacent to the root and includes the root epidermis and the mucilage. The outer ectorhizosphere extends from the rhizoplane into the core soil with a comparatively lower number of microorganisms [5].

As discussed earlier, soil microorganisms are a major source of antibiotics and over a thousand different antibiotics have been isolated from various microflora, predominantly from bacteria, actinomycetes, and fungi. Many of the medically important antibiotics are produced by the genus *Bacillus*, filamentous fungi, actinomycetes, and other bacteria [6] (Table **1**).

Table 1. Some clinically important antibiotics produced from soil microbes.

Antibiotic	Source of Organism	Spectrum of Activity
Penicillin	*Penicillium chrysogenum*	Gram-positive bacteria
Cephalosporin	*Cephalosporium acremonium*	Broad spectrum
Bacitracin	*Bacillus subtilis, B. licheniformis*	Gram-positive bacteria
Polymyxin	*B. polymyxa*	Gram-positive and Gram-negative bacteria
Streptomycin	*Streptomyces griseus*	Gram-negative bacteria

(Table 1) cont.....

Antibiotic	Source of Organism	Spectrum of Activity
Kanamycin	*Streptomyces kanamyceticus*	Gram-positive bacteria, Gram-negative bacteria, and mycobacteria
Tetracycline	*Streptomyces rimosus*	Broad spectrum
Vancomycin	*Streptomyces orientalis*	Gram-positive bacteria
Neomycin	*Streptomyces fradiae*	Broad spectrum
Erythromycin	*Streptomyces erythreus*	Gram-positive bacteria
Amphotericin B	*Streptomyces nodosus*	Fungi
Trichomycin	*Streptomyces hachijoensis*	Fungi
Fusidic acid	*Acremonium fusidioides*	*Staphylococci* and Gram-negative bacteria
Cochliodinol	*Chaetomium cochlioides*	Fungi and bacteria
Polymyxin	*Bacillus polymyxa*	Gram-negative bacteria
Gramicidin	*Bacillus brevis*	Gram-positive bacteria
Zwittermicin	*Bacillus cereus*	Gram-positive, Gram-negative, and prokaryotic microorganism

ACTINOMYCETES - A POTENTIAL SOURCE OF ANTIBIOTICS

Actinomycetes are best known for their ability to produce antibiotics. Although most species in this group produce bioactive compounds, few selective forms are reported to synthesize the maximum types of antimicrobials. The best-known among them is *Streptomyces*, followed by other related actinomycetes, which include *Micromonospora, Actinomadura, Actinoplanes, Nocardia, Streptosporangium, Streptoverticillium, Thermoactinomyces*, etc.

Streptomyces are a group of Gram-positive bacteria that grow in different soil environments, and their shape resembles filamentous fungi. The typical characteristic feature *of Streptomyces* includes the formation of a layer of hyphae that differentiates into a chain of spores. They are predominantly found in soil and dead organic vegetation. Most *Streptomyces* spp. produce a characteristic earthy fragrance due to the production of a volatile metabolite called geosmin (Fig. **1**).

Apart from the ability to synthesize antibiotics, *Streptomyces* also has the capacity to produce other bioactive secondary metabolites such as antifungals, antivirals, antitumorals, anti-inflammatories, immunosuppressives, *etc*. The antibiotic production of these organisms is species-specific, and these metabolites are essential for Streptomyces sp. to compete with another group of organisms, even within the same genera [6] (Table **2**).

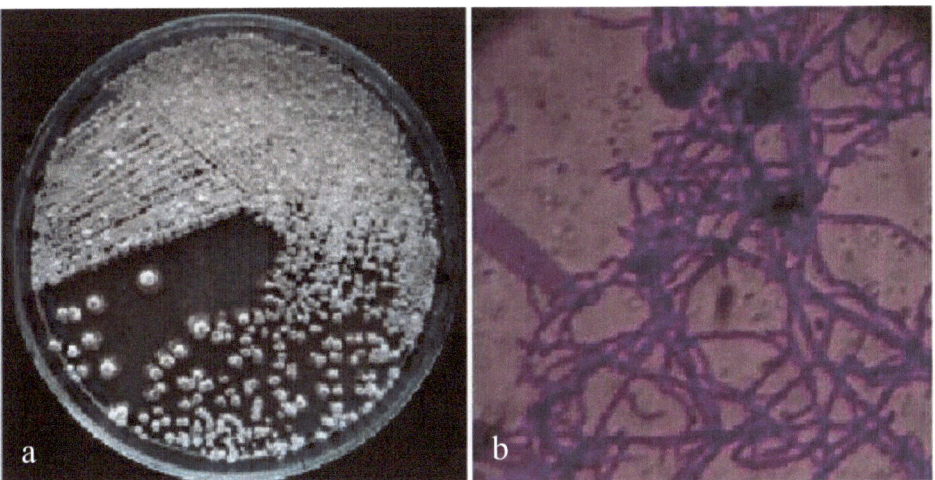

Fig. (1). (a) *Streptomyces* sp. grown on starch casein agar (SCA) and **(b)** Microscopic view of *Streptomyces* sp.

Table 2. Major classes of antibiotics produced by *Streptomyces* spp.

S. noursei	Macrolide, Polyenes, and Nystatin	Antifungal
S. nodosus	Amphotericin B	Antifungal
S. natalensis	Natamycin	Antifungal
S. griseus	Aminoglycosides	Antibacterial
S. fradiae	Neomycin	Antibacterial
S. kanamyceticus	Kanamycin	Antibacterial
S. erythraeus	Erythromycin	Antibacterial
S. rimosus	Tetracycline	Antibacterial
S. venezuelae	Chloramphenicol	Antibacterial
S. orientalis	Vancomycin	Antibacterial
S. cattleya	Thienamycin	Antibacterial
S. verticillus	Bleomycin	Anti-tumoral
S. lavendulae	Mitomycin	Anti-tumoral

ANTIBIOTIC PRODUCING BACTERIA

The most predominant bacteria found in soil are aerobic, Gram-positive, endospore-forming, and chemolithotrophic *Bacillus* spp. A number of members of the Bacillaceae family have the capacity to produce an array of antibiotics in an insoluble form and are widely preferred for the commercial production of various antibiotics.

Bacillus species produce different classes of antibiotics that share a full range of antimicrobial activity. Bacitracin, produced by *Bacillus licheniformis*, is a mixture of at least five polypeptides. It is active against many Gram-positive organisms, such as *Staphylococci*, *Streptococci*, *Corynebacterium*, and *Clostridia*. Gramicidin, which is produced by *Bacillus brevis*, is a linear polypeptide that inhibits phosphate group ATPase of other species. *Bacillus* generally produces polypeptide-type bacteriocin, and these antibiotics generally affect Gram-positive bacteria. The other antibiotic producers of this genus are *B. brevis* (*e.g.*, gramicidin and tyrothricin), *B. cereus* (*e.g.*, cerexin and zwittermicin), *B. circulans* (*e.g.*, circulin), *B. laterosporus* (*e.g.*, laterosporin), *B. licheniformis* (*e.g.*, bacitracin), *B. polymyxa* (*e.g.*, polymyxin and colistin), *B. pumilus* (*e.g.*, pumulin), and *B. subtilis* (*e.g.*, polymyxin, difficidin, subtilin, and mycobacillin) [7].

Another predominant group of microorganisms in the soil is *Pseudomonas* spp., which can produce an array of antimicrobial agents. The secondary metabolites released from various species of *Pseudomonas* are very effective against many of the phytopathogens and play a very important role as bioinoculants in agricultural biology. They are Gram-negative, aerobic bacilli with a single polar flagellum (Table 3).

FUNGI AS A SOURCE OF ANTIBIOTICS

Fungi are a potential source for the production of various pharmaceuticals. The first antibiotic was discovered from a group of soil fungi, *Penicillium notatum*, by Alexander Fleming in 1928. Many groups of fungi are known for the production of antibiotics, and they are widely used for industrial production (Table 4).

Table 3. Major groups of *Pseudomanas* spp. with antimicrobial property against various phytopathogens.

Organism	Antibiotic/Antimicrobial Agents	Target Phytopathogens
P. fluorescens	Phenazine-1-carboxylic acid	Broad spectrum- pathogenic fungi
P. fluorescens and *P. aureofaciens*	Phenazines	Broad spectrum- pathogenic bacteria & fungi
P. chlororaphis	Phenazine-1-carboxamide	Broad spectrum- pathogenic fungi
P. cepacia	Pyrrolnitrin	*Bipolaris maydis*

Table 4. Major groups of antibiotic producing fungi.

Fungi	Antibiotic	Target Group
P. notatum	Penicillin	Gram-positive bacteria
P. chrysogenum	Penicillin and isopenicillin	Gram-positive bacteria
Aspergillus proliferans	Proliferin	*Mycobacterium* spp.
Mucor remannianus	Ramycin	Bacteria
Psalliota campestris	Campestrin	Bacteria
Ustilago maydis	Ustilagic acid	Fungi
Aspergillus niger	Jawaharene	Bacteria and tumors
Gibberella baccata	Baccatin	Bacteria and tumors
Cephalosporium sp.	Cephalosporin	Gram-positive bacteria

CLASSIFICATION OF ANTIBIOTICS

Due to the developments in molecular chemistry, many antimicrobial substances are also chemically synthesized, like sulfa drugs. Drugs used in infection control are classified into two groups: chemically synthesized synthetic drugs and those produced by microbes - antibiotics. The antibiotics can be further classified based on their structure and the mechanism of action exerted on their target organisms [1].

Classification Based on Their Structure

Penicillin-Derived Antibiotics

They have a beta-lactam ring in their chemical structure, *e.g.*, semisynthetic penicillin, cephalosporins, ampicillin, *etc.*

Amino Glycosidase Antibiotics

These include the group with amino sugars in glycosidic linkages, *e.g.*, streptomycin, neomycin, kanamycin, gentamicin, paromomycin, tobramycin, and amikacin.

Macrolide Antibiotics

They possess a macrocyclic lactone ring attached to sugar residues, *e.g.*, erythromycin, spiramycin, and oleandomycin.

Tetracycline Antibiotics

These are derivatives of polycyclic napthacene carboxamide, *e.g.*, tetracycline, chlortetracycline, oxytetracycline, demeclocycline, and minocycline.

Peptide Antibiotics

This group of antibiotics has peptide-linked amino acids, *e.g.*, bacitracin, gramicidin, and polymyxins.

Antifungal Antibiotics

These are categorized based on polyenes containing rings with conjugated double bond systems, *e.g.*, nystatin and amphotericin B. Other antifungal antibiotics without a polyene double bond system include fluorocytosine, clotrimazole, and griseofulvin.

Chloramphenicol Antibiotics

It is a class by itself, a nitrobenzene derivative of dichloroacetic acid.

Other Groups of Antibiotics

These have varied structures and include cycloserine, fusidic acid, novobiocin, prasinomycin, streptomycin, and vancomycin.

Classification Based on Their Mechanism of Action

Inhibition of Cell Wall Synthesis

The groups of antimicrobial drugs that interfere with cell wall synthesis are the β-lactam antibiotics, which are bactericidal. They interfere with the peptidoglycan biosynthesis and, in turn, cell wall synthesis and kill the bacteria.

The major group of cell wall inhibiting antibiotics includes penicillin, aminopenicillins, ampicillin, amoxicillin, methicillin, oxacillin, ploxacillin, carbenicillin, cephalosporins, vancomycin, carbapenems, bacitracin, *etc*.

Protein Synthesis Inhibitors

These are groups of antibiotics that inhibit prokaryotic translation in its various stages, including initiation, elongation, translocation, termination, and proofreading. The following are some examples:

- Linezolid prevents the formation of the initiation complex.
- Tetracycline and tigecycline prevent the binding of aminoacyl tRNA to the ribosomes.
- Aminoglycosides interfere with the proofreading process and reduce the efficiency of the translation process.
- Chloramphenicol inhibits the elongation process by binding to 50 s ribosomal subunits.
- Macrolides, clindamycin, and aminoglycosides have evidence of inhibition of ribosomal translocation.
- Streptogramins cause the premature release of the peptide chain.

Inhibitors of Nucleic Acid Synthesis

Certain antibiotics interfere with nucleic acid and arrest the replication and transcription process of the central dogma. Quinolones bind with the topoisomerases, especially DNA gyrase, and inhibit DNA synthesis during replication. The fluoro-quinolones are second-generation quinolones that include levofloxacin, norfloxacin, and ciprofloxacin. They inhibit topoisomerase (DNA gyrase) enzymes, which inhibit the relaxation of supercoiled DNA and promote the breakage of double-stranded DNA. Rifampicin blocks the initiation of RNA synthesis by specifically inhibiting bacterial RNA polymerase and arrests the transcription process. Some groups of antibiotics interfere with RNA synthesis, such as doxorubicin, which is not only specific to prokaryotes but interferes with mammalian systems too. These are most often used as anticancer drugs that attack malignant cells as well as normal cells.

Other Biomolecule Inhibitors

Isoniazid, a group of antimicrobial agents effective against a group of microorganisms, synthesizes mycolic acid for its metabolism, like *Mycobacterium* sp., by inhibiting mycolic acid. Trimethoprim and sulfamethoxazole inhibit folic acid synthesis, essential for nucleic acid biosynthesis in microorganisms.

ISOLATION OF ANTIBIOTIC-PRODUCING ORGANISMS FROM SOIL

Crowded Plate Technique

This is one of the simplest and most efficient methods for identifying and isolating antibiotic-producing microorganisms from the soil. To perform the same, suitable soil samples should be collected and serially diluted to reduce the microbial load for optimum isolation. These diluted samples are plated on nutrient agar media and incubated. The plates showing approximately 300 colonies have to be selected for the screening of antibiotic-producing microorganisms. Further, the plates should be carefully observed for any clear zones around the colonies produced in the nutrient agar media, which are considered inhibitory zones due to the release of antimicrobial substances by a specific group of microorganisms. These colonies should be selected, sub-cultured, and purified by streaking in selective media. These organisms are then subjected to strain improvement by physical, chemical, or recombinant methods to create high efficiency in producing the desired antibiotics. To detect the specificity of the antibiotics produced by the organism, the isolates are inoculated in a medium containing different test organisms and incubated overnight. The clear zone produced in different media indicates the efficiency of the organism in producing the desired antibiotic. The diameter of the inhibitory zones will give an approximate idea of the amount of antibiotics released, which should be further subjected to antibiotic assay by the tube dilution method. By various media formulation tests, a suitable fermentation media has to be designed for further industrial production of antibiotics [1, 8] (Fig. 2).

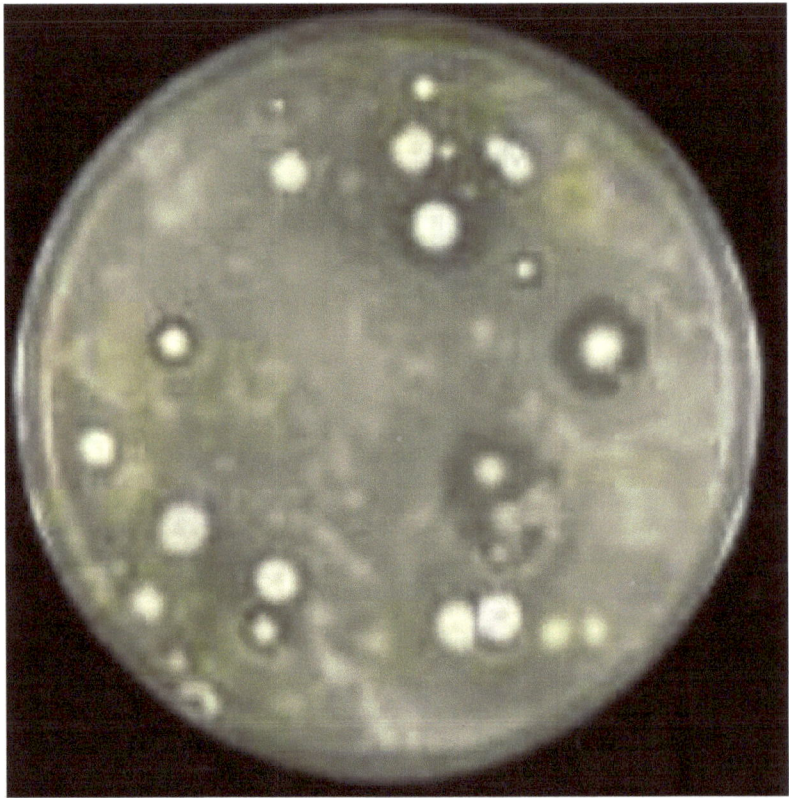

Fig. (2). Crowded plate technique showing growth of soil microflora on nutrient agar media.

A GENERALIZED PROTOCOL FOR THE PRODUCTION OF ANTIBIOTICS

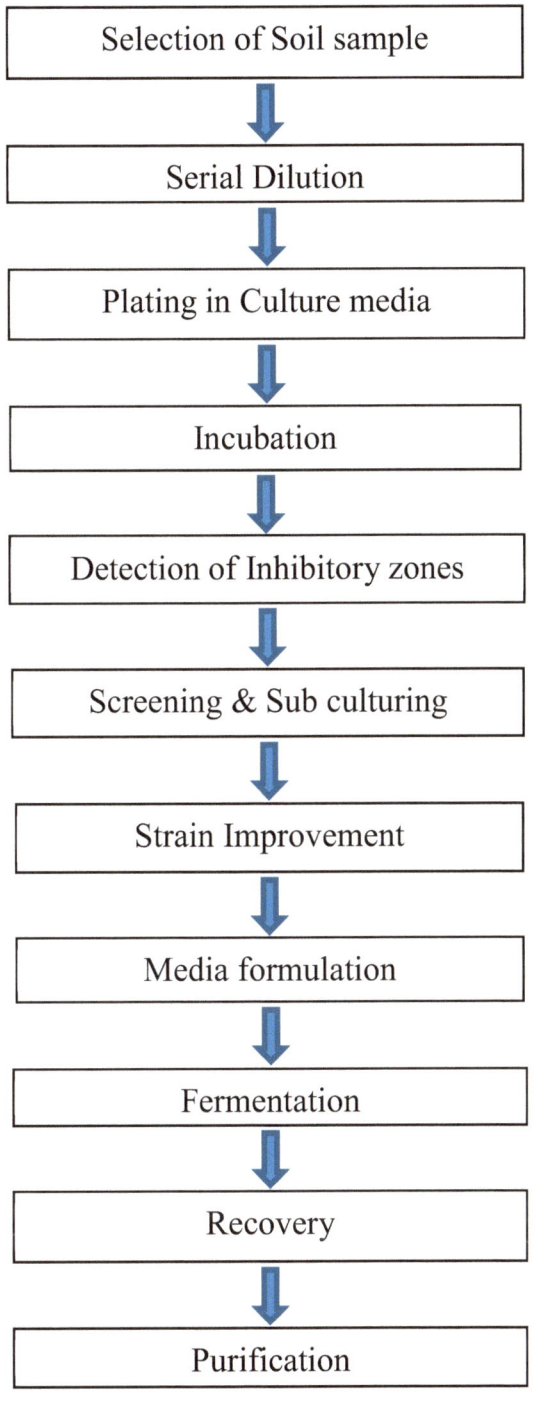

CONCLUDING REMARKS

Numerous groups of microorganisms are known to produce a large variety of antimicrobial substances, which can be extracted *in vitro* and can be applied against various deadly pathogenic infections and diseases in plants, vertebrates, and invertebrates. The major groups of microorganisms known to produce antibiotics include bacteria, fungi, and actinomycetes. The bacterial genera *Bacillus* and *Streptomyces*, along with the fungal genera *Penicillium* and *Cephalosporium*, are commonly found in soil. The genus *Streptomyces* is the most prolific antibiotic producer in soil. The synthesis of these antimicrobial substances serves as a natural defense mechanism for the microorganisms to better survive in their desired substratum. Soil serves as the major habitat for these microbes due to its heterogeneous nature. In a particular condition, soil acts as the perfect incubator and nutrient medium for the abundant growth and multiplication of these microorganisms with the highest metabolic rate. The rise in population and the need for improved aquatic and agricultural resources created a huge demand for antibiotics to keep the ecosystem healthy. The various biotechnological approaches made it possible to successfully isolate, cultivate, and extract these antibiotics for the welfare of human beings.

REFERENCES

[1] Patel AH, Ed. Industrial Microbiology. India: Trinity Press, Laxmi Publications 2011; p. 512.

[2] Gottlieb D. The production and role of antibiotics in soil. J Antibiot (Tokyo) 1976; 29: 987-1000.
[http://dx.doi.org/10.7164/antibiotics.29.987] [PMID: 994333]

[3] Geetanjali JP. Antibiotic production by rhizospheric soil microflora-a review. IJPSR 2016; 7: 4304-14.

[4] Sayyed RZ, Ilyas N, Tabassum B, *et al.* Plausible Role of Plant Growth-Promoting *Rhizobacteria* in Future Climatic Scenario. In: Sobti R, Arora N, Kothari R, Eds. Environmental Biotechnology: For Sustainable Future. Singapore: Springer 2019; pp. 175-97.
[http://dx.doi.org/10.1007/978-981-10-7284-0_7]

[5] Sandhya MVS, Ramyakrishna E, Divya P, *et al.* Isolation of antibiotic producing bacteria from soil. Int J Appl Biol Pharm Technol 2015; 6: 46-51.

[6] Manasa M, Poornima G, Abhipsa V, *et al.* Antimicrobial and Antioxidant Potential of *Streptomyces* sp. RAMPP- 065 isolated from Kudremukh soil, Karnataka, India. Sci Technol Arts Res J 2012; 1: 39-44.
[http://dx.doi.org/10.4314/star.v1i3.98798]

[7] Tortora GJ, Funke BR, Case CL, Eds. Microbiology: An Introduction. 11th ed. Pearson 2014; p. 960.

[8] Sethi S, Kumar R, Gupta S. Antibiotic production by microbes isolated from soil. IJPSR 2013; 4: 2967-73.
[http://dx.doi.org/10.13040/IJPSR.0975-8232.4(8).2967-73]

CHAPTER 7

Role of Nonpathogenic Strains in Rhizosphere

Rana Muhammad Sabir Tariq[1,*], **Maheen Tariq**[1], **Sarah Ali**[2], **Shahan Aziz**[2] and **Jam Ghulam Mustafa**[2]

[1] *Punjab College Sambrial, Wazirabad Road, University Road, Sambrial, Sialkot, Pakistan.*
[2] *Department of Agriculture, Agribusiness Management, University of Karachi, Karachi, Pakistan.*

Abstract: As the world's population is increasing rapidly, there is an urgent need to increase crop production. To achieve this goal, an eco-friendly alternative to chemical fertilizers and pesticides is required. Several types of microbes have been identified inhabiting the plant rhizosphere, such as nitrogen-fixing bacteria, plant growth-promoting rhizobacteria, fungi, proteobacteria, mycoparasitic and mycorrhizal fungi. These microorganisms not only influence the growth and development of plants but also suppress pathogenic microbes near plant roots through several different mechanisms. Non-symbiotic microbes play a crucial role in the biogeochemical cycling of organic and inorganic phosphorus (P) near the root zone *via* solubilization and mineralization of P from total soil phosphorus. Additionally, some non-pathogenic microbes have also been reported to induce systemic resistance in plants, which is phenotypically similar to pathogen-induced systemic acquired resistance (SAR). The present review summarizes the latest knowledge on the role of non-pathogenic strains of microbiomes residing in the rhizosphere and their commercial applications.

Keywords: Arbuscular mycorrhizal fungi, Non pathogenic, Rhizhobacteria.

INTRODUCTION

When we talk about the effects of the rhizosphere on plants and other organisms, the possibilities are nearly endless. For the sake of study, we are concerned here with only the beneficial aspects, otherwise known as non-pathogenic effects. It has been known for more than 300 years that endophytic microbes reside in plant roots. However, only recently has the value of these microbes been fully understood (Fig. **1**). Endophytes in plant roots improve crop yield and environmental buffering. Symbiotic interaction among plants and microbes was first reported by Malpighi in 1697. He described the formation of galls or nodules composed of both bacteria (Rhizobiaceae) and plant cells on the roots of plants as the mechanism to fix atmospheric nitrogen (N_2) and make ammonia (NH_3) availa-

---

[*] **Corresponding author Rana Muhammad Sabir Tariq:** Punjab College Sambrial, Wazirabad Road, University Road, Sambrial, Sialkot, Pakistan; E-mail: principal.sambrial@pgc.edu.pk

ble for leguminous plants [1]. In 1988, after almost two decades, plant roots were reported to be colonized by fungi called arbuscular mycorrhizal fungi (AMF). These fungi increase the productivity of plants by symbiotic means [2]. Soil-borne fungi *Trichoderma* were shown to have the ability to control pathogenic fungi of agricultural crops in the 1920s and 1930s [3]. Various other microbes are associated with roots. However, some true plant symbionts (Rhizobiaceae, *Trichoderma* strains, and AMF) have demonstrated the ability to serve as components of enhanced plant holobiomes [4].

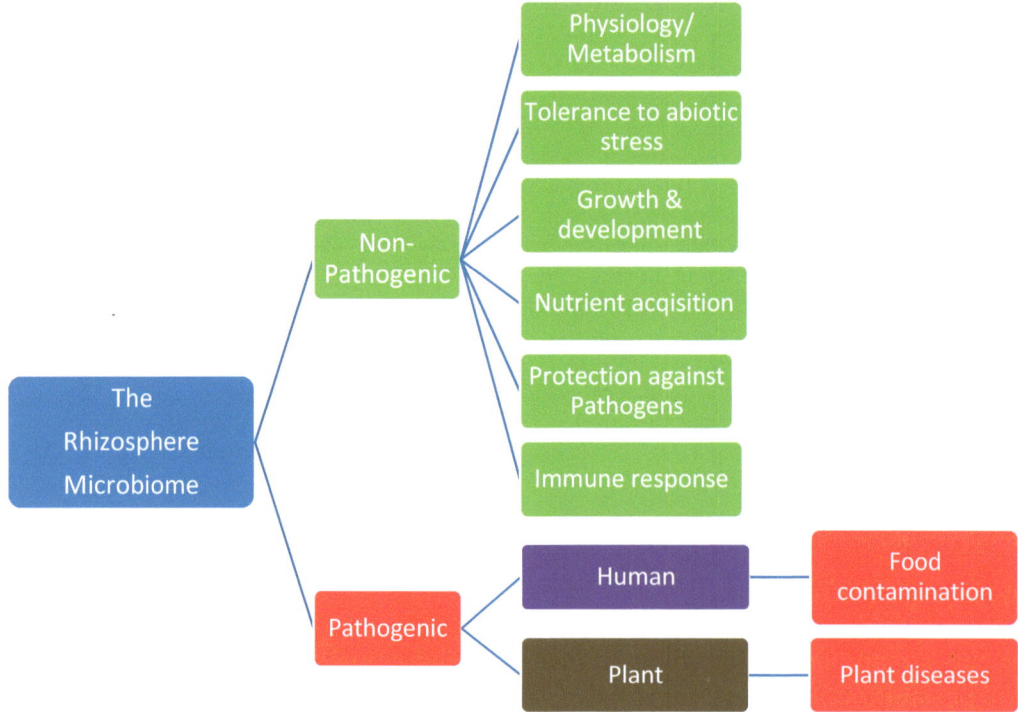

Fig. (1). Various effects of rhizosphere microbiome pathogenic (disease causing) as well as non-pathogenic (non-disease causing) strains.

IMPORTANCE OF NON-PATHOGENIC STRAINS OF RHIZOSPHERE

Rhizobiaceae

Bacteria in the Rhizobiaceae family are well known as nitrogen-fixers. However, two other groups of bacteria, comprising cyanobacteria and *Frankia*, can also fix atmospheric nitrogen in symbiosis with plants. Cyanobacteria can fix atmospheric nitrogen in a wide range of plants [5]. Bacteria in the Frankiaceace family fix di-nitrogen from the atmosphere like other symbiotic fixers of nitrogen. *Frankia* is capable of nodulating about 24 genera related to 8 varied angiosperm families [6].

Leguminous plants produce complex galls or nodules around the nitrogen-fixing cells referred to as bacteroids. These nodules are occupied with a protein containing iron called leghemoglobin that removes oxygen, consequently facilitating the hypoxic environment essential for the bacteroids to fix atmospheric nitrogen, reducing N_2 to NH_3 [7]. The symbiotic interactions of bacteria and plants that result in gall formation are highly specific, and only certain bacterial species or strains can colonize a specific host plant [8].

Trichoderma

The genus *Trichoderma* contains several species and strains isolated from forests and agricultural soil that can be easily cultured *in vitro*. *Trichoderma* sporulation is usually green and some strains possess coconut odour due to a biologically active compound called 6-pentyl-α-pyrone, which is volatile in nature [9]. Recently, *Trichoderma* isolates have been recognized as being capable of acting as endophytic plant symbionts. The strains become endophytic in roots. However, major variations in gene expression appear in shoots. These variations modify plant physiology and may result in the enhancement of photosynthetic efficiency, nitrogen uptake, and biotic and abiotic stress resistance. Generally, the overall result of these effects is excellent for plant growth and yield [10].

Arbuscular Mycorrhizal Fungi (AMF)

This association of host plant and fungus roots evolved around 400 million years ago. AMF are obligate fungi and require host plants for their growth and survival. Fungi form a complex with the roots of most terrestrial plants and involve chemical signalling pathways [11]. After the AM fungi enter the host root cortical cells, a prepenetration apparatus leads the fungi to an appropriate cell determined by the host. Once inside the host cell, the fungus forms lobed structures called arbuscules (sometimes vesicles) that are present between the host cell wall and cell membrane [12]. AM fungi are reported to induce the expression of ammonium and phosphorus transporter proteins, especially in nutrient-deficient soil conditions [13]. AMF also interacts with bacteria in the soil, which leads to significant effects in agriculture. In this interaction, soil bacteria bind with the fungi and degrade the fungal cell wall by releasing volatile compounds. This process affects the AMF's gene expression, which results in improved performance and yield [14].

BENEFICIAL ROLE OF NON-PATHOGENIC STRAINS IN RHIZOSPHERE

Studies in the past few years have shown the adverse effects of agrochemicals used to ensure the high yield of agricultural crops. Agrochemicals, generally pesticides, are divided into four major classes: fungicides, insecticides, herbicides, and rodenticides [15]. These pesticides can be further branched into carbamates, organophosphorus, organochlorines, neonicotinoids, and pyrethroids according to their chemical arrangements and modes of action [16]. While ensuring a high yield of crops, the overuse of agrochemicals has also become the leading cause of polluting soil, air and water [17]. Agrochemicals induce resistance in pest species and have a negative impact on beneficial insects, aquatic animals, and human life [18, 19]. Therefore, to reduce the harmful effects of synthetic pesticides, an efficient alternative is necessary to manage different pathogen species on agricultural crops. Biological control emerges as one of the best alternatives to agrochemicals. It refers to the management of plant pest populations by naturally occurring living organisms. Biological control agents of plant pathogens, called antagonistic microorganisms, belong to various groups of fungi and bacteria, while plant pests include plant pathogens, weeds, and insects [20]. The global biological pesticide market is anticipated to reach $4,556.37 million by 2019, at an annual growth rate of 15.30% from 2014 to 2019 [21].

Soilborne fungal pathogens are the most destructive pests of agro-ecosystems. The ban of the effective fumigant, methyl bromide, against soilborne pathogens and the occurrence of fungicide resistance in pathogenic fungal strains are the primary reasons for finding an alternative to control plant diseases [22]. The microbes are not only helpful for nutrient uptake as symbionts but also have the potential as biocontrol agents. *Trichoderma* species can establish themselves in a wide range of soil types globally due to their natural resistance to various toxic compounds, such as fungicides, insecticides, and herbicides [23]. They can persist for a longer time on plant propagules, making them suitable as a successful bio-control agent as the establishment of an ample population on the target site is the key limiting factor for the application of any biocontrol agent [24]. The presence of *Trichoderma* in soil controls the population of other pathogenic microorganisms, consequently making it a superior biocontrol agent [25].

Several species of *Rhizobium* inhibit the development of various soilborne pathogens, such as *Rhizobium* sp. RS12 is effective against *Fusarium solani* and *Macrophomina phaseolina*, causing disease in chickpeas [26]. Arbuscular mycorrhizal fungi (*Glomus* spp.) are reported to have the potential to inhibit the growth of *Rhizoctonia solani*, a causal agent for damping-off disease on cucumber [27]. Arbuscular mycorrhizal fungi also showed promising results when used with the endophytic fungal strain *Epicoccum nigrum* ASU11 against pathogenic bacteria *Pectobacterium carotovora*, responsible for the blackleg disease of potatoes.

Rhizosphere microorganisms have various effects on plants' health, growth, and disease. They can directly or indirectly affect the productivity and composition of flora [28 - 31]. As a result, the diversity and richness of microbial species are heavily influenced by surface diversity and productivity [32, 33]. Wagg and coworkers [34] proposed that the variation in the underground atmosphere helps the plant survive the variables of the environment while being diverse and productive. Researchers have shown that soil and rhizosphere can act as environmental indicators (bioindicators), determining the quality of soil, as they are sensitive to changes in their surrounding ecosystem, including perturbation and ecological stress.

We will be discussing the impact of non-pathogenic strains in the rhizosphere on the health, growth, and immunity of plants and other organisms as well [35]. Good rhizospheres with non-pathogenic strains promote rapid plant growth while also making them immune to pathogens through a variety of mechanisms [36]. Several factors, supported by rhizobacteria belonging to various classes like proteobacteria, firmicutes (*Bacillus* and *Pseudomonas*), Deuteromycetes (*Gliocladium* and *Trichoderma*) and the order Sebacinales (*Piriformospora*) [37], are root growth, biofertilization, abiotic stress, disease control mechanism, and rhizoremediation [38].

Fortunately, studies are being conducted to obtain more information on the matter regarding different soils and the effects of rhizosphere inhabitants and their effects on 'rare' or 'unusual' microbes like Planctomycetes [39, 40].

EFFECTS OF NON-PATHOGENIC RHIZOSPHERE STRAINS ON PLANTS' HEALTH

Non-pathogenic strains of microbes in the rhizosphere have a strong ability to change the nutritional status of plants [41]. A well-studied example is a nitrogen-fixing rhizosphere and mycorrhizal fungi, which help the plant with phosphorus uptake, a critical element in plant growth [42]. Mycorrhizal fungi are important for mineral and nutrient translocation from the ground to the plant body [43], for generating stable aggregates and physical structure (of soil) [44, 45], and for confining infectious agents [46, 47]. Subsequently, *Bradyrhizobium*, *Rhizobium*, and numerous nitrogen-fixing bacteria living inside rhizosphere colonies have been detected [52]. A highly genetically diverse symbiotic species has been found in the western Amazon, in the cowpea rhizosphere [52]. According to 16S rRNA gene sequencing and glasshouse experiments, *Burkholderia*, *Achromobacter*, *Bradyrhizobium*, and *Rhizobium* species are highly modified and thus able to form nodulates in cowpea, increasing the efficiency of nitrogen-fixation in plants [53]. Irrespective of the elaborated research on the project for the relocation of N_2

fixing and specified legume symbionts with the help of rhizobia for the plant, from the agricultural point of view, it has not been attained so far. Geurts and colleagues [54] recently reported that studying the fundamental differences between superficially similar reactions in cells persuaded by mycorrhizal fungi and *Rhizobium* will aid in achieving the "old dream".

Rhizosphere microorganisms can also increase the uptake of micro-elements like iron in the soil. Iron is present in abundance but in alkaline or neutral conditions and is found in a non-soluble state, which is not necessarily available for microbial species development. Because of the paucity of readily available iron in various microbiomes, along with the harmfulness of elevated amounts of iron, by secretion of siderophores, bacteria regulate their intracellular iron concentrations [55 - 58].

Two strategies are being specifically followed by the host plants to overcome the limitation of iron 1) Increased solubilization of ions, inorganic in nature, in the rhizosphere; and 2) Compensation of phytosiderophores content by first removal action, followed by back transportation through a definite uptake system into the root tissue [59]. Rice uses similar strategies to keep its iron levels stable [59, 60]. The utilization of iron chelates involving microbial siderophores has been proposed as a potent source for iron acquisition by various studies [61]. Fluorescent pseudomonads are the species that have been reported to promote iron uptake in dicotyledonous and graminaceous plants [62]. A well-known rhizobacterium also acts as an effective plant machinery for iron acquisition [63]. Most rhizobacterial species acquire energy through organic compound assimilation. The limited availability of degradable organic compounds limits bacterial growth in soil [64 - 66]. Most bacterial colonies in the rhizosphere, particularly mineral weathering bacteria [67, 68], contribute significantly to plant nutrition requirements and normal plant maturation in nutrient-deficient soil [69, 70].

An attack of soilborne pathogens on plant roots is also defended by microbial communities in the rhizosphere. The process of antibiosis is mainly used against these pathogens besides competition for essential elements, parasitism [71], structured opposition [72], and other virulence interference affected by quorum sensation [73]. Microsites, nutrient rhizobacteria, and rhizosphere fungi are the active agents responsible for the production of antimicrobial metabolic entities [74, 75]. *Trichoderma* species have been found to produce a considerable number of antimicrobial compounds. Antibiotic compounds produced by various bacterial and fungal biocontrol agents exhibit different degrees of antimicrobial activity. For example, agrocin 84 produced by *Agrobacterium* presents antibiotic action towards various related genera, while non-proteinaceous antibiotics show wide-spectral results [76]. Volatile organic compounds (VOCs) are the most important mediators of plant growth and interactions between plants and microorganisms [77]. However, a small proportion of VOCs are produced as metabolites but play a vital role in long-distance communication, better solubility, and normal plant growth [78].

The influence of abiotic circumstances on VOC production was shown by Weise and coworkers [79]. They also showed the disparity in the form of spectral graphs and data figures of the volatile compounds produced by bacterial species in solid agar and broth culture. The available pool of root exudates influences the production of volatiles. These volatile compounds exhibit strong antifungal activity when supplemented with amino acids [80].

Recently, Chernin and coworkers [81] proposed that metabolites from bacteria suppress the process of transcription of lactone synthase genes, thus efficiently impeding the quorum sensing of various bacteria. Dimethyl-sulfide has been recognized as a potent interferent. VOCs also play a role in comprehensive plant control [82] and plant magnification promotion [83, 84]. The rhizosphere microbial community also contributes to plant immunity modulation [85]. Jasmonic acid and ethylene are productive phytohormones for the regulation of systemic resistance responses [86]. Often, some bacteria undertake the salicylic acid pathway to perform regulatory activity [87]. *Bacillus cereus* AR156, a rhizobacterium, is reported to adopt both the strategic paths [88]. Finally, quorum-sensitive molecules can also stimulate various plant responses, which involve the actuation of various defensive genes, *e.g.*, Pdf1.2, WRKY22, MPK6, MPK3, *etc*.

In contrast, Cartieaux and coworkers [89] noticed that strains of bacteria that utilize the JA/ET pathways result in very few changes in transcription in *Arabidopsis* [90]. However, other bacterial species that use the SA pathway lead to prominent metabolic and transcriptional changes. The integration of transcription profiles and metabolic changes performed by Weston and coworkers [91, 92] on 2 strains of *P. fluorescens* showed that rhizobacteria exert profound effects on physiology, immunity, and metabolic properties of plants and aggravate the synthesis of unknown and secondary metabolites [93].

The rhizosphere microbiome develops the possibilities for the host plants to survive stress. *Achromobacter piechaudii* ARV8 significantly increases the biomass of pepper and tomato when exposed to drought stress [94, 95] as well as during flooding to stabilise plant growth. Jha and coworkers [96] discovered a few diazotrophic halotolerant bacteria from *Salicornia brachiate*. The isolates have been recognized as *Cronobacter sakazakii*, *Zhihengliuella* sp., *Haererehalobacter* sp., *Brevibacterium casei*, *Halomonas* sp., *Rhizobium radiobacter*, *Vibrio* sp., *Pseudomonas* spp., *Mesorhizobium* sp., and *Brachybacterium saurashtrense*. Berg and coworkers [97] reviewed the mode of action of rhizobacteria with a productive effect on (1) Osmolyte accumulation for water homeostasis, (2) Source-sink relationships modulation for plant energetics, (3) Alteration of host physiology, physical barrier modification, reduction in toxic ion accumulation and regulation of root toxic ion uptake, and (4) Hormonal signalling alteration for crop salt tolerance. Low temperatures have a profound effect on nitrogen fixation and nodule formation, but the high arctic native legumes exhibit comparable rates for both the phenomena as in temperate climates. *Serratia proteamaculans* vitalizes soybean growth at 15°C [98, 99].

Katiyar and coworkers [100] nominated various mutants, which are cold-tolerant strains of *P. fluorescens*, because of the phosphorus solubilization capability that promotes plant growth at low temperatures. They found 2 cold-tolerant mutants, which efficiently solubilize phosphorus at 10°C more efficiently than other wild types. Trivedi and coworkers [101] identified two mutants of 115, which were more competent than wild-type strains of *P. corrugata* in the solubilization of phosphorus at 4 to 28°C. Both studies did not mention the genes corresponding to the solubilization of phosphorus and cold tolerance. Different abiotic factors that may adequately affect plant growth are the concentration of toxic compounds and pH. Soil having a low pH is the major challenge in most production systems utilized worldwide.

Raudales and coworkers [102] investigated pH stress and discovered that foliar lesions stimulated on corn cultivating in low-pH soil were significantly reduced with plants treated with DAPG (2,4-diacetylphloroglucinol)-producing *P. fluorescens* strain. It was discovered that DAPG contributes to pathogen control as well as acts to enhance abiotic stress parameters. Various pollutants have led to studies showing that there are more effective methods for bioremediation than incineration and excavation [103]. A combination of bioaugmentation and phytoremediation, rhizoremediation, is a promising approach to tidying up polluted sites.

Conclusively, rhizosphere microbiome members, which can relieve biotic and abiotic stress on plants, provide a biologically sound substitute for genetic engineering as well as plant breeding. Nevertheless, the effective application of microbial inoculants is still in its early stages because of numerous constraints involving variable effectiveness throughout the environment and various plants, restricted shelf-life, and various processes in different countries to solve these issues. More fundamental understanding is required of how beneficial rhizosphere microorganisms interact with plants, what changes occur in the plant, and how useful microorganisms influence population dynamics as well as pathogenic microorganism virulence.

INDUSTRIALIZATION OF NON-PATHOGENIC STRAINS OF RHIZOSPHERE

In 1979, the United States Environmental Protection Agency [EPA] registered the first bacterial biocontrol agent named *Agrobacterium radiobacter* strain K84 for the management of crown gall. After ten years, the first fungal biocontrol agent, *Trichoderma harzianum* ATCC 20476, was registered by the EPA for plant disease management. Over the last few years, a large number of microbial biocontrol products have made the successful transition from lab to market. These products, however, account for only 1% of agricultural management strategies [104].

It is well-known that biocontrol agents require specific conditions to perform efficiently. One of the chief elements of a successful biocontrol product is formulation. Failures caused by the irrational use of effective products are usually taken as proof that these products are unreliable and non-effective as compared to chemical pesticides [105].

Commercialization of biocontrol agents requires a large array of steps comprising identification, isolation of pure cultures, characterization, designing of formulation, mass production, analyzing the efficacy of the product, registration, and then marketing. All of these steps necessitate a large number of inputs, making biocontrol agent commercialization an extremely costly process [106].

CONCLUDING REMARKS

To achieve the goal of zero hunger with an eco-friendly approach, there is a need for an alternative to chemical fertilizers and pesticides. The rhizosphere is a region around plant roots harboured by thousands of non-pathogenic microbes that add up to the protection and productivity of crop plants. General species include nitrogen-fixing bacteria, PGPR, mycoparasitic, and mycorrhizal fungi. These microbes have great potential to be explored as industrial biofertilizers and biopesticides. Various studies have proven this phenomenon already. There is a need to augment these microbes on an industrial scale. Commercial production of these microbes will surely replace the big pesticide industries.

REFERENCES

[1] Harman GE, Uphoff N. Symbiotic root-endophytic soil microbes improve crop productivity and provide environmental benefits. Scientifica (Cairo) 2019; 2019: 9106395.
[http://dx.doi.org/10.1155/2019/9106395] [PMID: 31065398]

[2] Berch SM, Massicotte HB, Tackaberry LE. Re-publication of a translation of 'The vegetative organs of *Monotropa hypopitys* L.' published by F. Kamienski in 1882, with an update on *Monotropa* mycorrhizas. Mycorrhiza 2005; 15: 323-32.
[http://dx.doi.org/10.1007/s00572-004-0334-1] [PMID: 15549481]

[3] Anjum MZ, Hayat S, Ghazanfar MU, *et al.* Does seed priming with *Trichoderma* isolates have any impact on germination and seedling vigor of wheat. Int J Botany Stud 2020; 5: 65-8.
[https://www.botanyjournals.com/assets/archives/2020/vol5issue2/5-1-38-125.pdf]

[4] Hacquard S. Disentangling the factors shaping microbiota composition across the plant holobiont. New Phytol 2016; 209: 454-7.
[http://dx.doi.org/10.1111/nph.13760] [PMID: 26763678]

[5] Lindström K, Mousavi SA. Effectiveness of nitrogen fixation in rhizobia. Microb Biotechnol 2020; 13: 1314-35.
[http://dx.doi.org/10.1111/1751-7915.13517] [PMID: 31797528]

[6] Satyanarayana T, Johri BN, Prakash A, Eds. Microorganisms in Sustainable Agriculture and Biotechnology. Netherlands: Springer 2012; p. 829.
[http://dx.doi.org/10.1007/978-94-007-2214-9]

[7] Jones KM, Kobayashi H, Davies BW, *et al.* How rhizobial symbionts invade plants: The *Sinorhizobium–Medicago* model. Nat Rev Microbiol 2007; 5: 619-33.
[http://dx.doi.org/10.1038/nrmicro1705] [PMID: 17632573]

[8] Andrews M, Andrews ME. Specificity in legume-rhizobia symbioses. Int J Mol Sci 2017; 18: 705.
[http://dx.doi.org/10.3390/ijms18040705] [PMID: 28346361]

[9] Brotman Y, Kapuganti JG, Viterbo A. *Trichoderma*. Curr Biol 2010; 20: R390-1.
[http://dx.doi.org/10.1016/j.cub.2010.02.042] [PMID: 20462476]

[10] Harman GE, Herrera-Estrella AH, Horwitz BA, *et al.* Special issue: *Trichoderma*-from basic biology to biotechnology. Microbiology (Reading) 2012; 158: 1-2.
[http://dx.doi.org/10.1099/mic.0.056424-0] [PMID: 22210803]

[11] Roth R, Paszkowski U. Plant carbon nourishment of arbuscular mycorrhizal fungi. Curr Opin Plant Biol 2017; 39: 50-6.
[http://dx.doi.org/10.1016/j.pbi.2017.05.008] [PMID: 28601651]

[12] Smith SE, Smith FA. Fresh perspectives on the roles of arbuscular mycorrhizal fungi in plant nutrition and growth. Mycologia 2012; 104: 1-13.
[http://dx.doi.org/10.3852/11-229] [PMID: 21933929]

[13] Shaikh AM, Sharma I, Sharma V, *et al.* Arbuscular mycorrhizal fungi: A sustainable tool for agriculture. IJRAR 2019; 6: 661-79. [http://ijrar.org/viewfull.php?&_id=IJRAR19K8465]

[14] Miransari M. Interactions between arbuscular mycorrhizal fungi and soil bacteria. Appl Microbiol Biotechnol 2011; 89: 917-30.
[http://dx.doi.org/10.1007/s00253-010-3004-6] [PMID: 21104242]

[15] Rossetti MF, Stoker C, Ramos JG. Agrochemicals and neurogenesis. Mol Cell Endocrinol 2020; 510: 110820.
[http://dx.doi.org/10.1016/j.mce.2020.110820] [PMID: 32315720]

[16] Ren XM, Kuo Y, Blumberg B. Agrochemicals and obesity. Mol Cell Endocrinol 2020; 515: 110926.
[http://dx.doi.org/10.1016/j.mce.2020.110926] [PMID: 32619583]

[17] Sellare J, Meemken EM, Qaim M. Fair trade, agrochemical input use, and effects on human health and the environment. Ecol Econ 2020; 176: 106718.
[http://dx.doi.org/10.1016/j.ecolecon.2020.106718]

[18] Mamun MSA, Ahmed M. Prospect of indigenous plant extracts in tea pest management. Int J Agric Res Innov Technol 2013; 1: 16-23.

[http://dx.doi.org/10.3329/ijarit.v1i1-2.13924]

[19] Raza M, Hussain F, Lee JY, *et al*. Groundwater status in Pakistan: A review of contamination, health risks, and potential needs. Crit Rev Environ Sci Technol 2017; 47: 1713-62.
[http://dx.doi.org/10.1080/10643389.2017.1400852]

[20] Ghazanfar MU, Raza M, Raza W, *et al*. *Trichoderma* as potential biocontrol agent, its exploitation in agriculture: A review. Plant Protec 2018; 2: 109-35. [https://esciencepress.net/journals/index.php/PP/article/view/3142/1571]

[21] Singh S, Kumar R, Yadav S, *et al*. Effect of bio-control agents on soil borne pathogens: A review. J Pharmacogn Phytochem 2018; 7: 406-11. [https://www.phytojournal.com/archives/2018/vol7issue3/PartF/7-2-600-894.pdf]

[22] Bonanomi G, Antignani V, Pane C, *et al*. Suppression of soilborne fungal diseases with organic amendments. J Plant Pathol 2007; 89: 311-24.

[23] Khalid SA. *Trichoderma* as biological control weapon against soil borne plant pathogens. Afr J Biotechnol 2017; 16: 2299-306.
[http://dx.doi.org/10.5897/AJB2017.16270]

[24] Kumar G, Maharshi A, Patel J, *et al*. *Trichoderma*: A potential fungal antagonist to control plant diseases. SATSA Mukhapatra Annu Tech Issue 2017; 21: 206-18.

[25] Adnan M, Islam W, Shabbir A, *et al*. Plant defense against fungal pathogens by antagonistic fungi with *Trichoderma* in focus. Microb Pathog 2019; 129: 7-18.
[http://dx.doi.org/10.1016/j.micpath.2019.01.042] [PMID: 30710672]

[26] Das K, Prasanna R, Saxena AK. Rhizobia: A potential biocontrol agent for soilborne fungal pathogens. Folia Microbiol (Praha) 2017; 62: 425-35.
[http://dx.doi.org/10.1007/s12223-017-0513-z] [PMID: 28285373]

[27] Aljawasim BD, Khaeim HM, Manshood MA. Assessment of arbuscular mycorrhizal fungi (*Glomus* spp.) as potential biocontrol agents against damping-off disease *Rhizoctonia solani* on cucumber. JCP 2020; 9: 141-7. [http://jcp.modares.ac.ir/article-3-33473-en.html]

[28] Schnitzer SA, Klironomos JN, Lambers JHR, *et al*. Soil microbes drive the classic plant diversity–productivity pattern. Ecology 2011; 92: 296-303.
[http://dx.doi.org/10.1890/10-0773.1] [PMID: 21618909]

[29] van der Heijden MGA, Boller T, Wiemken A, *et al*. Different arbuscular mycorrhizal fungal species are potential determinants of plant community structure. Ecology 1998; 79: 2082-91.
[http://dx.doi.org/10.1890/0012-9658(1998)079[2082:DAMFSA]2.0.CO;2]

[30] Van Der Heijden MGA, Bakker R, Verwaal J, *et al*. Symbiotic bacteria as a determinant of plant community structure and plant productivity in dune grassland. FEMS Microbiol Ecol 2006; 56: 178-87.
[http://dx.doi.org/10.1111/j.1574-6941.2006.00086.x] [PMID: 16629748]

[31] van der Heijden MGA, Bardgett RD, van Straalen NM. The unseen majority: soil microbes as drivers of plant diversity and productivity in terrestrial ecosystems. Ecol Lett 2008; 11: 296-310.
[http://dx.doi.org/10.1111/j.1461-0248.2007.01139.x] [PMID: 18047587]

[32] Hooper DU, Chapin FS III, Ewel JJ, *et al*. Effects of biodiversity on ecosystem functioning: A consensus of current knowledge. Ecol Monogr 2005; 75: 3-35.
[http://dx.doi.org/10.1890/04-0922]

[33] Lau JA, Lennon JT. Evolutionary ecology of plant–microbe interactions: soil microbial structure alters selection on plant traits. New Phytol 2011; 192: 215-24.
[http://dx.doi.org/10.1111/j.1469-8137.2011.03790.x] [PMID: 21658184]

[34] Wagg C, Jansa J, Schmid B, *et al*. Belowground biodiversity effects of plant symbionts support aboveground productivity. Ecol Lett 2011; 14: 1001-9.

[http://dx.doi.org/10.1111/j.1461-0248.2011.01666.x] [PMID: 21790936]

[35] Lugtenberg B, Kamilova F. Plant-growth-promoting rhizobacteria. Annu Rev Microbiol 2009; 63: 541-56.
[http://dx.doi.org/10.1146/annurev.micro.62.081307.162918] [PMID: 19575558]

[36] Raaijmakers JM, Paulitz TC, Steinberg C, et al. The rhizosphere: A playground and battlefield for soilborne pathogens and beneficial microorganisms. Plant Soil 2009; 321: 341-61.
[http://dx.doi.org/10.1007/s11104-008-9568-6]

[37] Kogel KH, Franken P, Hückelhoven R. Endophyte or parasite – what decides? Curr Opin Plant Biol 2006; 9: 358-63.
[http://dx.doi.org/10.1016/j.pbi.2006.05.001] [PMID: 16713330]

[38] Qiang X, Weiss M, Kogel KH, et al. Piriformospora indica - a mutualistic basidiomycete with an exceptionally large plant host range. Mol Plant Pathol 2012; 13: 508-18.
[http://dx.doi.org/10.1111/j.1364-3703.2011.00764.x] [PMID: 22111580]

[39] Hol WHG, de Boer W, Termorshuizen AJ, et al. Reduction of rare soil microbes modifies plant-herbivore interactions. Ecol Lett 2010; 13: 292-301.
[http://dx.doi.org/10.1111/j.1461-0248.2009.01424.x] [PMID: 20070364]

[40] Jogler C, Waldmann J, Huang X, et al. Identification of proteins likely to be involved in morphogenesis, cell division, and signal transduction in Planctomycetes by comparative genomics. J Bacteriol 2012; 194: 6419-30.
[http://dx.doi.org/10.1128/JB.01325-12] [PMID: 23002222]

[41] Richardson AE, Barea JM, McNeill AM, et al. Acquisition of phosphorus and nitrogen in the rhizosphere and plant growth promotion by microorganisms. Plant Soil 2009; 321: 305-39.
[http://dx.doi.org/10.1007/s11104-009-9895-2]

[42] Hawkins HJ, Johansen A, George E. Uptake and transport of organic and inorganic nitrogen by arbuscular mycorrhizal fungi. Plant Soil 2000; 226: 275-85.
[http://dx.doi.org/10.1023/A:1026500810385]

[43] Gianinazzi S, Gollotte A, Binet MN, et al. Agroecology: The key role of arbuscular mycorrhizas in ecosystem services. Mycorrhiza 2010; 20: 519-30.
[http://dx.doi.org/10.1007/s00572-010-0333-3] [PMID: 20697748]

[44] Miller RM, Jastrow JD. Mycorrhizal Fungi Influence Soil Structure. In: Kapulnik Y, Douds DD, Eds. Arbuscular Mycorrhizas: Physiology and Function. Dordrecht: Springer 2000; pp. 3-18.
[http://dx.doi.org/10.1007/978-94-017-0776-3_1]

[45] Degens BP, Sparling GP, Abbott LK. Increasing the length of hyphae in a sandy soil increases the amount of water-stable aggregates. Appl Soil Ecol 1996; 3: 149-59.
[http://dx.doi.org/10.1016/0929-1393(95)00074-7]

[46] Pozo MJ, Azcón-Aguilar C. Unraveling mycorrhiza-induced resistance. Curr Opin Plant Biol 2007; 10: 393-8.
[http://dx.doi.org/10.1016/j.pbi.2007.05.004] [PMID: 17658291]

[47] Whipps JM. Ecological and Biotechnological Considerations in Enhancing Disease Biocontrol. In: Vurro M, Gressel J, Butt T, et al., Eds., Enhancing Biocontrol Agents and Handling Risks. Ohmsha, USA: IOS Press 2001; pp. 43-51.

[48] Brundrett MC. Coevolution of roots and mycorrhizas of land plants. New Phytol 2002; 154: 275-304.
[http://dx.doi.org/10.1046/j.1469-8137.2002.00397.x] [PMID: 33873429]

[49] Smith SE, Read DJ, Eds. Mycorrhizal Symbiosis. New York: Academic Press 1997; p. 800.

[50] Kapulnik Y, Douds DD Jr, Eds. Arbuscular Mycorrhizas: Physiology and Function. Dordrecht: Springer 2000; p. 372.
[http://dx.doi.org/10.1007/978-94-017-0776-3]

[51] Varma A, Bertold H, Eds. Mycorrhiza: Structure, Function, Molecular Biology and Biotechnology. Berlin, Heidelberg: Springer-Verlag 1999; p. 704.
[http://dx.doi.org/10.1007/978-3-662-03779-9]

[52] Zehr JP, Jenkins BD, Short SM, *et al.* Nitrogenase gene diversity and microbial community structure: A cross-system comparison. Environ Microbiol 2003; 5: 539-54.
[http://dx.doi.org/10.1046/j.1462-2920.2003.00451.x] [PMID: 12823187]

[53] Azarias Guimarães A, Duque Jaramillo PM, Simão Abrahão Nóbrega R, *et al.* Genetic and symbiotic diversity of nitrogen-fixing bacteria isolated from agricultural soils in the western Amazon by using cowpea as the trap plant. Appl Environ Microbiol 2012; 78: 6726-33.
[http://dx.doi.org/10.1128/AEM.01303-12] [PMID: 22798370]

[54] Geurts R, Lillo A, Bisseling T. Exploiting an ancient signalling machinery to enjoy a nitrogen fixing symbiosis. Curr Opin Plant Biol 2012; 15: 438-43.
[http://dx.doi.org/10.1016/j.pbi.2012.04.004] [PMID: 22633856]

[55] Lindsay WL, Schwab AP. The chemistry of iron in soils and its availability to plants. J Plant Nutr 1982; 5: 821-40.
[http://dx.doi.org/10.1080/01904168209363012]

[56] Andrews SC, Robinson AK, Rodríguez-Quiñones F. Bacterial iron homeostasis. FEMS Microbiol Rev 2003; 27: 215-37.
[http://dx.doi.org/10.1016/S0168-6445(03)00055-X] [PMID: 12829269]

[57] Buckling A, Harrison F, Vos M, *et al.* Siderophore-mediated cooperation and virulence in *Pseudomonas aeruginosa*. FEMS Microbiol Ecol 2007; 62: 135-41.
[http://dx.doi.org/10.1111/j.1574-6941.2007.00388.x] [PMID: 17919300]

[58] Hider RC, Kong X. Chemistry and biology of siderophores. Nat Prod Rep 2010; 27: 637-57.
[http://dx.doi.org/10.1039/b906679a] [PMID: 20376388]

[59] Walker EL, Connolly EL. Time to pump iron: iron-deficiency-signaling mechanisms of higher plants. Curr Opin Plant Biol 2008; 11: 530-5.
[http://dx.doi.org/10.1016/j.pbi.2008.06.013] [PMID: 18722804]

[60] Lemanceau P, Bauer P, Kraemer S, *et al.* Iron dynamics in the rhizosphere as a case study for analyzing interactions between soils, plants and microbes. Plant Soil 2009; 321: 513-35.
[http://dx.doi.org/10.1007/s11104-009-0039-5]

[61] Marschner H, Römheld V. Strategies of plants for acquisition of iron. Plant Soil 1994; 165: 261-74.
[http://dx.doi.org/10.1007/BF00008069]

[62] Shirley M, Avoscan L, Bernaud E, *et al.* Comparison of iron acquisition from Fe–pyoverdine by strategy I and strategy II plants. Botany 2011; 89: 731-5.
[http://dx.doi.org/10.1139/b11-054]

[63] Zhang H, Sun Y, Xie X, *et al.* A soil bacterium regulates plant acquisition of iron *via* deficiency-inducible mechanisms. Plant J 2009; 58: 568-77.
[http://dx.doi.org/10.1111/j.1365-313X.2009.03803.x] [PMID: 19154225]

[64] Aldén L, Demoling F, Bååth E. Rapid method of determining factors limiting bacterial growth in soil. Appl Environ Microbiol 2001; 67: 1830-8.
[http://dx.doi.org/10.1128/AEM.67.4.1830-1838.2001] [PMID: 11282640]

[65] Demoling F, Figueroa D, Bååth E. Comparison of factors limiting bacterial growth in different soils. Soil Biol Biochem 2007; 39: 2485-95.
[http://dx.doi.org/10.1016/j.soilbio.2007.05.002]

[66] Rousk J, Bååth E. Fungal and bacterial growth in soil with plant materials of different C/N ratios. FEMS Microbiol Ecol 2007; 62: 258-67.
[http://dx.doi.org/10.1111/j.1574-6941.2007.00398.x] [PMID: 17991019]

[67] Puente ME, Bashan Y, Li CY, *et al.* Microbial populations and activities in the rhizoplane of rock-weathering desert plants. I. Root colonization and weathering of igneous rocks. Plant Biol 2004; 6: 629-42.
[http://dx.doi.org/10.1055/s-2004-821100] [PMID: 15375735]

[68] Calvaruso C, Turpault MP, Leclerc E, *et al.* Impact of ectomycorrhizosphere on the functional diversity of soil bacterial and fungal communities from a forest stand in relation to nutrient mobilization processes. Microb Ecol 2007; 54: 567-77.
[http://dx.doi.org/10.1007/s00248-007-9260-z] [PMID: 17546519]

[69] Leveau JHJ, Uroz S, de Boer W. The bacterial genus *Collimonas*: Mycophagy, weathering and other adaptive solutions to life in oligotrophic soil environments. Environ Microbiol 2010; 12: 281-92.
[http://dx.doi.org/10.1111/j.1462-2920.2009.02010.x] [PMID: 19638176]

[70] Mapelli F, Marasco R, Balloi A, *et al.* Mineral–microbe interactions: Biotechnological potential of bioweathering. J Biotechnol 2012; 157: 473-81.
[http://dx.doi.org/10.1016/j.jbiotec.2011.11.013] [PMID: 22138043]

[71] Druzhinina IS, Seidl-Seiboth V, Herrera-Estrella A, *et al. Trichoderma*: The genomics of opportunistic success. Nat Rev Microbiol 2011; 9: 749-59.
[http://dx.doi.org/10.1038/nrmicro2637] [PMID: 21921934]

[72] Conrath U. Systemic acquired resistance. Plant Signal Behav 2006; 1: 179-84.
[http://dx.doi.org/10.4161/psb.1.4.3221] [PMID: 19521483]

[73] Lin YH, Xu JL, Hu J, *et al.* Acyl-homoserine lactone acylase from *Ralstonia* strain XJ12B represents a novel and potent class of quorum-quenching enzymes. Mol Microbiol 2003; 47: 849-60.
[http://dx.doi.org/10.1046/j.1365-2958.2003.03351.x] [PMID: 12535081]

[74] Hoffmeister D, Keller NP. Natural products of filamentous fungi: Enzymes, genes, and their regulation. Nat Prod Rep 2007; 24: 393-416.
[http://dx.doi.org/10.1039/B603084J] [PMID: 17390002]

[75] Harman GE, Howell CR, Viterbo A, *et al. Trichoderma* species - opportunistic, avirulent plant symbionts. Nat Rev Microbiol 2004; 2: 43-56.
[http://dx.doi.org/10.1038/nrmicro797] [PMID: 15035008]

[76] Gross H, Loper JE. Genomics of secondary metabolite production by *Pseudomonas* spp. Nat Prod Rep 2009; 26: 1408-46.
[http://dx.doi.org/10.1039/b817075b] [PMID: 19844639]

[77] Bailly A, Weisskopf L. The modulating effect of bacterial volatiles on plant growth. Plant Signal Behav 2012; 7: 79-85.
[http://dx.doi.org/10.4161/psb.7.1.18418] [PMID: 22301973]

[78] Wheatley RE. The consequences of volatile organic compound mediated bacterial and fungal interactions. Antonie van Leeuwenhoek 2002; 81: 357-64.
[http://dx.doi.org/10.1023/A:1020592802234] [PMID: 12448734]

[79] Weise T, Kai M, Gummesson A, *et al.* Volatile organic compounds produced by the phytopathogenic bacterium *Xanthomonas campestris* pv. *vesicatoria* 85-10. Beilstein J Org Chem 2012; 8: 579-96.
[http://dx.doi.org/10.3762/bjoc.8.65] [PMID: 22563356]

[80] Schrey SD, Schellhammer M, Ecke M, *et al.* Mycorrhiza helper bacterium *Streptomyces* AcH 505 induces differential gene expression in the ectomycorrhizal fungus *Amanita muscaria*. New Phytol 2005; 168: 205-16.
[http://dx.doi.org/10.1111/j.1469-8137.2005.01518.x] [PMID: 16159334]

[81] Chernin L, Toklikishvili N, Ovadis M, *et al.* Quorum-sensing quenching by rhizobacterial volatiles. Environ Microbiol Rep 2011; 3: 698-704.
[http://dx.doi.org/10.1111/j.1758-2229.2011.00284.x] [PMID: 23761359]

[82] Ryu CM, Farag MA, Hu CH, *et al.* Bacterial volatiles promote growth in *Arabidopsis*. Proc Natl Acad Sci USA 2003; 100: 4927-32.
[http://dx.doi.org/10.1073/pnas.0730845100] [PMID: 12684534]

[83] Blom D, Fabbri C, Eberl L, *et al.* Volatile-mediated killing of *Arabidopsis thaliana* by bacteria is mainly due to hydrogen cyanide. Appl Environ Microbiol 2011; 77: 1000-8.
[http://dx.doi.org/10.1128/AEM.01968-10] [PMID: 21115704]

[84] Blom D, Fabbri C, Connor EC, *et al.* Production of plant growth modulating volatiles is widespread among rhizosphere bacteria and strongly depends on culture conditions. Environ Microbiol 2011; 13: 3047-58.
[http://dx.doi.org/10.1111/j.1462-2920.2011.02582.x] [PMID: 21933319]

[85] De Vleesschauwer D, Höfte M. Rhizobacteria-induced systemic resistance. Adv Bot Res 2009; 51: 223-81.
[http://dx.doi.org/10.1016/S0065-2296(09)51006-3]

[86] Zamioudis C, Pieterse CMJ. Modulation of host immunity by beneficial microbes. Mol Plant Microbe Interact 2012; 25: 139-50.
[http://dx.doi.org/10.1094/MPMI-06-11-0179] [PMID: 21995763]

[87] Maurhofer M, Keel C, Schnider U, *et al.* Influence of enhanced antibiotic production in *Pseudomonas fluorescens* strain CHA0 on its disease suppressive capacity. Phytopathology 1992; 82: 190-5.
[http://dx.doi.org/10.1094/Phyto-82-190]

[88] Niu DD, Liu HX, Jiang CH, *et al.* The plant growth-promoting rhizobacterium *Bacillus cereus* AR156 induces systemic resistance in *Arabidopsis thaliana* by simultaneously activating salicylate- and jasmonate/ethylene-dependent signaling pathways. Mol Plant Microbe Interact 2011; 24: 533-42.
[http://dx.doi.org/10.1094/MPMI-09-10-0213] [PMID: 21198361]

[89] Cartieaux F, Contesto C, Gallou A, *et al.* Simultaneous interaction of *Arabidopsis thaliana* with *Bradyrhizobium* Sp. strain ORS278 and *Pseudomonas syringae* pv. *tomato* DC3000 leads to complex transcriptome changes. Mol Plant Microbe Interact 2008; 21: 244-59.
[http://dx.doi.org/10.1094/MPMI-21-2-0244] [PMID: 18184068]

[90] van de Mortel JE, de Vos RCH, Dekkers E, *et al.* Metabolic and transcriptomic changes induced in *Arabidopsis* by the rhizobacterium *Pseudomonas fluorescens* SS101. Plant Physiol 2012; 160: 2173-88.
[http://dx.doi.org/10.1104/pp.112.207324] [PMID: 23073694]

[91] Weston DJ, Pelletier DA, Morrell-Falvey JL, *et al. Pseudomonas fluorescens* induces strain-dependent and strain-independent host plant responses in defense networks, primary metabolism, photosynthesis, and fitness. Mol Plant Microbe Interact 2012; 25: 765-78.
[http://dx.doi.org/10.1094/MPMI-09-11-0253] [PMID: 22375709]

[92] Weston LA, Ryan PR, Watt M. Mechanisms for cellular transport and release of allelochemicals from plant roots into the rhizosphere. J Exp Bot 2012; 63: 3445-54.
[http://dx.doi.org/10.1093/jxb/ers054] [PMID: 22378954]

[93] Jorquera MA, Shaharoona B, Nadeem SM, *et al.* Plant growth-promoting rhizobacteria associated with ancient clones of creosote bush (*Larrea tridentata*). Microb Ecol 2012; 64: 1008-17.
[http://dx.doi.org/10.1007/s00248-012-0071-5] [PMID: 22639075]

[94] Mayak S, Tirosh T, Glick BR. Plant growth-promoting bacteria that confer resistance to water stress in tomatoes and peppers. Plant Sci 2004; 166: 525-30.
[http://dx.doi.org/10.1016/j.plantsci.2003.10.025]

[95] Mayak S, Tirosh T, Glick BR. Plant growth-promoting bacteria confer resistance in tomato plants to salt stress. Plant Physiol Biochem 2004; 42: 565-72.
[http://dx.doi.org/10.1016/j.plaphy.2004.05.009] [PMID: 15246071]

[96] Jha B, Gontia I, Hartmann A. The roots of the halophyte *Salicornia brachiata* are a source of new

halotolerant diazotrophic bacteria with plant growth-promoting potential. Plant Soil 2012; 356: 265-77.
[http://dx.doi.org/10.1007/s11104-011-0877-9]

[97] Berg G, Grube M, Schloter M, *et al.* Unraveling the plant microbiome: Looking back and future perspectives. Front Microbiol 2014; 5: 148.
[http://dx.doi.org/10.3389/fmicb.2014.00148] [PMID: 24926286]

[98] Zhang F, Lynch DH, Smith DL. Impact of low root temperatures in soybean [*Glycine max.* (L.) Merr.] on nodulation and nitrogen fixation. Environ Exp Bot 1995; 35: 279-85.
[http://dx.doi.org/10.1016/0098-8472(95)00017-7]

[99] Zhang F, Dashti N, Hynes RK, *et al.* Plant growth promoting rhizobacteria and soybean [*Glycine max* (L.) Merr.] nodulation and nitrogen fixation at suboptimal root zone temperatures. Ann Bot (Lond) 1996; 77: 453-60.
[http://dx.doi.org/10.1006/anbo.1996.0055]

[100] Katiyar V, Goel R. Solubilization of inorganic phosphate and plant growth promotion by cold tolerant mutants of *Pseudomonas fluorescens*. Microbiol Res 2003; 158: 163-8.
[http://dx.doi.org/10.1078/0944-5013-00188] [PMID: 12906389]

[101] Trivedi P, Sa T. *Pseudomonas corrugate* (NRRL B-30409) mutants increased phosphate solubilization, organic acid production, and plant growth at lower temperatures. Curr Microbiol 2008; 56: 140-4.
[http://dx.doi.org/10.1007/s00284-007-9058-8] [PMID: 18026795]

[102] Raudales RE, Stone E, McSpadden Gardener BB. Seed treatment with 2,4-diacetylphloroglucino--producing pseudomonads improves crop health in low-pH soils by altering patterns of nutrient uptake. Phytopathology 2009; 99: 506-11.
[http://dx.doi.org/10.1094/PHYTO-99-5-0506] [PMID: 19351246]

[103] Kuiper I, Lagendijk EL, Bloemberg GV, *et al.* Rhizoremediation: A beneficial plant-microbe interaction. Mol Plant Microbe Interact 2004; 17: 6-15.
[http://dx.doi.org/10.1094/MPMI.2004.17.1.6] [PMID: 14714863]

[104] Junaid JM, Dar NA, Bhat TA, *et al.* Commercial biocontrol agents and their mechanism of action in the management of plant pathogens. Int J Modern Plant Anim Sci 2013; 1: 39-57.

[105] Leggett M, Leland J, Kellar K, *et al.* Formulation of microbial biocontrol agents – an industrial perspective. Can J Plant Pathol 2011; 33: 101-7.
[http://dx.doi.org/10.1080/07060661.2011.563050]

[106] Thambugala KM, Daranagama DA, Phillips AJL, *et al.* Fungi *vs.* fungi in biocontrol: An overview of fungal antagonists applied against fungal plant pathogens. Front Cell Infect Microbiol 2020; 10: 604923.
[http://dx.doi.org/10.3389/fcimb.2020.604923] [PMID: 33330142]

CHAPTER 8

Bioremediation Industry: A Microbial Perspective

Pooja Singh[1,*]

[1] Department of Botany, DDU Gorakhpur University, Gorakhpur, UP, India

Abstract: Bioremediation of environmental pollutants and contaminants in soil is an emerging technology, which will gain relevance and importance in the near future. Microbiological bioremediation is not only cost-effective but also environmentally sustainable, as it does not cause undesirable effects like toxic byproducts or residues, requires heavy infrastructure, has on-site application, and is the least hazardous to human health. With new biotechnological tools, the microbes can be designed to have desirable effects for the bioremediation of more toxic wastes. However, the free release of genetically modified microbes for this purpose is still under risk assessment. This is an effective method to use indigenous microflora and harness their biodegradation properties to remove unwanted contaminants from soil, water bodies, underground water aquifers, ocean spills, *etc.* Currently, they are mostly used for cleaning oil spills and removing petroleum products and heavy metals from soil. Both *in situ* and *ex situ* methods are employed, where microbes can be used in varied ways. Much work is going on to explore and enhance the properties of microbes, especially bacteria, to be used as agents for contaminant removal from our environment. Global bioremediation is an emerging market that is slowly growing and will become a multibillion-dollar market worldwide in days to come. The current review tries to view the subject with microbes in perspective; their role in bioremediation; mode of action; technologies used; and their use for sustainable cleanup of the environment.

Keywords: Bioremediation, Biotransformation, Microflora, Pollutants, Pollutants.

INTRODUCTION

With the advent of the modern era with a rampant rate of development, anthropogenic activities have released several harmful chemicals and exposed the earth, air, soil, and water to some of the most destructive compounds. Pollution of our environment and its consequences are not new to us, and now several steps are being taken by humanity to negate and remove contaminants from the environment. Microbes are employed for the removal of pollutants by biological means. Bioremediation is one such process that is gaining importance.

---

[*] **Corresponding author Pooja Singh:** Department of Botany, DDU Gorakhpur University, UP, India; E-mail: pooja.ddu@gmail.com

Shampi Jain, Ashutosh Gupta and Neeraj Verma
All rights reserved-© 2023 Bentham Science Publishers

BIOREMEDIATION

It is the process of chemical removal by degradation/immobilization of a contaminant/pollutant from the environment by specific microbes. These microbes act singly, in sequence or in a consortium to achieve and complete the degradation process. Bioremediation is a safe option with the possibility of removing or rendering various harmless contaminants using biological entities. With the current advancement in microbial technology, bioremediation is an emerging industry where an integrated approach is sought to achieve the solution to the problem, *i.e.*, the removal of contaminants. The bioremediation process involves several microbial processes, some of which are: (a) Biodegradation is the natural modification of compounds by microbial metabolic pathways, such as mineralization of organic substrates; and (b) Biotransformation is the modification of a pollutant's chemical configuration or constituent to make it less toxic through the use of microbes.

The concept of using microbes to remove harmful chemicals is not new; they have always been part of domestic, agricultural, and industrial processes. About 50 years ago, Raymond and coworkers reported bioremediation in the work titled "Beneficial stimulation of bacterial activity in groundwater containing petroleum products" [1]. Bioremediation was invented by George Robinson in the 1960s for oil spill cleanup in California. Since then, it has been widely used for the cleanup of petrochemical spills, sewage, and leach fields, as well as for odour and pest control. It is widely used in the American continent and some of the most famous examples of bioremediation (microbial) are the Exxon Valdez oil spill in Alaska (1989), the crude oil spill in Minnesota (1979), sewage effluent cleanup, Cape Cod in Massachusetts and removal of chlorinated solvents in New Jersey [2].

POTENTIAL POLLUTANTS FOR BIOREMEDIATION

Organic Pollutants

A large range of chemicals have now been recognized as being biodegradable with the help of microbes. Chlorinated solvents, like trichloroethylene and perchloroethylene, are industrial-grade solvents used in textile, chemical, and allied industries. Some of the examples are polychlorinated biphenyls (4-chlorobiphenyl and 4,4-dichlorobiphenyl) produced in powerhouses and in the electrical manufacturing industry, chlorinated phenols (pentachlorophenol) from the timber industry, landfills, benzene and allied compounds (toluene, ethylbenzene, and xylene) obtained from the petroleum industry, paints, released from ports and airports, chemical manufacturing byproducts, and polyaromatic hydrocarbons (PAH) (naphthalene, anthracene, fluorine, pyrene, and benzopyrene) that are petroleum byproducts found in refineries, coke plants, landfills, tar production, power plants, gas plants, *etc.* Pesticides originate from agricultural runoff and contaminate soil, ground, and surface water. Some pesticides, where the bioremediation process is studied are atrazine, carbaryl, carbofuran, diazinon, glyphosate, parathion, propham, and 2-4 D [3].

The very first superbug (*Pseudomonas putida*) for bioremediation of oil spills was created in the 1970s when Anand Mohan Chakrabarty and co-workers [4] reported the development of a new strain of bacterium by the transfer of plasmids and named it superbug, which could biodegrade a number of toxic organic chemicals like octane, hexane, xylene, toluene, camphor, and naphthalene. The United States granted this patent, making it the first genetically engineered microorganism to be patented. This superbug was then used to clean an oil spill in Texas in 1990. Many bacteria are now reported and are used commercially to harness their ability to biodegrade organic compounds, *viz.*, *Dechloromonas aromatica* (benzoate, chlorobenzoate, and toluene), *Nitrosomonas* sp. (nitrogenous pollutants), *Deinococcus radiodurans* (solvents, heavy metals, and radioactive), *Methylibium petroleiphilum* (methyl tert-butyl ether), and *Alcanivorax borkumensis* (various hydrocarbons) (Table 1).

Table 1. Bioremediation of contaminants by bacteria.

Pollutant	Bacteria Bioremediating Agent	Reference
Polyaromatic hydrocarbons (PAH)	*Achromobacter* sp., *Arthrobacter* sp., *Bacillus* sp., *Mycobacterium* sp., *Burkholderia* sp., *Pseudomonas* sp., *Rhodococcus* sp., *Stenotrophomonas maltophilia*, *Sphingomonas* sp., *Xanthomonas* sp., etc.	[5, 6]
Benzene and allied compounds	*Pseudomonas putida*, *Acinetobacter johnsonii*, and *R. erythropolis*	[7]
Polychlorinated biphenyls	*Arthrobacter* sp., *Pseudomonas* sp., *Azotobacter* sp., etc.	[8, 9]
Pesticides	*Flavobacterium* sp., *Arthrobacter* sp., *Azotobacter* sp., *Burkholderia* sp., and *Pseudomonas* sp.	[10]

Inorganic Pollutants

These include several heavy metals, which come from industrial and other human activities like mining, tanneries, paints, battery wastes, pesticides, fertilizers, air emissions, fuel burning, and natural processes like weathering of minerals, erosion, volcanic activity, and forest fires. Heavy metals, which are prime candidates for the bioremediation process, are arsenic, cadmium, chromium, copper, mercury, nickel, and lead. Heavy metals are absorbed by microbes at cellular binding sites on the cell wall. Their extracellular polymers form complexes with heavy metals and immobilize them by various mechanisms, like changing their oxidation states and facilitating their removal. Bacteria couple the oxidation of alcohols, simple organic acids or aromatic compounds with the oxidation of heavy metals like Mn and Fe. Many bacteria have been reported to bioaccumulate and modulate chemically several heavy metals (Table 2). Many genetically engineered bacteria are now being explored for bioremediation of heavy metals. A few examples are as follows:

- Arsenic bioremediation by *Escherichia coli* strains (induced to produce metalloregulatory protein ArsR) [11].
- *Methylococcus* capsulatus removes chromium (IV) (expression of the CrR gene) [12].
- Mercury bioaccumulation by genetically engineered *coli*. JM109 (Hg^{2+} transporter gene expression) [13].
- Enhanced accumulation of cadmium in bioengineered *Mesorhizobium huakuii* [introduction of the *Arabidopsis thaliana* gene for phytochelatin synthase (PCS_{At})] [14].

Table 2. Bioremediation of heavy metals by bacteria.

Bacterial Bioremediating Agents	Heavy Metal	Reference
Bacillus sp. (endophyte)	Cu, Cd, and Pb (uptake upto 84%)	[15]
Enterobacter sp., *Stenotrophomonas*, *Comamonas*, and *Ochrobactrum*	Cu, Cd, Co, and Cr from activated sewage sludge	[16]
Bacillus carotarum, B. cereus, B. lentus, and *B. licheniformis*	Pb	[17]
Escherichia coli, Salmonella typhi, Bacillus licheniformis, and *Pseudomonas fluorescens*	Cd, Pb, and Zn (from textile industry)	[18]
Serratia sp.	Ni and Cd in polluted soil	[19]
Paenibacillus sp. (endophyte)	Cu, Zn, Pb, and As	[20]
Sulfate-reducing bacteria	Ba, Cd, Co, Cu, Fe, Mg, Mo, Zn, Hg, and Pb from saline water	[21]
Exiguobacterium profundum, Pseudomonas putida, and *Bacillus marisflavi*	Uranium sequestration by biofilm formation	[22]

MYCOREMEDIATION

Fungi also play a promising role in the biodegradation process and cycling of elements. Many possess complex biochemical machinery, which enables them to biotransform harmful compounds. Filamentous fungi (molds) and macrofungi both have shown bioremediation properties, especially saprotrophic and biotrophic fungi. The white rot fungi (WRF) have the best role in biodegradation. During mineralization, these can completely degrade lignin and cellulose biopolymers. They produce some exclusive enzymes (which help in this process), like lignin peroxidases, manganese-dependent peroxidases, versatile peroxidases, laccases, glutathione transferases, peroxide generating oxidases, and dehydrogenases. The most utilizable WRF in bioremediation are *Phanerochaete chrysosporium*, *Pleurotus ostreatus*, *Trametes versicolor*, *Bjerkandera adusta*, *Lentinula edodes*, *Irpex lacteus*, *Agaricus bisporus*, *Pleurotus tuber-regium*, and *Pleurotus pulmonarius*. Mycoremediation over the years has gained the attention of environmentalists and many fungal species are reported to degrade a wide variety of chemicals (Table 3).

Fungi have been proved useful bioremediating agents in harsh environments, like extreme cold or radioactive environments, where other methods have failed. Many fungi are hyperaccumulators and accumulate toxins inside their fruiting bodies, which can be removed later. They have been traditionally used in mycofiltration in wastewater treatment and the removal of the industrial, distillery, and pharmaceutical wastes in nontoxic ways. Fungi are used as inoculums with suitable media in methods like *ex situ* biopiles along with other technologies like bioventing.

Table 3. Mycoremediation by Fungi.

Pollutant	Degrading Fungal Species
Poly cyclic aromatic hydrocarbon, monoaromatic hydrocarbon, and chlorinated hydrocarbon	*Phanerochaete chrysosporium*, *Aspergillus* sp., *Penicillium* sp., *Rhizopus* sp., *Fusarium*, and *Cladosporium* sp.
Heavy metals	*Agaricus* sp., *Amanita* sp., *Cortinarius* sp., *Boletus* sp., *Leccinum* sp., *Suillus* sp., and *Phellinus* sp.
Crude oil	*Rhizopus* sp., *Paecilomyces* sp., *Alternaria* sp., *Mucor* sp., *Gliocladium* sp., *Aspergillus* sp., and *Fusarium* sp.
Phenolic compounds	*Trametes* sp., *Lentinus* sp., *Pleurotus* sp., and *Ganoderma* sp.
Pesticides like endosulphan, imazalil, and chlorpyrifos	Several white rot fungi
Industrial dyes	Several white rot fungi

BIOLOGICAL PROCESSES INVOLVED IN BIOREMEDIATION

There are several biological processes that help in bioremediation by microbes. Some of them are shown here (Fig. **1**). Biosorption, bioleaching, bioaccumulation, biomineralization, bioleaching, and biotransformation are the major processes. Biosorption is one of the major processes, which involves several other processes, where microbes adsorb heavy metals and other toxic substances (Fig. **2**).

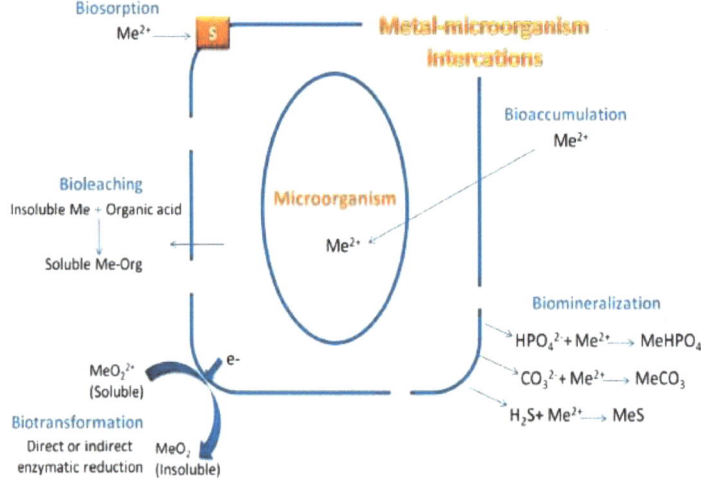

Fig. (1). Microbial processes involved in bioremediation [23].

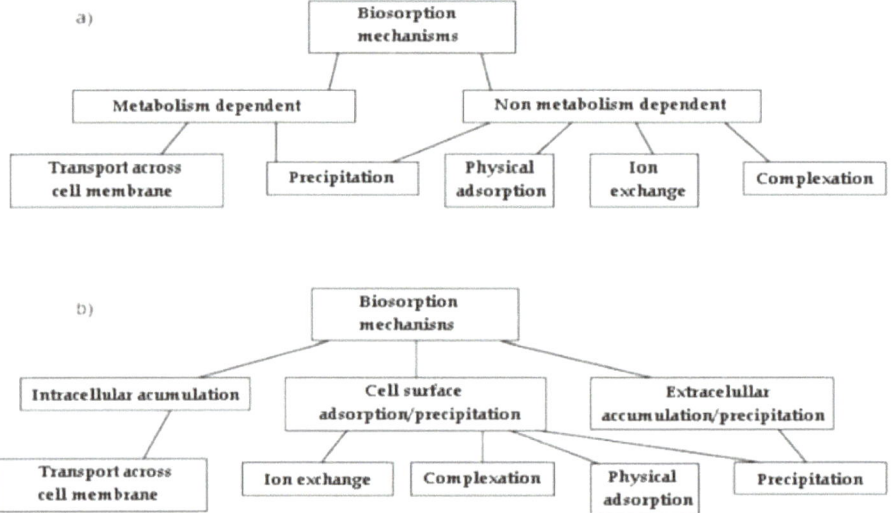

Fig. (2). Biosorption mechanisms in microbes [24].

Viable microbes undertake the transportation of metals across the membrane for intracellular accumulation. This may be a result of microbe defense against metal toxicity. In non-metabolism-dependent biosorption, metal ions physically interact at the membrane surface with some functional groups, *viz.*, in physical adsorption, ion exchange, and chemical adsorption. Microbial membranes are rich in metal-binding functional groups like carboxyl sulphate, phosphate and amino groups. During precipitation, metal uptake takes place in solution on the cell surface. Microorganisms produce precipitation-accelerating substances in the metabolic pathway, while in the non-metabolic pathway there is a simple membrane-metal interaction. Different improved strains are produced by recombinant DNA technology to enhance biosorptive properties. Transport across a membrane uses the same machinery for ions' transportation of contaminants as that of other ions. This transport may be both metabolically and non-metabolically dependent. Physical adsorption is accomplished by Van der Waals forces. Metals are also adsorbed by ion exchange, where they are exchanged with similarly charged counter ions.

Metals are also removed by a process of complexation, where metals interact with membrane-based chelating agents, polysaccharides, proteins, *etc.* Several organic acids produced by microbes also form metallo-organic complexes and aid their removal by leaching. Various groups of siderophores (having phenolate, catecholate, or hydroxamate binding groups) and special metal chelating compounds are capable of capturing metals from soil. Many bacteria produce glycolipids and other biosurfactants, which form emulsions in water. They are produced internally metabolically and are less toxic. Microbes internalize and break down the metallic contaminants through several biochemical pathways. Microbes transform the toxic state of metals like Hg, As, Cd, Se, and Pb into less toxic forms by biomethylation. Many factors affect microbial activity in the environment, like temperature, soil pH, and humidity. The type and concentration of contaminants also play a major role. The consortium of bacteria available and the nutrient content of soil also affect bacterial metabolism, which indirectly or directly affects the bioremediation process.

USES OF BIOREMEDIATION TECHNOLOGY

Several technologies are now being explored, tested, and used to harness the bioremediation potential of microbes. The microbial population to be used can be isolated from nature itself based on the need. The main groups of microbes used for the purpose can be grouped into aerobic, anaerobic, ligninolytic fungi, and methylotrophs. The main environmental factors that affect the process in soil are the nutrient level of soil, including trace elements, adequate soil moisture, soil pH, the oxygen content of the soil, soil temperature, and pollutant concentration. There are two types of bioremediation strategies used in water and soil:

In Situ Bioremediation

It involves bioremediating at the affected site itself for treatment and cleanup without the disturbance of any kind of excavation. This is applied to both soil and water treatment. It is comparatively cheaper, more cost-effective and does not cause the spread of pollutants or contaminants to other areas. Important *in situ* strategies in soil treatment are biostimulation, bioventing, biosparaging, and bioaugmentation (Fig. **3**).

Biostimulation

Creating an optimum environment for a specific native microbial community on purpose for harnessing their biodegradative properties is essential. This is achieved by providing suitable changes in nutrients (N, P, O, and C), pH, temperature, moisture, aeration, trace elements, *etc*. Several technologies are used for biostimulation, and it is still an emerging field. Stimulants in the form of additives are added to the subsurface by injection wells. It is mostly used for oil spills, herbicide removal, and groundwater treatment.

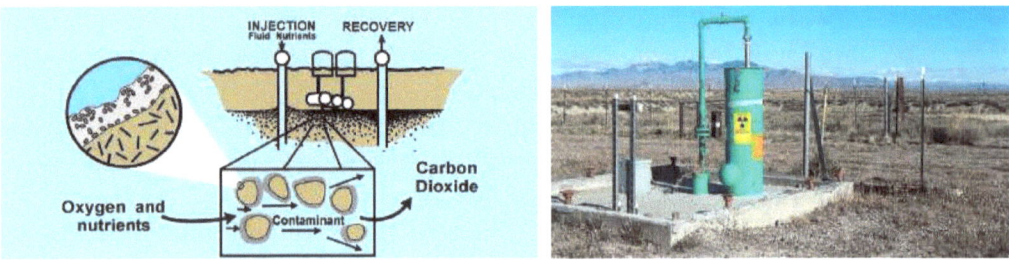

Fig. (3). Biostimulation by injection well in the subsurface (Source: http://rt-bi.nl/social-responsibility/ biodegradation/bioremediation/)

Bioaugmentation

It involves the addition of microbes indigenously or exogenously to contaminated sites. Genetic bioaugmentation is based on the spread of specific degradative catabolic genes, located on plasmids, into native microbial populations (Fig. **4**).

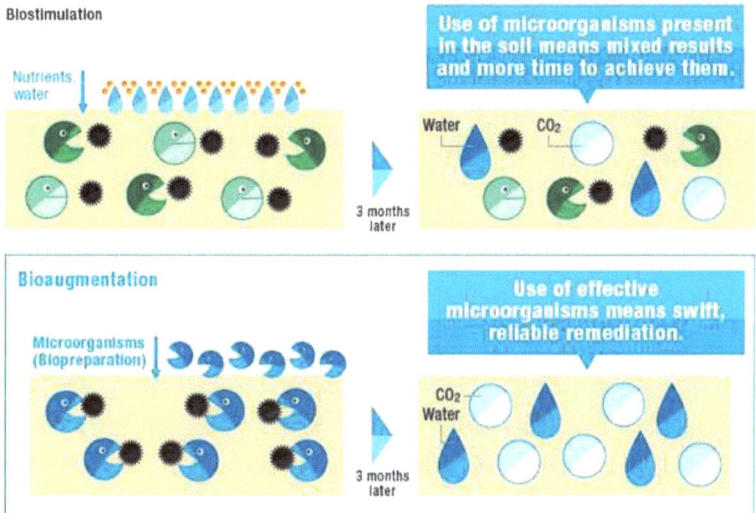

Fig. (4). Biostimulation and bioaugmentation (Source: https://www.bri.co.jp/english/teq/index.html)

Bioventing

Here, oxygen is supplied to the contaminated sites, probably unsaturated through a process of extraction and injection well. Oxygen/nutrient flow is highly regulated so that it allows only biodegradation and prevents volatilization. It is the most common *in situ* method, which is used for deep contaminated sites and may take a longer time (Fig. **5**).

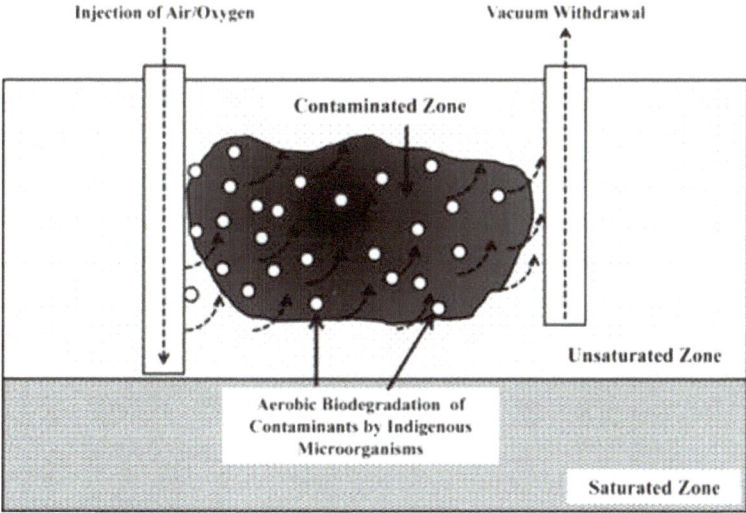

Fig. (5). Bioventing of contaminated soil [25].

Biosparaging

Here, the air is injected under pressure under the water table to accelerate biodegradation by native flora. It eases contact with soil and groundwater. It is a cheap method and is a highly effective *in situ* method without disturbing the surface. It promotes volatilization and biodegradation of dissolved organic components in the underground aquifer. Vapour extraction well in the subsurface, then creates a vacuum to direct the flow of liberated vapours and monitoring wells (Fig. **6**).

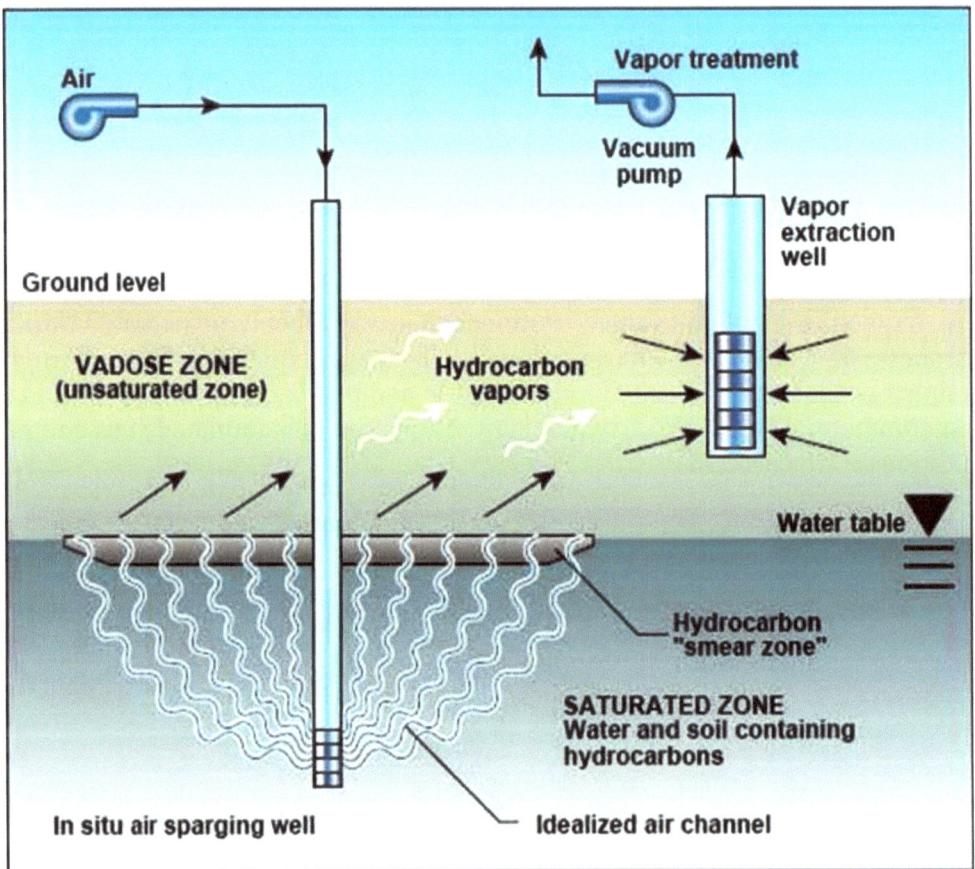

Fig. (6). Biosparaging (Source: https://soilvaporintrusion.wordpress.com/remedial-strategies-for-soil-v-por-intrusion/)

Ex Situ Bioremediation

It involves the removal of pollutants from the actual site by transportation, *etc.*, for treatment elsewhere (off-site). Some of the techniques used here are as follows:

Land Farming

This technique is used for the treatment of pollutants in soil. The affected soil is excavated to a depth of 10-40 cm and spread on prepared inert beds, made up of plastic, tiles, *etc*. Then the soil is inoculated with a relevant bioremediating microbe or used with native microflora. This process accelerates natural degradation aerobically. Regular tilling is done till desired results are attained. The process is simple, cost - effective, and widely used. The process also improves the physical and chemical properties of soil (Fig. 7).

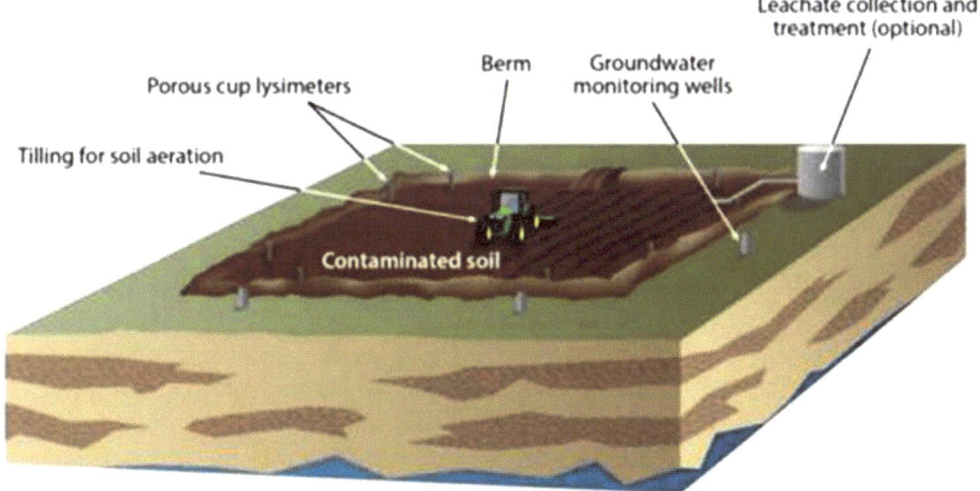

Fig. (7). Land farming [26].

Composting

This process involves mixing affected soil with natural substances like animal manure, leaf litter, fruit or vegetable wastes, sawdust, *etc*., to enrich the soil for the microbial degradation process. These are kept in large heaps to facilitate degradation by enrichment and increased temperatures. Heaps are regularly turned and watered to maintain moisture and air to accelerate the process of degradation by microbes.

Biopiles

This involves a combined approach of land farming and composting methods. The contaminated soil is excavated, piled, and typically constructed in the treatment area. The treatment area consists of a leachate collection and aeration system. The piles are covered, and there is minimal loss of contaminants outside by leaching and volatilization. This method provides a favourable environment for both aerobic and anaerobic microbes to biodegrade the contaminants. This method is widely used to treat soil contaminated with petroleum products (Fig. **8**).

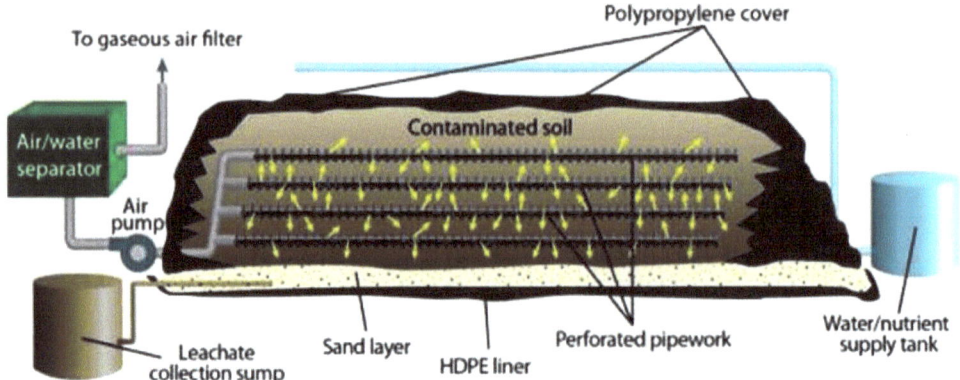

Source: UK EA 2002

Fig. (8). Schematic view of biopile [27]

Bioreactors

This is used for the treatment of contaminated soil and water. It comprises an engineered container/bioreactor system, in which contaminated soil, sludge, water, sediments, *etc.*, are processed. A slurry bioreactor may be defined as a containment vessel and apparatus that forms a three-phase (solid, liquid, and gas) mixing condition to increase the bioremediation rate of soil-bound and water-soluble pollutants as a mixture of contaminated soil and biomass (usually indigenous microorganisms) capable of degrading target contaminants. The system is controlled, so the process can be accelerated desirably (Fig. **9**).

BENEFITS OF BIOREMEDIATION

It is a natural system that uses microbial power to remediate pollutants from the natural environment. The end products are almost non-toxic, unlike other processes. It does not involve elaborate set up or infrastructure to process. There is no toxic residue, so there is no problem with disposal after the process is complete. On-site remediation helps in the non-transfer of contaminants to new areas. It poses minimal hazards to human health with the use of indigenous microbes in most cases. It is a less expensive way of removing hazardous chemicals from the environment. Despite its indisputable effectiveness, the process also has some drawbacks. Many compounds cannot be bioremediated by microbes due to a lack of relevant enzymes or microbial biochemistry.

Fig. (9). Schematic representation of bioslurry [26].

BIOREMEDIATION BY GENETICALLY ENGINEERED MICROORGANISMS

Genetically engineered microorganisms (GEMs) can be beneficially used for the treatment of contaminated soil, groundwater, and activated sludge. They can exhibit desirable, need-based, and enhanced degrading capabilities, and have a wide spectrum of capabilities for several groups of diverse chemicals. Although the idea is lucrative and effective, its release into the natural environment needs thorough analysis like risk assessment, ecological impact involved and biosafety.

There are at least four principal approaches to GEM development for bioremediation applications. These include:

- Study of specific enzymes involved and their activity and affinity
- The metabolic pathway involved and its regulation
- Development and improvisation of a relevant and effective bioprocess; its monitoring, control, and risk assessment
- Applications of bioaffinity bioreporter sensors for chemical sensing, toxicity reduction, and endpoint analysis

For example, genes responsible for the degradation of environmental pollutants like toluene, chlorobenzene acids, and other halogenated pesticides and toxic wastes have been identified. Several groups of plasmids have been identified in various groups of bacteria, which have degradative properties for different compounds. Some of these are XYL (for xylene and toluene), OCT (for octane,hexane, and decane), CAM (for camphor), NAH (for naphthalene), TOL (for toluene), SAL (for salicylate), NIC (for nicotine, nicotinate), pRA500 (for 3, 5-xylenol), pCIT1 (for aniline), pEG (for styrene), pCS1 (for parathion), pJP2 (for 2,4-dichlorophenoxyacetic acid), pWR1 (for 3-chlorobenzoic acid) *etc.* Several species of *Pseudomonas* are genetically modified to treat several contaminants like dichlorobenzoate, PCB, benzene, toluene, and biphenyls. *P. putida* was the first bacteria to be developed into a superbug for cleaning oil spills.

CONCLUDING REMARKS

Bioremediation is an emerging technology. It will expand its scope and use. Environmental degradation is increasing, and we are facing the hazards of technological development and rampant misuse of natural resources. Every form of resource, like water, oceans, and soil on which human lives depend, has been contaminated by recalcitrant compounds, which can be managed in an eco-friendly way by microbiological bioremediation. It leaves minimal hazardous byproducts and has now been recognized as a new clean-up technology. The global bioremediation market is a multimillion-dollar industry that is going to expand globally on an unprecedented scale, especially in agriculture, aquaculture, mining, chemical industries, oil and gas industries, and municipal and ocean spills. Currently, North America dominates this market, with expanding corporations in the Asia-Pacific regions. Increasing urbanization and industrialization have posed a serious threat to the environment, which will eventually affect human life and survival. The world has awakened to the bioremedial power of microbes. Recombinant DNA technology is further enhancing the desirable properties of the microbes to increase their spectrum and efficacy. Much work is being done and further research has been going on, and we will see the results in the near future as microbes of the future with cleaning superpowers. In the coming years, it will be one of the most emerging technologies that will pave the way to sustainable development and environmental healing.

REFERENCES

[1] Litchfield C. Thirty years and counting: Bioremediation in its prime? Bioscience 2005; 55: 273-9.
[http://dx.doi.org/10.1641/0006-3568(2005)055[0273:TYACBI]2.0.CO;2]

[2] Chapelle FH. Bioremediation: Nature's Way to a Cleaner Environment US Geological Survey Fact Sheet. US Department of the Interior 1997; FS-054-95: p. 2. [http://water.usgs.gov/wid/html/bioremed.html#HDR1]

[3] Vidali M. Bioremediation. An overview. Pure Appl Chem 2001; 73: 1163-72.
[http://dx.doi.org/10.1351/pac200173071163]

[4] Chakrabarty AM, Mylroie JR, Friello DA, *et al.* Transformation of *Pseudomonas putida* and *Escherichia coli* with plasmid-linked drug-resistance factor DNA. Proc Natl Acad Sci USA 1975; 72: 3647-51.
[http://dx.doi.org/10.1073/pnas.72.9.3647]

[5] Liu SH, Zeng GM, Niu QY, *et al.* Bioremediation mechanisms of combined pollution of PAHs and heavy metals by bacteria and fungi: A mini review. Bioresour Technol 2017; 224: 25-33.
[http://dx.doi.org/10.1016/j.biortech.2016.11.095]

[6] Gupte A, Tripathi A, Patel H, *et al.* Bioremediation of polycyclic aromatic hydrocarbon (PAHs): A perspective. Open Biotechnol J 2016; 10: 363-78.
[http://dx.doi.org/10.2174/1874070701610010363]

[7] Genovese M, Denaro R, Cappello S, *et al.* Bioremediation of benzene, toluene, ethylbenzene, xylenes-contaminated soil: A biopile pilot experiment. J Appl Microbiol 2008; 105: 1694-702.
[http://dx.doi.org/10.1111/j.1365-2672.2008.03897.x] [PMID: 19149767]

[8] Robinson GK, Lenn MJ. The bioremediation of polychlorinated biphenyls (PCBs): Problems and perspectives. Biotechnol Genet Eng Rev 1994; 12: 139-88.
[http://dx.doi.org/10.1080/02648725.1994.10647911] [PMID: 7727027]

[9] Ohtsubo Y, Kudo T, Tsuda M, *et al.* Strategies for bioremediation of polychlorinated biphenyls. Appl Microbiol Biotechnol 2004; 65: 250-58.
[http://dx.doi.org/10.1007/s00253-004-1654-y]

[10] Uqab B, Mudasir S, Nazir R. Review on bioremediation of pesticides. J Bioremed Biodeg 2016; 7: 343.
[http://dx.doi.org/10.4172/2155-6199.1000343]

[11] Kostal J, Yang R, Wu CH, *et al.* Enhanced arsenic accumulation in engineered bacterial cells expressing ArsR. Appl Environ Microbiol 2004; 70: 4582-87.
[http://dx.doi.org/10.1128/AEM.70.8.4582-4587.2004]

[12] Al Hasin A, Gurman SJ, Murphy LM, *et al.* Remediation of chromium (VI) by a methane-oxidizing bacterium. Environ Sci Technol 2010; 44: 400-05.
[http://dx.doi.org/10.1021/es901723c]

[13] Zhao XW, Zhou MH, Li QB, *et al.* Simultaneous mercury bioaccumulation and cell propagation by genetically engineered *Escherichia coli.* Proc Biochem 2005; 40: 1611-16.
[http://dx.doi.org/10.1016/j.procbio.2004.06.014]

[14] Sriprang R, Hayashi M, Ono H, *et al.* Enhanced accumulation of Cd^{2+} by a *Mesorhizobium* sp. transformed with a gene from *Arabidopsis thaliana* coding for phytochelatin synthase. Appl Environ Microbiol 2003; 69: 1791-1796.
[http://dx.doi.org/10.1128/AEM.69.3.1791-1796.2003]

[15] Guo H, Luo S, Chen L, *et al.* Bioremediation of heavy metals by growing hyperaccumulaor endophytic bacterium *Bacillus* sp. L14. Biore Technol 2010; 101: 8599-605.
[http://dx.doi.org/10.1016/j.biortech.2010.06.085]

[16] Bestawy EE, Helmy S, Hussien H, *et al.* Bioremediation of heavy metal-contaminated effluent using optimized activated sludge bacteria. Appl Water Sci 2013; 3: 181-92.
[http://dx.doi.org/10.1007/s13201-012-0071-0]

[17] Gupta MK, Kumari K, Shrivastava A, *et al.* Bioremediation of heavy metal polluted environment using resistant bacteria. J Environ Res Develop 2014; 8: 883-889.

[18] Basha SA, Rajaganesh K. Microbial bioremediation of heavy metals from textile industry dye effluents using isolated bacterial strains. Int J Curr Microbiol Appl Sci 2014; 3: 785-94.
[https://www.ijcmas.com/vol-3-5/S.Ameer%20Basha%20and%20K.Rajaganesh.pdf]

[19] Li X, Dong S, Yao Y, *et al.* Inoculation of bacteria for the bioremediation of heavy metals contaminated soil by *Agrocybe aegerita*. RSC Advances 2016; 6: 65816-24.
[http://dx.doi.org/10.1039/C6RA11767H]

[20] Govarthanan M, Mythili R, Selvankumar T, *et al.* Bioremediation of heavy metals using an endophytic bacterium *Paenibacillus* sp. RM isolated from the roots of *Tridax procumbens*. 3 Biotech 2016; 6: 242.

[http://dx.doi.org/10.1007/s13205-016-0560-1]

[21] Kerkar S, Ranjan Das K. Bioremediation of Heavy Metals from Saline Water Using Hypersaline Dissimilatory Sulfate-Reducing Bacteria. In: Naik M, Dubey S, Eds. Marine Pollution and Microbial Remediation. Singapore: Springer 2017; pp. 15-28.
[http://dx.doi.org/10.1007/978-981-10-1044-6_2]

[22] Manobala T, Shukla SK, Rao TS, *et al.* Uranium sequestration by biofilm-forming bacteria isolated from marine sediment collected from Southern coastal region of India. Int Biodeter Biodegrad 2019; 145: 104809.
[http://dx.doi.org/10.1016/j.ibiod.2019.104809]

[23] Lovley DR, Lloyd JR. Microbes with a mettle for bioremediation. Nat Biotechnol 2000; 18: 600-01.
[http://dx.doi.org/10.1038/76433]

[24] Veglio F, Beolchini F. Removal of metals by biosorption: A review. Hydrometallurgy 1997; 44: 301-16.
[http://dx.doi.org/10.1016/S0304-386X(96)00059-X]

[25] Brown LD, Ulrich AC. Bioremediation of Oil Spills on Land. In: Fingas M, Ed. Handbook of Oil Spill Science and Technology. John Wiley & Sons, Inc 2014; pp 395-406.
[http://dx.doi.org/10.1016/B978-0-12-809413-6.00012-6]

[26] US EPA. How to Evaluate Alternative Cleanup Technologies for Underground Storage Tank Sites: A Guide for Corrective Action Plan Reviewers, Chapter V: Landfarming. Land and Emergency Management 5401R. United States Environmental Protection Agency, 1994; p. 27.
[https://www.epa.gov/sites/production/files/2014-03/documents/tum_ch5.pdf]

[27] UK EA. Remedial Treatment Action Data Sheet on Landfarming, Data Sheet No. DS-03. United Kingdom Environment Agency, Bristol, 2002; [www.environment-agency.gov.uk/research/planning/40381.aspx]

CHAPTER 9

Alleviation of Salinity Stress by Microbes

Sampat Nehra[1,*,#], Raj Kumar Gothwal[1,2,#], Alok Kumar Varshney[1,2], Pooran Singh Solanki[1,2], Poonam Meena[3], P.C. Trivedi[3] and P. Ghosh[1]

[1] Birla Institute of Scientific Research, Statue Circle, Jaipur, Rajasthan, India

[2] Department of Bioengineering and Biotechnology, Birla Institute of Technology, Mesra, Ranchi, Jaipur Campus, Jaipur, Rajasthan, India

[3] Deptartment of Botany, University of Rajasthan, Jaipur, Rajasthan, India

Abstract: Agricultural production is majorly hampered by the negative impact of both biotic and abiotic stress in most developing countries. Among abiotic stresses, soil salinity is a major problem, affecting crop production and responsible for limiting the growth and productivity of plants in different areas of the world due to increasing use of poor quality of water, flooding, over-irrigation, seepage, silting, and a rising water table. In agriculture, salt-tolerant rhizospheric/endophytic microorganisms play an important role in helping alleviate abiotic stresses in plants. Under plant-microbe interactions, plant root-associated microbes, including endophytes, closely interact and cooperate with plants, and mediate important physiological and metabolic processes, thereby enhancing the plant's tolerance to salinity stress. Several mechanisms have been developed for microbial alleviation of salinity stress in plants, including the production of phytohormones, improving plant nutrient status, production of ACC deaminase, salt exclusion, and enhancing resistance to drought in plant cells. A wide range of micro-organisms are available that have diverse mechanisms for salt stress alleviation in plants. Future research needs to be directed towards field evaluation for the validation of the potential microbes.

Keywords: Mechanism of stress alleviation, Plant-microbe interaction, Soil salinity, Salt stress, Salt stress alleviation.

INTRODUCTION

Soil salinity is not a new phenomenon. This has been described for centuries, where salinity and humanity have lived one aside from the other. A better example is the publication '*Salt and silt in ancient Mesopotamian agriculture'* that reported the history of salinization in Mesopotamia where three episodes, *i.e.*, the earliest and most serious one, affected Southern Iraq from 2400 BC until almost

---

[*] **Corresponding author Sampat Nehra:** Birla Institute of Scientific Research, Statue Circle, Jaipur, Rajasthan, India; E-mail: nehrasampat@gmail.com
[#] Sampat Nehra and Raj Kumar Gothwal have equal contribution

1700 BC; a milder episode in central Iraq happened between 1200 and 900 BC; and in the east of Baghdad (which became salinized after 1200 AD) have been described [1]. The salinization of soil is a serious problem and is steadily increasing in many parts of the world, in particular in arid and semi-arid areas [2]. The main causes of soil salinization that have been reported are flooding, over-irrigation, seepage, silting, and a rising water table.

According to the FAO Land and Nutrition Management Service [3], saline soils occupy 7% of the Earth's land surface [4] and increased salinization of arable land has resulted in 50% land loss by the middle of the 21st century [5] (Table **1** and **2**). In 2008, about 77 million hectares out of 1-5 billion hectares of cultivated land around the world was affected by excess salt content [6]. Approximately 12 million hectares of land suffers from various kinds of salt-related afflictions. Over 2.1 million hectares of salt-affected land is located in the country's key breadbasket in the north [7]. Recent precise reports on the global extent of salt-affected soils are not recorded. Many countries have assessed their soils and soil salinization at the national level, such as the United Arab Emirates, Kuwait, the Middle East, Australia, *etc.* In consideration of the current extent of salt-affected soils, the damage to salt-induced land in 2013 was $441 per hectare, and annual economic losses were about $27 billion [1]. Salt stress presents an increasing threat to plant agriculture [8]. Among the various sources of soil salinity, irrigation combined with poor drainage is the most serious, because it represents losses of once productive agricultural land.

Salt Stress Damage to Plants

Combating soil pollution to feed the world's growing population is very important and needs attention. Several environmental factors adversely affect plant growth and the final crop yield. Drought, salinity, nutrient imbalances (toxicities and deficiencies) and temperature extremes are some of the major environmental barriers to crop production. Adverse climatic conditions are among the principal limiting abiotic factors for the decline in agricultural production [9]. Of the world's 5.2 billion ha of dryland agriculture, 3.6 billion ha is affected by the problems of erosion, degradation, and salinity [10].

Globally, soil salinity has become a major issue for plant growth and has reduced agricultural yield [11]. It is one of the major abiotic stresses faced by crops and has been reported to affect about 950×10^6 ha of land worldwide [12]. By salt accumulation, the amount of world agricultural land destroyed each year is estimated to be 10 million ha [13]. Due to soil salinity, the annual cost of land degradation in irrigated lands could be US $27.3 billion worldwide due to a loss in crop production [14]. The land area under ever-increasing salinization has reached almost 34 million irrigated hectares [15].

In plants, salt stress conditions cause various physiological and metabolic changes such as photosynthesis, nutritional imbalance, inhibition of water uptake, a decrease in growth and seed germination. In the natural environment, microorganisms colonize plants. Under plant-microbe interaction, plant root-associated microbes, including endophytes, closely interact, cooperate with plants, and mediate important physiological and metabolic processes, thereby enhancing the plant's tolerance to salinity stress [11].

Many plants can tolerate salinity to an optimal level, and yield decreases as salinity increases [16]. It has been recognized that a crop's sensitivity to salinity varies depending on the growth stage of the plant [17]. Most annual plants are tolerant at germination but are sensitive during emergence and early vegetative development [18]. A mature plant is more tolerant to salts, particularly during later stages of development. Most of the plants are tolerant during germination. Though salinity stress delays the germination process, there may be no difference in the percentage of germinated seeds [18].

However, higher salt concentrations have been shown to reduce seed germination in sorghum [18], cotton [19], tomato [20], and globe artichoke [21]. An increase in salinity levels significantly reduced the rate and percentage of seed germination of alfalfa [22].

Table1. Variation in salinity levels in the world, in million hectares (Mha) [3].

Regions	Total Area (Mha)	Saline Soils	Percentage	Sodic Soils	Percentage
Africa	1899.1	38.7	2.0	33.5	1.8
Asia, the Pacific & Australia	3107.2	195.1	6.3	248.6	8.0
Europe	2010.8	6.7	0.3	72.7	3.6
Latin America	2038.6	60.5	3.0	50.9	2.5
Near East	1801.9	91.5	5.1	14.1	0.8
North America	1923.7	4.6	0.2	14.5	0.8
Total	12781.3	397.1	3.1	434.3	3.4

Table 2. In drylands, salt-affected soils by continents [23, 24].

Continents	Total Salt Affected Area (mha)	Salt Affected Area (mha) Saline soils	Salt Affected Area (mha) Sodic soils
Africa	209.6	122.9	86.7

Continents	Total Salt Affected Area (mha)	Salt Affected Area (mha)	Salt Affected Area (mha)
		Saline soils	Sodic soils
Australia	357.6	17.6	340
North America	15.8	6.2	9.6
Central America	112	61	51
South America	84.1	82.3	1.8
North and Central Asia	211.7	91.5	120.2
South Asia	84.1	82.3	1.8
Southeast Asia	20	-	20.0
Total	1031.2	397	434

Salinity also affects both vegetative and reproductive development. Wheat [25], sorghum [26], and cowpea [27] crops were the most sensitive during the vegetative and early reproductive stages, while less sensitive during flowering and least sensitive during the seed filling stage. According to Yadav and coworkers [28], salinity may decrease shoot growth more than that of root [29], and can decrease the number of florets per ear, increase sterility, and affect the flowering time and maturity in both wheat [30] and rice [31].

The symptoms of damage by salt stress are generally growth inhibition, accelerated development, senescence, and death due to prolonged exposure. Growth inhibition is the primary step leading to other symptoms. Programmed cell death may also occur under severe salinity stress.

SALT TOLERANT BACTERIA

Soil has a vast diversity of microorganisms, which include different groups of bacteria, fungi, and archaea. Some microbes have been demonstrated for their inherent capability to tolerate different salt concentrations and to promote plant growth as well. Microorganisms that can grow in the absence or in the presence of salt are known as salt tolerant or halotolerant, while those that can grow above approximately 15% (w/v) NaCl (2.5 M) are called extremely halotolerant [32]. These salt-tolerant plant-beneficial microbes are extremely important in agriculture and have been shown to improve crop productivity in arid and semiarid environments [33]. Such microorganisms belong to all three domains of life, namely archaea, bacteria, and eukarya. Halophiles include a number of microorganisms, such as moderately halophilic aerobic bacteria, cyanobacteria, sulphur-oxidizing bacteria, heterotrophic bacteria, anaerobic bacteria, archaea, protozoa, fungi, algae, multicellular eukaryotes, *etc.* They can grow over an extended range of salt concentrations (3-15% NaCl w/v and above), unlike the truly halophilic archaea, whose growth is restricted to high salt concentrations [34].

Under saline conditions, the bacterial genera *Pseudomonas, Bacillus, Agrobacterium, Klebsiella, Streptomyces, Enterobacter*, and *Achromobacter* are reported with better results in improving the productivity of different crops [35, 36]. The diazotrophic salt-tolerant bacterial strains of *Pseudomonas, Agrobacterium, Klebsiella,* and *Ochrobactrum* isolated from the roots of a halophytic plant, *Arthrocnemum indicum*, were reported with salinity tolerance ranging from 4 to 8% NaCl and promoted the productivity of peanuts in saline conditions [37]. An alkaliphilic bacterium, *Planococcus rifietoensis*, has been reported to improve the growth and yield of wheat crops under salinity stress [38].

These microbes have developed certain osmoadaptive mechanisms for their protection from salt stress. These osmoadaptive mechanisms include the accumulation of compatible osmolytes in their cytosol-like sugars (*e.g.*, trehalose), polyols (*e.g.*, glycerol and glucosyl glycerol), free amino acids (*e.g.*, proline and glutamate) and their (*e.g.*, glycine betaine, carnitine and dimethylsulfoniopropionate), sulfate esters (*e.g.*, choline-*o*-sulfate), *N*-acetylated diamino acids and small peptides (*e.g.*, N-acetylornithine and N-acetyl glutaminyl glutamine amide), by either endogenous *de novo* synthesis or uptake mediated by specific carriers [39, 40].

Characterization of Salt Tolerant Bacteria

To mitigate salt stress conditions, beneficial microorganisms could play a significant role once unique properties of microorganisms such as tolerance to extremities, their ubiquity, and genetic diversity are understood [41]. Several reports are available on the existence and functional diversity of agriculturally important microorganisms in stressed conditions [42]. Several plant growth-promoting rhizobacteria (PGPR) such as *Rhizobium, Bradyrhizobium, Azotobacter, Azospirillum, Pseudomonas*, and *Bacillus* have been isolated from salt-affected areas of India [43]. *Pseudomonas alcaligenes* and *P. pseudoalcaligenes* have been found to dominate the saline sites. Other bacteria, namely *Alcaligenes xylosoxidans, Ochrobactrum anthropi, Serratia marcescens*, and *Pseudomonas aeruginosa*, have been reported from saline soils [43].

Several reports are available on the isolation and characterization of PGPRs from salt-affected areas. Shultana and coworkers [41] isolated *Rhizobacteria* from a saline-infested zone of the wheat rhizosphere and identified them as mainly belonging to *Bacillus* and *Bacillus*-derived genera. These were tested for plant growth promoting (PGP) traits at higher salt concentrations [41]. By providing a viable alternative to inorganic fertilisers and pesticides, PGPRs could effectively colonise plant roots and maintain soil fertility [44]. A very high percentage of these have indole-3-acetic acid (IAA) production, phosphorus solubilization, and siderophore production abilities. Only a few isolates were capable of producing gibberellin and 1-aminocyclopropane-1-carboxylate (ACC) deaminase.

For salinity mitigation, screening of locally-isolated salt-tolerant PGPR is difficult to ensure effectiveness, and indigenous strains are more efficient in boosting plant resistance to salinity stress compared to PGPR originated from the non-saline ecosystem [45, 46]. Siddikee and coworkers [47] isolated halotolerant bacterial strains from both the soil of barren fields and the rhizosphere of six naturally growing halophytic plants. They belonged to 10 different bacterial genera, namely Bacillus, *Brevibacterium, Planococcus, Zhihengliuella, Halomonas, Exiguobacterium, Oceanimonas, Corynebacterium, Arthrobacter,* and *Micrococcus* on the basis of their 16S rRNA gene sequencing analysis. Under *in vitro* conditions, these were characterized for multiple plant growth promoting traits, such as the production of IAA, nitrogen fixation, phosphorus (P) and zinc (Zn) solubilization, thiosulfate (S_2O_3) oxidation, production of ammonia (NH_3), and the production of extracellular hydrolytic enzymes (protease, chitinase, pectinase, cellulase, and lipase). Many isolates were discovered to have multiple plant-growth promoting traits as well as 1-aminocyclopropane-1-carboxylic acid (ACC) deaminase activity.

The role of PGPRs in increasing the growth of different crops under salt stress conditions has been reported earlier [48, 49]. On the basis of 16S rRNA gene sequences, novel P-solubilizing salt-tolerant bacteria, namely *Kushneria* sp., have also been reported from salt-affected soils in China, South Korea, and Algeria [50], *Burkholderia vietnamiensis* [51] and *Azospirillum brasilense* [52], respectively. *A. brasilense* was also reported to produce IAA under saline conditions. The stress hormone abscisic acid (AA) has been reported to be produced by two P-solubilizing salt-tolerant bacteria isolated from the Khewra salt range in Pakistan [53] and two P-solubilizing bacteria isolated from the tomato rhizosphere [54]. Three ACC deaminase-producing halotolerant bacterial strains, RS16, RS656, and RS111, were isolated from South Korea and were identified as belonging to *Brevibacterium iodinun, Bacillus licheniformis,* and *Zhihengliuella alba,* respectively [55]. Several mechanisms have been correlated with these beneficial microbes for salt stress mitigation, such as by retaining an appropriate Na^+/K^+ ratio through the secretion of extracellular polymeric substances called exopolysaccharides (EPS), which explains their survivability under unfavourable soil conditions [56, 57]. Several bacterial genera, such as *Pseudomonas, Bacillus, Enterobacter, Planococcus, Burkholderia, Halomonas,* and *Microbacterium,* can produce EPS in salt stress conditions [41, 58].

Soil Tolerant Bacteria in Salt Stress Alleviation of Plants

Soil salinity in arid regions is a significant limit for crop growth. Many technologies have contributed to improved salt tolerance in plants, and it has been reported that a number of microorganisms have been reported to reduce salt stress in plants.

In arid regions, soil salinity is an important limiting factor for cultivating crops. Many technologies have contributed to improving salt tolerance in plants, and various microorganisms have also been found to alleviate salt stress in plants (Table 2). Plants are colonized by both external and internal microorganisms in their natural environment. Some microorganisms, particularly beneficial bacteria and fungi, can improve plant performance under stress conditions and, consequently, enhance yield [49]. The role of microorganisms in plant growth promotion, nutrient management, and disease control is well known. These beneficial microorganisms multiply and colonize the rhizosphere and endorhizosphere of plants and promote the growth of the plants through various direct and indirect mechanisms [41, 46]. The bacteria inhabiting the rhizosphere may influence seedling emergence, root development, plant vigour and health, thus reducing or enhancing the crop yield either by directly stimulating plant growth by production of plant growth regulators, solubilization of minerals, fixation of atmospheric nitrogen, or indirectly by reducing the damage from plant pathogens [59].

Recently, bacteria such as *Rhizobium*, *Bacillus*, *Pseudomonas*, *Burkholderia*, *Achromobacter*, *Azospirillum*, *etc.*, have been found to provide tolerance to host plants tested under salt stress environments. Under saline conditions, salt-tolerant PGPRs have been found to improve tomato, pepper, canola, bean, soybean, and lettuce growth [32, 41, 60, 61]. The use of these microorganisms can alleviate stresses in agriculture, thus opening a new and emerging application of microorganisms (Table 3).

Table 3. Alleviation of salt stress in plants by plant growth-promoting rhizobacteria.

Bacteria	Crop	Reference
Pseudomonas pseudoalcaligenes	Rice (*Oryza sativa*)	[62]
Bacillus pumilus		
Bacillus megaterium	Maize (*Zea mays*)	[63]
Azospirillum brasilense	Barley (*Hordeum vulgare*)	[64]
Pseudomonas mendocina	Lettuce (*Lactuca sativa* L. cv. Tafalla)	[65, 66]
Azospirillum spp.	Pea (*Phaseolus vulgaris*)	[67]

(Table 3) cont.....

Bacteria	Crop	Reference
Bacillus subtilis	*Arabidopsis thaliana*	[68]
Pseudomonas syringae	Maize (*Zea mays*)	[69, 70]
Pseudomonas fluorescens		
Enterobacter aerogenes		
P. fluorescens	Groundnut (*Arachis hypogaea*)	[71]
Azospirillum spp.	Lettuce (*Lactuca sativa*)	[60, 66]
Achromobacter piechaudii	Tomato (*Lycopersicon esculentum*)	[72]
Aeromonas hydrophila and *A. caviae*	Wheat (*Triticum aestivum*)	[73, 74]
Bacillus insolitus		
Bacillus sp.		
Azospirillum spp.	Maize (*Z. mays*)	[75]
A. brasilense	Chickpeas (*Cicer arietinum*) and Faba beans (*Vicia faba*)	

Inoculation with bacteria could mitigate the effects of salt stress in different plants. *Azospirillum* inoculated seeds of lettuce (*Lactuca sativa* cv. Mantecosa) showed better germination rates and vegetative growth than non-inoculated control plants when exposed to NaCl [76]. Siddikee and coworkers [47] inoculated the halotolerant bacterial strains of *Brevibacterium epidermidis*, *Micrococcus yunnanensis*, and *Bacillus aryabhattai* to ameliorate salt stress (150 mM NaCl) in canola plants and reported a more than 40% increase in root length and dry weight as compared to uninoculated plants. When used as inoculants under saline conditions, isolates 2P and 1P isolated from the rhizosphere soil of *Lactuca dissecta* and *Chrysopogon aucheri*, respectively, promoted the growth of chickpea plants [53]. Siddikee and coworkers [55] reported a significant improvement in plant growth and root/shoot dry weight ratio in salt stress in the plants inoculated with salt-tolerant *Brevibacterium iodinum*, *Bacillus licheniformis*, and *Zhihengliuela alba*. Inoculation of the osmotolerant *Azospirillum brasilense* strain in wheat could improve the growth under salt stress conditions [52]. *Pseudomonas putida* inoculation in cotton seedlings can protect against salt stress and promote growth [77], and salt stress can affect germination rate, plant height, fresh weight, and dry weight not only in cotton but also in wheat plants [78, 79]. Salt-sensitive wheat cultivars have been shown to exhibit a better response to inoculation [80]. Inoculation of *Staphylococcus kloosii* and *Kocuria erythromyxa* in radish grown under salt stress conditions not only boosted seedling emergence and root/shoot dry weight but also increased chlorophyll content, leaf relative water content, and decreased electrolyte leakage [81]. *Azospirillum brasilense* also significantly improved the growth and yield of barley under a saline environment [64] and resulted in an increase in chlorophyll content, photosynthesis, and accumulation of K, P, Mg, Ca, and Fe in the shoots and roots.

MECHANISM OF SALT STRESS ALLEVIATION

Among abiotic stresses, salinity is very important and is responsible for limiting the growth and productivity of plants in different areas of the world due to the increasing use of very poor quality water for irrigation and soil salinization. Salt stress is the major reason for cellular dehydration by lowering turgor pressure in plant cells. To nullify the stress effect, plants employ different mechanisms to tolerate salt stress [82]. Several mechanisms have been proposed to explain microbial elicited stress tolerance in plants, which include plant hormone production, enhancing nutrient acquisition, exopolysaccharide synthesis, and ACC deaminase production, improving rhizospheric and soil conditions [83], and altering the physiological and biochemical properties of the host [84].

Phytohormones Synthesis

The effects of phytohormone-producing bacteria and plant growth regulators on germination and seedling growth of crops under saline conditions were studied worldwide, and it was concluded that growth regulators considerably alleviated salinity. Root colonizing bacteria that produce phytohormones can be used to alleviate salt stress from soil salinity [85]. In the processes of plant growth, phytohormones, *e.g.*, the production of IAA, auxins, cytokinins, gibberellins, and salicylic acid (SA), plays a significant role. The repressive effect of salinity on germination could be related to a decline in endogenous levels of plant growth hormones or phytohormones [86], resulting in reduced plant growth. These hormones can be synthesized not only by the plants themselves but also by their associated microorganisms, such as free-living denitrifier *Azospirillum* spp [85]. The phytohormones have a beneficial effect on the lateral root growth of *Arabidopsis* under salt stress conditions. It is also suggested that root-colonizing bacteria, which produce phytohormones when bound to the seed coat of a developing seed, may stimulate plant growth and prevent the deleterious effects of stresses from the environment [87].

IAA Production

In plants, the high concentration of salts in saline soils imposes osmotic, ionic, oxidative, and water stress. Biological solutions (hormone production; IAA production) can be the most reliable and sustainable approach to ensure food security and limit the use of agrochemicals [88]. Bacteria belonging to the genera *Azospirillum*, *Pseudomonas*, *Xanthomonas*, and *Rhizobium* as well as *Alcaligenes faecalis, Enterobacter cloacae, Acetobacter diazotrophicus,* and *Bradyrhizobium japonicum* have been reported to produce auxins, which help in stimulating plant growth [89]. Salt-tolerant IAA-producing bacterial strains *P. aurantiaca* TSAU22 and *P. extremorientalis* TSAU20 alleviated quite successfully the reductive effect of salt stress on the percentage of germination (up to 79%), probably through their ability to produce IAA [78]. They were able to produce IAA in saline conditions as well. Inoculation of wheat with *Pseudomonas* spp. stimulated plant growth by reducing plant uptake of toxic ions and increasing the auxin content [90].

ABA Production

ABA is now established as a widely occurring and important plant growth regulator, an important regulator of plant growth, including embryo and seed development, seedling establishment, vegetative and reproductive growth, as well as promoting seed dormancy [91]. ABA plays a pathosystem-dependent role in plant pathogen responses and has the potential to antagonize the germination-promoting effects of the plant hormone gibberellin, regulate guard cells, and regulate stress-responsive gene expression when water is scarce [92]. Salinity increases ABA concentrations in roots, xylem sap and shoots [93], concomitant with decreased transpiration rates, to help the plant develop tolerance to the stress. Several PGPR isolated from weed species grown in saline soil (2.3 dSm^{-1}) produced multiple phytohormones *in vitro*, including ABA, and increased the growth of salinized (20 dSm^{-1}) soybean seedlings [94]. Some mycorrhizal fungi accumulate substantial quantities of ABA [95], which may be responsible for the increased root ABA concentrations of some mycorrhizal plants [96]. However, contradictory results were reported by Yao and coworkers [77] in cotton. Immersing cotton (*Gossypium hirsutum*) seeds in suspensions of *Pseudomonas putida* Rs-198 before planting increased seedling biomass accumulation by 10% in saline soil and prevented any salinity-induced ABA accumulation in seedlings. In contrast, untreated seeds showed a 33% increase in foliar ABA concentration. Similarly, *Glomus intraradices* BEG121 reduced the concentration of ABA in lettuce roots in both well-watered and dry soil [97]. In both these cases, it was speculated that mycorrhizal effects on root ABA concentration may be related to increased hyphal water capture, resulting in improved water uptake by the plant. ABA induces both drought stress and salinity stress and upregulates osmotic stress responsive genes. Drought and salt stress result in osmotic imbalance and, thus, salt and drought stress tolerance mechanisms aim at restoring cellular homeostasis.

These mechanisms are adaptive responses that are responsible for creating stress tolerance or avoiding stress conditions.

Cytokinin Production

Cytokinins act as nodes in abiotic stress signalling pathways [98]. Cytokinin is a multifaceted plant hormone that has a major role in stress responses and plays an important role in diverse plant growth and development processes. Salinity decreased shoot growth and root, xylem, and shoot cytokinin concentrations of tomato plants. Plants grafted onto a rootstock constitutively expressing the *ipt* gene not only improved trans-zeatin concentrations in developing fruits but also fruit yield (30%) when grown under moderate salinity [99], indicating a beneficial effect of cytokinins in a plant grown under salt stress. Because cytokinin production appears to be a fairly common trait of PGPR and mycorrhizal fungi [100, 101], more emphasis should be placed on selecting microbial inoculants with high cytokinin production to potentially alleviate salt stress. Stimulation of shoot biomass of lettuce plants grown in drying soil by the cytokinin-producing PGPR *Bacillus subtilis* was observed [102]. Pavlu and coworkers [98] concluded from their studies that cytokinin metabolism and signalling play important roles in abiotic stress tolerance and that the manipulation of these processes in crops would be beneficial for sustainable agriculture.

Ethylene Production

The rhizobacteria (PGPRs) with ACC deaminase activity have a major role in promoting plant growth and development under stress conditions [103]. The increment in the synthesis of ethylene from its immediate precursor, ACC, secreted by plants' root exudates, has been recorded in almost all plants growing under stress conditions [104, 105]. The plant hormone ethylene, under adverse and stressful conditions, endogenously regulates plant homoeostasis, resulting in reduced root and shoot growth. Plant ACC is sequestered and degraded by bacterial cells to supply nitrogen and energy in the presence of ACC deaminase producing bacteria. Further, removing ACC resulted in the reduction of the adverse effect of ethylene, ameliorating plant stress and promoting plant growth [106].

Inoculation of the red pepper plant with three ACC deaminase-producing halotolerant bacteria, namely *Brevibacterium iodinum*, *Bacillus licheniformis*, and *Zhihengliuela alba*, reduced ethylene production by 53, 57, and 44%, respectively, and improved plant growth [54]. Another PGPR strain, *Achromobacter piechaudii* ARV8, which also produced ACC deaminase, conferred IST against salt stress in pepper and tomato [72]. In groundnut grown under saline field conditions, the plant growth-promoting effects of ACC deaminase-possessed *Pseudomonas fluorescens* TDK1 were more pronounced compared to strains lacking the enzyme [71]. *Pseudomonas putida* UW4 protected canola seedlings from growth inhibition by high levels of salt [107]. Siddikee and coworkers [47] have also confirmed that inoculation with 14 halotolerant bacterial strains ameliorates salt stress in canola plants through the reduction of ethylene production *via* ACC deaminase activity.

Inoculation of maize plants with *Pseudomonas fluorescens* containing ACC deaminase boosted root elongation and fresh weight significantly under saline conditions [108]. Inoculation with *Pseudomonas* spp. containing ACC-deaminase partially eliminated the effects of drought stress on growth, yield, and ripening of pea (*Pisum sativum*) [109]. Nadeem and coworkers [110] reported that rhizobacteria capable of producing ACC deaminase mitigate salt stress in wheat. Climate change-induced ethylene production in plants reduces plant growth and development significantly and, if not properly monitored, can result in plant death [111]. As a result, it is clear that an increase in ethylene production is an indicator of susceptibility to various stresses [112, 113].

Protective Compounds

In plants, various nitrogen-containing compounds (NCC) such as amino acids, amids, imino acids, proteins, quaternary ammonium compounds (QAC) and polyamines accumulate when exposed to salinity stress. In the saline environment, the specific NCC that accumulates varies with the plant species. Several studies correlated the accumulation of NCC with salt tolerance in plants [114]. A very high accumulation of cellular proline (up to 80% of the amino acid pool under stress and 5% under normal conditions) due to increased synthesis and decreased degradation under a variety of stress conditions such as salt and drought has been documented in many plant species [115].

Some PGPRs enhance the production of compatible osmolytes in inoculated plants. Microbial inoculation enhanced proline accumulation in the roots and provided tolerance to plants under salinity stress [116]. Increased production of proline along with decreased electrolyte leakage, maintenance of relative water content of leaves, and selective uptake of K^+ ions resulted in salt tolerance in *Zea mays* inoculated with *Rhizobium* and *Pseudomonas* [117]. Hayat and coworkers [118] reported that proline protects membranes and proteins from the effects of high concentrations of inorganic ions and temperature extremes and acts as a protein-compatible hydroxyl radical scavenger. Under environmental stresses, the accumulation of proline buffers cellular redox potential [119]. Under stress conditions, NCC has been proposed to perform several functions, including osmotic adjustment, protection of cellular macromolecules, the storage form of nitrogen, maintaining cellular pH, detoxification of the cells, and free radical scavenging [120, 121].

Antioxidative Enzymes

Salt stress affects many physiological processes in plants negatively. The generation of reactive oxygen species is promoted by salinity, and subsequently, oxidative damage of cellular components takes place [122]. Salt stress not only imposes osmotic stress and ion toxicity but also oxidative stress [123], which can stimulate the accumulation of reactive oxygen species (ROS) such as superoxide, hydrogen peroxide, hydroxyl radicals, and singlet oxygen [124]. Plant cells produce both antioxidant enzymes and non-enzymatic antioxidants to protect against stress conditions [125].

Several strains of PGPR can improve salinity by increasing the concentration of antioxidative enzymes like glutathione reductase and ascorbate peroxidase and their antioxidant activity in inoculated salt-stressed plants. Inoculation of *Bradyrhizobium japonicum* with *Serratia proteamaculans* in soybean has been reported to increase antioxidant activity and concentration of proline and malondialdehyde, and also enhance the activity of antioxidative enzymes (glutathione reductase and ascorbate peroxidase) [61]. Induction of antioxidant enzymes (catalase and total peroxidase) was involved in the alleviation of salinity stress in lettuce plants inoculated with PGPR strains [65]. Salinity decreased the dry weight of the shoots and roots for all lettuce plants. However, the plants inoculated with *P. mendocina* had significantly greater shoot biomass than the control plants at both medium and high salinity levels. Medicago plants inoculated with IAA-overproducing PGPR strains showed high antioxidant enzyme activity, which contributed to enhancing plant protection against salt stress [126]. A reduction in the activity of antioxidant enzymes was also observed in barley plants. Omar and coworkers [64] reported that, without inoculation, salinity led to a significant increase in catalase and peroxidase activities in salt-stressed leaves of two barley cultivars differing in salinity tolerance. Inoculation of the two cultivars with *Azospirillum brasilense* lowered the magnitude of increase and significantly ameliorated the deleterious effects of salinity, improving crop productivity. Salt stress tolerance in plants needs activation of antioxidative pathways to control plant cells' injurious effects [122].

Exopolysaccharides

EPS production is stimulated by salts that have a significant Na^+ removal capability from an aqueous solution. With the application of cyanobacterial EPS vigour index, seed germination and mobilization efficiency in the crops have improved, which indicates the significant role of EPS in salt stress alleviation [127]. Certain bacteria such as *Pseudomonas* can survive under stress conditions due to exopolysaccharide (EPS) production, which possesses water-holding and cementing properties. Therefore, it plays an important role in the formation and stabilization of soil aggregates, regulation of mineral nutrients, and water flow across plant roots through biofilm formation [128]. Bacterial EPSs have a great capacity for binding cations as well as trace elements [129]. Thus, increasing the population density of EPS-producing bacteria in the root zone would decrease the amount of Na^+ available for plant uptake and help in alleviating salt stress in plants growing in saline environments. Due to the higher proportion of the root zones covered with soil sheaths in EPS-producing rhizobacteria inoculated wheat roots, a reduced apoplasmic (passive) flow of Na^+ into the stele may restrict Na uptake [130]. The inoculation also substantially increased the dry matter yield of roots and shoots. EPS-producing rhizobacteria also increased the RS mass/root mass ratio as well as the population density of EPS bacteria on the rhizoplane, and both these parameters were significantly correlated with the content of water-insoluble saccharides in the RS.

Ion Homeostasis

The tendency of a cell or an organism to maintain an internal steady state, even in response to any environmental perturbation or stimulus tending to disturb normality, because of the coordination, can be called homeostasis. Especially, under controlled conditions, ions constantly flux in and out of cells, with net flux adjusted to accommodate cellular requirements, thus maintaining ionic homeostasis. The two most important ions that induce salt stress in plants are Na^+ and Cl^-. Sodium is a non-essential but beneficial element, whereas Cl^- is an essential phytomicronutrient. However, both are potentially toxic in excessive concentrations, triggering specific disorders and causing substantial damage to crops. Salt tolerance in glycophyte species is mostly related to the exclusion of these ions from the leaves, thereby avoiding or delaying toxic effects [131]. Hence, any contribution of the soil biota towards maintaining the homeostasis of toxic ions must benefit plant growth under salinity. Microbes can influence root uptake of toxic ions and nutrients by modifying host physiology (by regulating ion transporter expression and/or activity) or directly reducing the foliar accumulation of toxic ions (Na^+, Cl^-), and may even improve K^+/Na^+ ratios in beneficial plant/microbe interactions [132].

Inoculation of salt-stressed maize with ACC deaminase containing PGPRs increased plant N, R, and K uptake in salinized maize plants [69] and increased P, K^+, and Ca^{2+} uptake at the expense of Mg^{2+} and Na^+ uptake in salinized tomatoes [72], resulting in higher K^+/Na^+ ratios. High K^+/Na^+ ratios were also found in salt-stressed maize, in which selectivity for Na^+, K^+ and Ca^{++} was altered upon inoculation with *Azospirillum* [75]. Similarly, inoculation of pepper with *Bacillus* spp. TW4 led to relief from osmotic stress, which is often manifested as salinity (or drought) stress. Salt stress alleviation of *Lotus glaber* by *Glomus intraradices* BAFC 3108 has been related to decreased root and shoot N accumulation and enhanced root K^+ concentrations [133].

Salt tolerance can be a macroevolutionary self-destructive trait because it is gained frequently, but is also lost easily by reversal or extinction [134]. Despite the high ion concentration in the growth environment, AMF *Glomus intraradices* could selectively take up elements such as K^+, Mg^{2+}, and Ca^{2+}, while avoiding Na^+ uptake [135] to keep internal K^+/Na^+ and Ca^{2+}/Na^+ ratios within narrow limits. The fungus selectively excluded Nat but included Cl^- as indicated by the tissue concentration and distribution of these ions. These selective mechanisms for ion uptake could partially alleviate salinity stress in host plants by obtaining a significant proportion of the elemental uptake *via* the mycorrhizal fungi [135], resulting in often higher K^+/Na^+ ratios in mycorrhized plants [136]. It has been speculated that AMF could modify the expression of plant ion transporters, thus altering the uptake of both nutrients and toxic ions. Although overexpression of Na^+/H^+ and K^+/H^+ antiporters improves salt tolerance in plants [137], no changes in the expression of the antiporter genes LeNHX1 (Na^+/H^+) and LeNHX2 (K^+/H^+) of mycorrhized tomato plants were observed under salt stress [138]. Nevertheless, selective uptake of ions by mycorrhized plants indicates an important role of AMF under salinized soils.

Salt-tolerant species have emerged in at least more than a hundred different flowering plant species [139]. *Arabidopsis* plants exposed to bacterial volatile organic compounds from *Bacillus subtilis* decreased root transcriptional expression of a high-affinity K^+ transporter (AtHKT1) but upregulated it in the shoots. This resulted in decreased root Na^+ import and, simultaneously, facilitated Na^+ exclusion from the shoot by retrieving Na^+ from the xylem and also facilitated root-to-root Na recirculation [68]. As a result, there was reduced Na^+ accumulation in the plants and improved salt tolerance. Higher stomatal conductance and photosynthetic activities in PGPR-inoculated plants [140] resulted in lower toxic ion accumulation (Na^+ and Cl^-) and improved leaf K^+/Na^+, alleviating osmotic stress [141]. Soil microbes may also modulate other indirect and novel mechanisms to increase plant salt tolerance. Plant salt tolerance is improved by limiting Na^+ uptake through an unknown mechanism, improving photochemical efficiency, protecting cell membrane integrity, increasing antioxidant enzyme activity, and decreasing water loss [134]. AMF (*Glomus intraradices*) can accumulate and also enhance concentrations of silicon (a role in limiting Na^+ uptake) in the host plant [142].

ENHANCEMENT OF NUTRIENT UP-TAKE

According to Su and coworkers [143], soil salinization affects almost 10% of the land surface (950 million ha) globally, and this problem is expected to be exacerbated by the current environmental destruction. The mineral nutritional status of plants greatly affects their ability to adapt to adverse environmental conditions and, in particular, to abiotic stress. The impaired mineral nutrient status of plants exacerbates the adverse effects of abiotic stresses. An exogenous supply of macronutrients can help in the alleviation of the adverse effects of stress on plant growth [144]. In plants, potassium is an essential macronutrient that plays an important role in abiotic stress tolerance [145, 146]. AMF has been shown to have a positive influence on nutrient uptake by plants grown in salt-stress conditions by increasing their absorption and/or translocation [116]. The mycorrhizal dependency of the plant increases with increasing salt concentrations [147]. Improved plant growth and yield under saline conditions have also been linked to altered nutrient status in other host/AMF combinations, including *Glomus clarum* and mungbean (*Vigna radiata*) [148], *Glomus clarum* and pepper [149], and *Glomus etunicatum* and *Glomus mosseae* in wheat (*Triticum aestivum*) [150]. AMP promotes salinity tolerance in crop plants. AMF is capable of reducing the negative effects of salt stress by increasing antioxidant defense mechanisms in response to salinity stress conditions. AMF has significant potential to mitigate the salt-induced deleterious effects by maintaining the osmotic balance by regulating the Na^+ and K^+ ratio [151].

Nitrogen

The problem of salinity stress is caused by the excessive uptake of toxic ions. The most common ionic composition of saline soil is NaCl. Nitrogen is the most important nutrient for plant growth. Nitrogen, particularly in the form of nitrates, represents a stress signal that triggers the activation of antioxidant enzymes to safeguard the plants against salinity-induced oxidative damage [152]. Salinity interferes with nitrogen acquisition and utilization by the plants [153]. AMF-inoculated plants can assimilate nitrogen better. Giri and Mukerji [147] recorded a higher accumulation of N in shoots of mycorrhizal *Sesbania grandiflora* and *S. aegyptiaca* than in non-mycorrhizal control plants. Increased N uptake in an AM plant due to a change in N metabolism brought about by changes in the enzymes associated with N metabolism [154]. Improved N nutrition may help to reduce the toxic effects of Na ions by reducing their uptake, and this may indirectly help in maintaining the chlorophyll content of the plant [155].

Phosphorus

Plants grown under higher salinity often suffer from the toxicity of Na^+ and Cl^- ions, osmotic stress, nutritional disorders, and oxidative stress that result in a reduction in photosynthesis and inhibition of plant growth [156]. Plants are salt-sensitive or hypersensitive with characteristics of low resistance to salt stress [157]. After nitrogen, phosphorus is the second major nutrient for plant growth. Despite a large amount of total P, most of the soil P exists in plant-unavailable forms [158]. In an agricultural system, low P availability limits crop production, especially in regions where calcareous and alkaline soils largely prevail [159]. To cope with this constraint, plants have developed adaptive strategies, which include alterations in root morphology and architecture; acidification of root rhizosphere; secretion of acid phosphatases and organic acids from roots; induction of high-affinity P transport systems; changes in carbohydrate metabolism; and formation of symbiotic associations with mycorrhizal fungi [158, 160].

Under saline soil conditions, phosphate ions precipitate with Ca^{2+}, Mg^{2+}, and Zn^{2+} ions and become unavailable to plants [161]. Thus, there is decreased P accumulation in plants, resulting in P-deficiency [162]. Therefore, P solubilization or fertilization is necessary for plant growth, which may help mitigate salt stress by overcoming the P-binding capacity of the soil [163]. Phosphorus exists in two forms in soil, organic and inorganic phosphate, and possesses limited mobility in the soil [164]. Many PGPRs can solubilize insoluble inorganic phosphate compounds, thus making P available for uptake by the plant [163].

Higher P content was reported in mycorrhizal than in non-mycorrhizal *Acacia nilotica* and *Trifolium alexandrinum* plants in saline soils [136, 165]. The higher P concentrations in mycorrhizal plants under saline soil conditions suggest that AMF increased P uptake in plants under saline conditions. Mycorrhizal inoculation helps alleviate P deficiency by enhancing its uptake, facilitated by the extensive hyphae of the fungus, which increases the root surface area and thereby its nutrient absorption capacity as compared to non-mycorrhizal plants [95]. Improved P nutrition in AM-inoculated plants could have improved their growth rate, improved integrity and stability of the plasma membrane, enhanced salt tolerance, improved antioxidant status, and enhanced nodulation and nitrogen fixation in legumes [166]. Enhanced uptake of P by AMF in plants grown under saline conditions could have helped in reducing the negative effects of Na and C ions by maintaining a vacuolar membrane that prevented ions from interfering in metabolic pathways of growth [167]. Maintained membrane integrity facilitates compartmentalization within vacuoles and selective ion intake [168].

ROOT HYDRAULIC CONDUCTANCE

Salt stress disturbs plant growth and yield formation. Thus, plants respond to salinity by altering their physiological parameters to maintain their water balance. For example, the reduction in root hydraulic conductivity is one of the first responses of plants to the presence of salt to minimize water stress [169]. Both mycorrhizae and PGPR can improve root hydraulic conductance, resulting in increased water absorption capacity. This may lead to increased plant growth and subsequent dilution of toxic ion effects [170]. Improved hydraulic conductivity of the root at low water potential resulted in relatively higher water content in inoculated plants compared with uninoculated plants [6]. Under saline (dSm^{-1}) stress, mycorrhizal (*Glomus intraradices* BEG 123) bean (*Phaseolus vulgaris*) plants had greater osmotic root hydraulic conductance [171], which was not due to a concentrating effect of decreased osmotically driven sap flux, as this was approximately doubled in AM plants. Instead, it was proposed that mycorrhizal roots increase active solute transport as a mechanism to sustain water flow across the root. Regulation of root hydraulic properties by AM symbiosis was strongly correlated with the regulation of aquaporin (PvPIP2) protein abundance and phosphorylation state [171].

CONCLUDING REMARKS

Abiotic stresses, such as salinity, drought, temperature extremes, and heavy metals, are major factors that adversely affect crop productivity and sustainability worldwide. Abiotic stresses disturb plant growth and yield. Approximately 90% of arable lands are prone to one or more of these stresses and are responsible for up to 70% of the yield losses in major food crops. Rhizospheric/endophytic microorganisms play a very important role in helping alleviate abiotic stresses on plants. Several mechanisms have been suggested for microbial alleviation of salinity stress, including the production of phytohormones, improved plant nutrient status, production of ACC deaminase, salt exclusion, and enhanced resistance to drought in plant cells. Microbes can play an important role in salt stress alleviation in sensitive and moderately sensitive crops. A wide range of microorganisms are available, having diverse mechanisms for salt stress alleviation in plants. Future research needs to be directed towards field evaluation for the validation of the potential microbes.

REFERENCES

[1]   Shahid SA, Zaman M, Heng L, Eds. Soil Salinity: Historical Perspectives and a World Overview of the Problem. In: Guideline for Salinity Assessment, Mitigation and Adaptation Using Nuclear and Related Techniques. Cham: Springer 2018; pp. 43-53.
[http://dx.doi.org/10.1007/978-3-319-96190-3_2]

[2]   Al-Karaki GN. Nursery inoculation of tomato with arbuscular mycorrhizal fungi and subsequent

performance under irrigation with saline water. Sci Hortic (Amsterdam) 2006; 109: 1-7.
[http://dx.doi.org/10.1016/j.scienta.2006.02.019]

[3] FAO Land and nutrition management services 2008.[http://www.fao.org/soils-portal/soi--management/management-of-some-problem-soils/salt-affected-soils/more-informati-n-on-salt-affected-soils/en/]

[4] Ruiz-Lozano JM, Collados C, Barea JM, *et al*. Arbuscular mycorrhizal symbiosis can alleviate drought induced nodule senescence in soybean plants. New Phytol 2001; 151: 493-502.
[http://dx.doi.org/10.1046/j.0028-646x.2001.00196.x]

[5] Wang W, Vinocur B, Altman A. Plant responses to drought, salinity and extreme temperatures: Towards genetic engineering for stress tolerance. Planta 2003; 218: 1-14.
[http://dx.doi.org/10.1007/s00425-003-1105-5] [PMID: 14513379]

[6] Sheng M, Tang M, Chen H, *et al*. Influence of arbuscular mycorrhizae on photosynthesis and water status of maize plants under salt stress. Mycorrhiza 2008; 18: 287-96.
[http://dx.doi.org/10.1007/s00572-008-0180-7] [PMID: 18584217]

[7] Mandal AK, Sharma RC. Delination and characterization of waterlogged salt affected soils in IGNP using remote sensing and GIS. J Indian Soc Remote Sens 2011; 39: 39-50.
[http://dx.doi.org/10.1007/s12524-010-0051-5]

[8] Zörb C, Geilfus CM, Dietz KJ. Salinity and crop yield. Plant Biol 2019; 21: 31-8.
[http://dx.doi.org/10.1111/plb.12884] [PMID: 30059606]

[9] Grayson M. Agriculture and drought. Nature 2013; 501: S1.
[http://dx.doi.org/10.1038/501S1a] [PMID: 24067757]

[10] Riadh K, Wided M, Hans-Werner K, *et al*. Responses of halophytes to environmental stresses with special emphasis to salinity. Adv Bot Res 2010; 53: 117-45.
[http://dx.doi.org/10.1016/S0065-2296(10)53004-0]

[11] Narsing Rao MP, Dong ZY, Xiao M, *et al*. Effect of Salt Stress on Plants and Role of Microbes in Promoting Plant Growth under Salt Stress. In: Giri B, Varma A, Eds. Microorganisms in Saline Environments: Strategies and Functions. Soil Biology, 56. Cham: Springer 2019; pp. 423-35.
[http://dx.doi.org/10.1007/978-3-030-18975-4_18]

[12] Meena KK, Sorty AM, Bitla UM, *et al*. Abiotic stress responses and microbe-mediated mitigation in plants: The omics strategies. Front Plant Sci 2017; 8: 172.
[http://dx.doi.org/10.3389/fpls.2017.00172] [PMID: 28232845]

[13] Pimentel D, Berger B, Filiberto D, *et al*. Water resources: Agricultural and environmental issues. BioScience 2004; 54: 909-18.
[http://dx.doi.org/10.1641/0006-3568(2004)054[0909:WRAAEI]2.0.CO;2]

[14] Qadir M, Quillérou E, Nangia V, *et al*. Economics of salt-induced land degradation and restoration. Nat Resour Forum 2014; 38: 282-95.
[http://dx.doi.org/10.1111/1477-8947.12054]

[15] FAOSTAT statistical database 2012.[faostat.fao.org]

[16] Machado R, Serralheiro R. Soil salinity: Effect on vegetable crop growth. Management practices to prevent and mitigate soil salinization. Horticulturae 2017; 3: 30.
[http://dx.doi.org/10.3390/horticulturae3020030]

[17] Bernstein L, Hayward HE. Physiology of salt tolerance. Annu Rev Plant Physiol 1958; 9: 25-46.
[http://dx.doi.org/10.1146/annurev.pp.09.060158.000325]

[18] Dehnavi AR, Zahedi M, Ludwiczak A, *et al*. Effect of salinity on seed germination and seedling development of sorghum (*Sorghum bicolor* [L.] Moench) genotypes. Agronomy (Basel) 2020; 10: 859.
[http://dx.doi.org/10.3390/agronomy10060859]

[19] Kent LM, Lauchli A. Germination and seedling growth of cotton: Salinity-calcium interactions. Plant Cell Environ 1985; 8: 155-9.
[http://dx.doi.org/10.1111/j.1365-3040.1985.tb01223.x]

[20] Kaveh H, Nemati H, Farsi M, *et al.* How salinity affect germination and emergence of tomato lines. J Biol Environ Sci 2011; 5: 159-63. [https://dergipark.org.tr/en/pub/jbes/issue/38003/438816]

[21] Mauromicale G, Licandro P. Salinity and temperature effects on germination, emergence and seedling growth of globe artichoke. Agronomie 2002; 22: 443-50.
[http://dx.doi.org/10.1051/agro:2002011]

[22] Kafi M, Goldani M. Effect of water potential and type of osmoticum on seed germination of three crop species of wheat, sugarbeet, and chickpea. J Agric Sci Technol 2001; 15: 121-33.

[23] Barrow CJ. World atlas of desertification (United nations environment programme), edited by N. Middleton and D. S. G. Thomas. Edward Arnold, London, 1992. isbn 0 340 55512 2, £89.50 (hardback), ix + 69 pp. LDD 1992; 3: 249.
[http://dx.doi.org/10.1002/ldr.3400030407]

[24] FAO, ITPS. Status of the World's Soil Resources (SWSR) – Main Report. Rome: Food and Agriculture Organization of the United Nations and Intergovernmental Technical Panel on Soils, Rome, Italy. 2015; p. 650.[https://www.fao.org/3/i5199e/i5199e.pdf]

[25] Tareq MJ, Hossain MA, Mojakkir AM, *et al.* Effect of salinity on reproductive growth of wheat. Bangladesh J Seed Sci Tech 2011; 15: 111-116.

[26] Netondo GW, Onyango JC, Beck E. Sorghum and salinity: II. Gas exchange and chlorophyll fluorescence of sorghum under salt stress. Crop Sci 2004; 44: 806-11.
[http://dx.doi.org/10.2135/cropsci2004.8060]

[27] Troyo-Diéguez E, Murillo-Amador B. Effects of salinity on the germination and seedling characteristics of cowpea [*Vigna unguiculata* (L.) Walp.]. Aust J Exp Agric 2000; 40: 433-8.
[http://dx.doi.org/10.1071/EA99009]

[28] Yadav SP, Bharadwaj R, Nayak H, *et al.* Impact of salt stress on growth, productivity and physicochemical properties of plants: A review. Int J Chem Stud 2019; 7: 1793-8.
[https://www.chemijournal.com/archives/2019/vol7issue2/PartAD/7-2-319-755.pdf]

[29] Läuchli A, Epstein E. Plant Responses to Saline and Sodic Conditions. . In: Tanji KK, Ed. Agricultural Salinity Assessment and Management. Vol. 71. ASCE Manuals and Reports on Engineering Practice, American Society of Civil Engineers, New York, 1990; pp. 113-137

[30] Maas EV, Poss JA. Salt sensitivity of wheat at different growth stages. Irrig Sci 1989; 10: 29-40.

[31] Khatun S, Flowers TJ. Effects of salinity on seed set in rice. Plant Cell Environ 1995; 18: 61-7.
[http://dx.doi.org/10.1111/j.1365-3040.1995.tb00544.x]

[32] Egamberdieva D, Wirth S, Bellingrath-Kimura SD, *et al.* Salt-tolerant plant growth promoting rhizobacteria for enhancing crop productivity of saline soils. Front Microbiol 2019; 10: 2791.
[http://dx.doi.org/10.3389/fmicb.2019.02791] [PMID: 31921005]

[33] Niu X, Song L, Xiao Y, *et al.* Drought-tolerant plant growth-promoting rhizobacteria associated with foxtail millet in a semi-arid agroecosystem and their potential in alleviating drought stress. Front Microbiol 2018; 8: 2580.
[http://dx.doi.org/10.3389/fmicb.2017.02580] [PMID: 29379471]

[34] Litchfield CD, Gillevet PM. Microbial diversity and complexity in hypersaline environments: A preliminary assessment. J Ind Microbiol Biotechnol 2002; 28: 48-55.
[http://dx.doi.org/10.1038/sj/jim/7000175] [PMID: 11938471]

[35] Singh RP, Jha PN. A halotolerant bacterium *Bacillus licheniformis* HSW-16 augments induced systemic tolerance to salt stress in wheat plant (*Triticum aestivum*). Front Plant Sci 2016; 7: 1890.
[http://dx.doi.org/10.3389/fpls.2016.01890] [PMID: 28018415]

[36] Sarkar A, Ghosh PK, Pramanik K, *et al.* A halotolerant *Enterobacter* sp. displaying ACC deaminase activity promotes rice seedling growth under salt stress. Res Microbiol 2018; 169: 20-32.
[http://dx.doi.org/10.1016/j.resmic.2017.08.005] [PMID: 28893659]

[37] Sharma S, Kulkarni J, Jha B. Halotolerant rhizobacteria promote growth and enhance salinity tolerance in peanut. Front Microbiol 2016; 7: 1600.
[http://dx.doi.org/10.3389/fmicb.2016.01600] [PMID: 27790198]

[38] Rajput L, Imran A, Mubeen F, *et al.* Salt-tolerant PGPR strain *Planococcus rifietoensis* promotes the growth and yield of wheat (*Triticum aestivum* L.) cultivated in saline soil. Pak J Bot 2013; 45: 1955-62.

[39] Czech L, Hermann L, Stöveken N, *et al.* Role of the extremolytes ectoine and hydroxyectoine as stress protectants and nutrients: Genetics, phylogenomics, biochemistry, and structural analysis. Genes (Basel) 2018; 9: 177.
[http://dx.doi.org/10.3390/genes9040177] [PMID: 29565833]

[40] Gunde-Cimerman N, Plemenitaš A, Oren A. Strategies of adaptation of microorganisms of the three domains of life to high salt concentrations. FEMS Microbiol Rev 2018; 42: 353-75.
[http://dx.doi.org/10.1093/femsre/fuy009] [PMID: 29529204]

[41] Shultana R, Kee Zuan AT, Yusop MR, *et al.* Characterization of salt-tolerant plant growth-promoting rhizobacteria and the effect on growth and yield of saline-affected rice. PLoS One 2020; 15: e0238537.
[http://dx.doi.org/10.1371/journal.pone.0238537] [PMID: 32886707]

[42] Grover M, Ali SZ, Sandhya V, *et al.* Role of microorganisms in adaptation of agriculture crops to abiotic stresses. World J Microbiol Biotechnol 2011; 27: 1231-40.
[http://dx.doi.org/10.1007/s11274-010-0572-7]

[43] Upadhyay SK, Singh DP, Saikia R. Genetic diversity of plant growth promoting rhizobacteria isolated from rhizospheric soil of wheat under saline condition. Curr Microbiol 2009; 59: 489-96.
[http://dx.doi.org/10.1007/s00284-009-9464-1] [PMID: 19701667]

[44] Majeed A, Abbasi MK, Hameed S, *et al.* Isolation and characterization of plant growth-promoting rhizobacteria from wheat rhizosphere and their effect on plant growth promotion. Front Microbiol 2015; 6: 198.
[http://dx.doi.org/10.3389/fmicb.2015.00198] [PMID: 25852661]

[45] Etesami H, Beattie GA. Plant-Microbe Interactions in Adaptation of Agricultural Crops to Abiotic Stress Conditions. In: Kumar V, Kumar M, Sharma S, *et al.*, Eds. Probiotics and Plant Health. Singapore: Springer 2017; pp. 163-200.
[http://dx.doi.org/10.1007/978-981-10-3473-2_7]

[46] Egamberdieva D, Kucharova Z. Selection for root colonising bacteria stimulating wheat growth in saline soils. Biol Fertil Soils 2009; 45: 563-71.
[http://dx.doi.org/10.1007/s00374-009-0366-y]

[47] Siddikee MA, Chauhan PS, Anandham R, *et al.* Isolation, characterization, and use for plant growth promotion under salt stress, of ACC deaminase-producing halotolerant bacteria derived from coastal soil. J Microbiol Biotechnol 2010; 20: 1577-84.
[http://dx.doi.org/10.4014/jmb.1007.07011] [PMID: 21124065]

[48] Cardinale M, Ratering S, Suarez C, *et al.* Paradox of plant growth promotion potential of rhizobacteria and their actual promotion effect on growth of barley (*Hordeum vulgare* L.) under salt stress. Microbiol Res 2015; 181: 22-32.
[http://dx.doi.org/10.1016/j.micres.2015.08.002] [PMID: 26640049]

[49] Soldan R, Mapelli F, Crotti E, *et al.* Bacterial endophytes of mangrove propagules elicit early establishment of the natural host and promote growth of cereal crops under salt stress. Microbiol Res 2019; 223-225: 33-43.

[http://dx.doi.org/10.1016/j.micres.2019.03.008] [PMID: 31178049]

[50] Zhu F, Qu L, Hong X, *et al.* Isolation and characterization of a phosphate-solubilizing halophilic bacterium *Kushneria* sp. YCWA18 from Daqiao Saltern on the coast of Yellow Sea of China. Evid Based Complement Alternat Med 2011; 2011: 1-6.
[http://dx.doi.org/10.1155/2011/615032] [PMID: 21716683]

[51] Park KH, Lee OM, Jung HI, *et al.* Rapid solubilization of insoluble phosphate by a novel environmental stress-tolerant *Burkholderia vietnamiensis* M6 isolated from ginseng rhizospheric soil. Appl Microbiol Biotechnol 2010; 86: 947-55.
[http://dx.doi.org/10.1007/s00253-009-2388-7] [PMID: 20024543]

[52] Nabti E, Sahnoune M, Adjrad S, *et al.* A halophilic and osmotolerant *Azospirillum brasilense* strain from Algerian soil restores wheat growth under saline conditions. Eng Life Sci 2007; 7: 354-60.
[http://dx.doi.org/10.1002/elsc.200720201]

[53] Yasmin H, Bano A. Isolation and characterization of phosphate solubilizing bacteria from rhizosphere soil of weeds of Khewra salt range and Attock. Pak J Bot 2011; 43: 1663-8.

[54] Banerjee S. Stress induced phosphate solubilization by *Arthrobacter* sp. and *Bacillus* sp. isolated from tomato rhizosphere. Aust J Crop Sci 2010; 4: 378-69.

[55] Siddikee MA, Glick BR, Chauhan PS, *et al.* Enhancement of growth and salt tolerance of red pepper seedlings (*Capsicum annuum* L.) by regulating stress ethylene synthesis with halotolerant bacteria containing 1-aminocyclopropane-1-carboxylic acid deaminase activity. Plant Physiol Biochem 2011; 49: 427-34.
[http://dx.doi.org/10.1016/j.plaphy.2011.01.015] [PMID: 21300550]

[56] Singh RP, Jha PN. The multifarious PGPR *Serratia marcescens* CDP-13 augments induced systemic resistance and enhanced salinity tolerance of wheat (*Triticum aestivum* L.). PLoS One 2016; 11: e0155026.
[http://dx.doi.org/10.1371/journal.pone.0155026] [PMID: 27322827]

[57] Vurukonda SSKP, Vardharajula S, Shrivastava M, *et al.* Enhancement of drought stress tolerance in crops by plant growth promoting rhizobacteria. Microbiol Res 2016; 184: 13-24.
[http://dx.doi.org/10.1016/j.micres.2015.12.003] [PMID: 26856449]

[58] Qurashi AW, Sabri AN. Bacterial exopolysaccharide and biofilm formation stimulate chickpea growth and soil aggregation under salt stress. Braz J Microbiol 2012; 43: 1183-91.
[http://dx.doi.org/10.1590/S1517-83822012000300046] [PMID: 24031943]

[59] Kang SM, Joo GJ, Hamayun M, *et al.* Gibberellin production and phosphate solubilization by newly isolated strain of *Acinetobacter calcoaceticus* and its effect on plant growth. Biotechnol Lett 2009; 31: 277-81.
[http://dx.doi.org/10.1007/s10529-008-9867-2] [PMID: 18931973]

[60] Barassi CA, Ayrault G, Creus CM, *et al.* Seed inoculation with *Azospirillum* mitigates NaCl effects on lettuce. Sci Hortic (Amsterdam) 2006; 109: 8-14.
[http://dx.doi.org/10.1016/j.scienta.2006.02.025]

[61] Han HS, Lee KD. Physiological responses of soybean-inoculation of *Bradyrhizobium japonicum* with PGPR in saline soil conditions. Res J Agric Biol Sci 2005; 3: 216-21.

[62] Jha Y, Subramanian RB, Patel S. Combination of endophytic and rhizospheric plant growth promoting rhizobacteria in *Oryza sativa* shows higher accumulation of osmoprotectant against saline stress. Acta Physiol Plant 2011; 33: 797-802.
[http://dx.doi.org/10.1007/s11738-010-0604-9]

[63] Marulanda A, Barea J-M, Azcón R. Stimulation of plant growth and drought tolerance by native microorganisms (AM fungi and bacteria) from dry environments: Mechanisms related to bacterial effectiveness. J Plant Growth Regul 2009; 28: 115-24.
[http://dx.doi.org/10.1007/s00344-009-9079-6]

[64] Omar MNA, Osman MEH, Kasim WA, *et al.* Improvement of Salt Tolerance Mechanisms of Barley Cultivated Under Salt Stress Using *Azospirillum brasilense*. In: Ashraf M, Ozturk M, Athar H, Eds. Salinity and Water Stress: Improving Crop Efficiency. Tasks for Vegetation Sciences, vol 44. Dordrecht, The Netherlands: Springer 2009; 44: pp. 133-147.
[http://dx.doi.org/10.1007/978-1-4020-9065-3_15]

[65] Kohler J, Caravaca F, Roldán A. An AM fungus and a PGPR intensify the adverse effects of salinity on the stability of rhizosphere soil aggregates of *Lactuca sativa*. Soil Biol Biochem 2010; 42: 429-34.
[http://dx.doi.org/10.1016/j.soilbio.2009.11.021]

[66] Azarmi-Atajan F, Sayyari-Zohan MH. Alleviation of salt stress in lettuce (*Lactuca sativa* L.) by plant growth-promoting rhizobacteria. J Hortic Postharvest Res 2020; 3: 67-78.
[http://dx.doi.org/10.22077/jhpr.2020.3013.1114]

[67] Dardanelli MS, Fernández de Córdoba FJ, Espuny MR, *et al.* Effect of *Azospirillum brasilense* coinoculated with *Rhizobium* on *Phaseolus vulgaris* flavonoids and Nod factor production under salt stress. Soil Biol Biochem 2008; 40: 2713-21.
[http://dx.doi.org/10.1016/j.soilbio.2008.06.016]

[68] Zhang H, Kim MS, Sun Y, *et al.* Soil bacteria confer plant salt tolerance by tissue-specific regulation of the sodium transporter HKT1. Mol Plant Microbe Interact 2008; 21: 737-44.
[http://dx.doi.org/10.1094/MPMI-21-6-0737] [PMID: 18624638]

[69] Nadeem SM, Zahir ZA, Naveed M, *et al.* Preliminary investigations on inducing salt tolerance in maize through inoculation with rhizobacteria containing ACC deaminase activity. Can J Microbiol 2007; 53: 1141-9.
[http://dx.doi.org/10.1139/W07-081] [PMID: 18026206]

[70] Wu Z, Peng Y, Guo L, *et al.* Root colonization of encapsulated *Klebsiella oxytoca* Rs-5 on cotton plants and its promoting growth performance under salinity stress. Eur J Soil Biol 2014; 60: 81-7.
[http://dx.doi.org/10.1016/j.ejsobi.2013.11.008]

[71] Saravanakumar D, Samiyappan R. ACC deaminase from *Pseudomonas fluorescens* mediated saline resistance in groundnut (*Arachis hypogea*) plants. J Appl Microbiol 2007; 102: 1283-92.
[http://dx.doi.org/10.1111/j.1365-2672.2006.03179.x] [PMID: 17448163]

[72] Mayak S, Tirosh T, Glick BR. Plant growth-promoting bacteria confer resistance in tomato plants to salt stress. Plant Physiol Biochem 2004; 42: 565-72.
[http://dx.doi.org/10.1016/j.plaphy.2004.05.009] [PMID: 15246071]

[73] Nawaz MS, Arshad A, Rajput L, *et al.* Growth-stimulatory effect of quorum sensing signal molecule *N*-Acyl-homoserine lactone-producing multi-trait *Aeromonas* spp. on wheat genotypes under salt stress. Front Microbiol 2020; 11: 553621.
[http://dx.doi.org/10.3389/fmicb.2020.553621] [PMID: 33117303]

[74] Rajput L, Imran A, Mubeen F, *et al.* Wheat (*Triticum aestivum* L.) growth promotion by halo-tolerant PGPR-consortium. Soil Environ 2018; 37: 178-89.

[75] Hamdia MAES, Shaddad MAK, Doaa MM. Mechanisms of salt tolerance and interactive effects of *Azospirillum brasilense* inoculation on maize cultivars grown under salt stress conditions. Plant Growth Regul 2004; 44: 165-74.
[http://dx.doi.org/10.1023/B:GROW.0000049414.03099.9b]

[76] Fasciglione G, Casanovas EM, Yommi A, *et al. Azospirillum* improves lettuce growth and transplant under saline conditions. J Sci Food Agric 2012; 92: 2518-23.
[http://dx.doi.org/10.1002/jsfa.5661] [PMID: 22473714]

[77] Yao L, Wu Z, Zheng Y, *et al.* Growth promotion and protection against salt stress by *Pseudomonas putida* Rs-198 on cotton. Eur J Soil Biol 2010; 46: 49-54.
[http://dx.doi.org/10.1016/j.ejsobi.2009.11.002]

[78] Egamberdieva D. Alleviation of salt stress by plant growth regulators and IAA producing bacteria in

wheat. Acta Physiol Plant 2009; 31: 861-4.
[http://dx.doi.org/10.1007/s11738-009-0297-0]

[79] Ashraf M, Hasnain S, Berge O, *et al.* Inoculating wheat seedlings with exopolysaccharide-producing bacteria restricts sodium uptake and stimulates plant growth under salt stress. Biol Fertil Soils 2004; 40: 157-62.
[http://dx.doi.org/10.1007/s00374-004-0766-y]

[80] Afrasayab S, Faisal M, Hasnain S. Comparative study of wild and transformed salt tolerant bacterial strains on *Triticum aestivum* growth under salt stress. Braz J Microbiol 2010; 41: 946-55.
[http://dx.doi.org/10.1590/S1517-83822010000400013]

[81] Yildirim E, Turan M, Donmez MF. Mitigation of salt stress in radish (*Raphanus sativus* L.) by plant growth promoting rhizobacteria. Rom Biotechnol Lett 2008; 3: 3933-43.
[https://rombio.unibuc.ro/wp-content/uploads/2022/05/13-5-8.pdf]

[82] Zou YN, Wu QS, Huang YM, *et al.* Mycorrhizal-mediated lower proline accumulation in *Poncirus trifoliata* under water deficit derives from the integration of inhibition of proline synthesis with increase of proline degradation. PLoS One 2013; 8: e80568.
[http://dx.doi.org/10.1371/journal.pone.0080568] [PMID: 24260421]

[83] Ahemad M, Kibret M. Mechanisms and applications of plant growth promoting *rhizobacteria*: Current perspective. J King Saud Univ Sci 2014; 26: 1-20.
[http://dx.doi.org/10.1016/j.jksus.2013.05.001]

[84] Gupta B, Huang B. Mechanism of salinity tolerance in plants: Physiological, biochemical, and molecular characterization. Int J Genomics 2014; 2014: 1-18.
[http://dx.doi.org/10.1155/2014/701596] [PMID: 24804192]

[85] Mokrani S, Nabti E, Cruz C. Current advances in plant growth promoting bacteria alleviating salt stress for sustainable agriculture. Appl Sci (Basel) 2020; 10: 7025.
[http://dx.doi.org/10.3390/app10207025]

[86] Jini D, Joseph B. Physiological mechanism of salicyclic acid for alleviation of stress in rice. Rice Sci 2017; 24: 97-108.
[http://dx.doi.org/10.1016/j.rsci.2016.07.007]

[87] Shrivastava P, Kumar R. Soil salinity: A serious environmental issue and plant growth promoting bacteria as one of the tools for its alleviation. Saudi J Biol Sci 2015; 22: 123-31.
[http://dx.doi.org/10.1016/j.sjbs.2014.12.001] [PMID: 25737642]

[88] Arora NK, Fatima T, Mishra J, *et al.* Halo-tolerant plant growth promoting rhizobacteria for improving productivity and remediation of saline soils. J Adv Res 2020; 26: 69-82.
[http://dx.doi.org/10.1016/j.jare.2020.07.003] [PMID: 33133684]

[89] Manivannan M, Tholkappian P. Prevalence of *Azospirillum* isolates in tomato rhizosphere soils of coastal areas of Cuddalore district, Tamil Nadu. Int J Recent Sci Res 2013; 4: 1610-3.
[https://recentscientific.com/sites/default/files/Download_649.pdf]

[90] Egamberdieva D, Lugtenberg B. Use of Plant Growth-Promoting Rhizobacteria to Alleviate Salinity Stress in Plants. In: Miransari M, Eds. Use of Microbes for the Alleviation of Soil Stresses. New York, NY: Springer 2014; 1: pp. 73-96.
[http://dx.doi.org/10.1007/978-1-4614-9466-9_4]

[91] Fujii H, Zhu JK. *Arabidopsis* mutant deficient in 3 abscisic acid-activated protein kinases reveals critical roles in growth, reproduction, and stress. Proc Natl Acad Sci USA 2009; 106: 8380-5.
[http://dx.doi.org/10.1073/pnas.0903144106] [PMID: 19420218]

[92] Cutler SR, Rodriguez PL, Finkelstein RR, *et al.* Abscisic acid: Emergence of a core signaling network. Annu Rev Plant Biol 2010; 61: 651-79.
[http://dx.doi.org/10.1146/annurev-arplant-042809-112122] [PMID: 20192755]

[93] Albacete A, Ghanem ME, Martínez-Andújar C, *et al.* Hormonal changes in relation to biomass

partitioning and shoot growth impairment in salinized tomato (*Solanum lycopersicum* L.) plants. J Exp Bot 2008; 59: 4119-31.
[http://dx.doi.org/10.1093/jxb/ern251] [PMID: 19036841]

[94] Irum N, Asghari B, Ul-Hassan T. Isolation of phytohormones producing plant growth promoting rhizobacteria from weeds growing in Khewra salt range, Pakistan and their implication in providing salt tolerance to *Glycine max* L. Afr J Biotechnol 2009; 8: 5762-8.
[http://dx.doi.org/10.5897/AJB09.1176]

[95] Begum N, Qin C, Ahanger MA, *et al*. Role of arbuscular mycorrhizal fungi in plant growth regulation: Implications in abiotic stress tolerance. Front Plant Sci 2019; 10: 1068.
[http://dx.doi.org/10.3389/fpls.2019.01068] [PMID: 31608075]

[96] Jahromi F, Aroca R, Porcel R, *et al*. Influence of salinity on the *in vitro* development of *Glomus intraradices* and on the *in vivo* physiological and molecular responses of mycorrhizal lettuce plants. Microb Ecol 2008; 55: 45-53.
[http://dx.doi.org/10.1007/s00248-007-9249-7] [PMID: 17393053]

[97] Aroca R, Vernieri P, Ruiz-Lozano JM. Mycorrhizal and non-mycorrhizal *Lactuca sativa* plants exhibit contrasting responses to exogenous ABA during drought stress and recovery. J Exp Bot 2008; 59: 2029-41.
[http://dx.doi.org/10.1093/jxb/ern057] [PMID: 18469324]

[98] Pavlů J, Novák J, Koukalová V, *et al*. Cytokinin at the crossroads of abiotic stress signaling pathways. Int J Mol Sci 2018; 19: 2450.
[http://dx.doi.org/10.3390/ijms19082450] [PMID: 30126242]

[99] Ghanem ME, Albacete A, Smigocki AC, *et al*. Root-synthesized cytokinins improve shoot growth and fruit yield in salinized tomato (*Solanum lycopersicum* L.) plants. J Exp Bot 2011; 62: 125-40.
[http://dx.doi.org/10.1093/jxb/erq266] [PMID: 20959628]

[100] Dodd IC, Zinovkina NY, Safronova VI, *et al*. Rhizobacterial mediation of plant hormone status. Ann Appl Biol 2010; 157: 361-79.
[http://dx.doi.org/10.1111/j.1744-7348.2010.00439.x]

[101] Ludwig-Müller J. Hormonal Responses in Host Plants Triggered by Arbuscular Mycorrhizal Fungi. In: Koltai H, Kapulnik Y, Eds. Arbuscular Mycorrhizas: Physiology and Function. Dordrecht: Springer 2010; pp. 169-90.

[102] Arkhipova TN, Prinsen E, Veselov SU, *et al*. Cytokinin producing bacteria enhance plant growth in drying soil. Plant Soil 2007; 292: 305-15.
[http://dx.doi.org/10.1007/s11104-007-9233-5]

[103] Gupta S, Pandey S. ACC deaminase producing bacteria with multifarious plant growth promoting traits alleviates salinity stress in french bean (*Phaseolus vulgaris*) plants. Front Microbiol 2019; 10: 1506.
[http://dx.doi.org/10.3389/fmicb.2019.01506] [PMID: 31338077]

[104] Liu JL, Xie BM, Shi XH, *et al*. Effects of two plant growth-promoting rhizobacteria containing 1-aminocyclopropane-1-carboxylate deaminase on oat growth in petroleum-contaminated soil. Int J Environ Sci Technol 2015; 12: 3887-94.
[http://dx.doi.org/10.1007/s13762-015-0798-x]

[105] Abiri R, Shaharuddin NA, Maziah M, *et al*. Role of ethylene and the APETALA 2/ethylene response factor superfamily in rice under various abiotic and biotic stress conditions. Environ Exp Bot 2017; 134: 33-44.
[http://dx.doi.org/10.1016/j.envexpbot.2016.10.015]

[106] Glick BR, Todorovic B, Czarny J, *et al*. Promotion of plant growth by bacterial ACC deaminase. Crit Rev Plant Sci 2007; 26: 227-42.
[http://dx.doi.org/10.1080/07352680701572966]

[107] Cheng Z, Park E, Glick BR. 1-Aminocyclopropane-1-carboxylate deaminase from *Pseudomonas putida* UW4 facilitates the growth of canola in the presence of salt. Can J Microbiol 2007; 53: 912-8.
[http://dx.doi.org/10.1139/W07-050] [PMID: 17898846]

[108] Kausar R, Shahzad SM. Effect of ACC-deaminase containing rhizobacteria on growth promotion of maize under salinity stress. J Agric Soc Sci 2006; 2: 216-8. [https://www.fspublishers.org/published_papers/81545_..pdf]

[109] Arshad M, Shaharoona B, Mahmood T. Inoculation with *Pseudomonas* spp. containing ACC-deaminase partially eliminates the effects of drought stress on growth, yield, and ripening of pea (*Pisum sativum* L.). Pedosphere 2008; 18: 611-20.
[http://dx.doi.org/10.1016/S1002-0160(08)60055-7]

[110] Nadeem SM, Zahir ZA, Naveed M, *et al.* Rhizobacteria capable of producing ACC-deaminase may mitigate salt stress in wheat. Soil Sci Soc Am J 2010; 74: 533-42.
[http://dx.doi.org/10.2136/sssaj2008.0240]

[111] Dubois M, Van den Broeck L, Inzé D. The pivotal role of ethylene in plant growth. Trends Plant Sci 2018; 23: 311-23.
[http://dx.doi.org/10.1016/j.tplants.2018.01.003] [PMID: 29428350]

[112] Glick BR. Bacteria with ACC deaminase can promote plant growth and help to feed the world. Microbiol Res 2014; 169: 30-9.
[http://dx.doi.org/10.1016/j.micres.2013.09.009] [PMID: 24095256]

[113] Müller M, Munné-Bosch S. Ethylene response factors: A key regulatory hub in hormone and stress signalling. Plant Physiol 2015; 169: 32-41.
[http://dx.doi.org/10.1104/pp.15.00677] [PMID: 26103991]

[114] Parida AK, Das AB. Salt tolerance and salinity effects on plants: A review. Ecotoxicol Environ Saf 2005; 60: 324-49.
[http://dx.doi.org/10.1016/j.ecoenv.2004.06.010] [PMID: 15590011]

[115] Szabados L, Savouré A. Proline: A multifunctional amino acid. Trends Plant Sci 2010; 15: 89-97.
[http://dx.doi.org/10.1016/j.tplants.2009.11.009] [PMID: 20036181]

[116] Sharifi M, Ghorbanli M, Ebrahimzadeh H. Improved growth of salinity-stressed soybean after inoculation with salt pre-treated mycorrhizal fungi. J Plant Physiol 2007; 164: 1144-51.
[http://dx.doi.org/10.1016/j.jplph.2006.06.016] [PMID: 16919369]

[117] Bano A, Fatima M. Salt tolerance in *Zea mays* (L). following inoculation with *Rhizobium* and *Pseudomonas*. Biol Fertil Soils 2009; 45: 405-13.
[http://dx.doi.org/10.1007/s00374-008-0344-9]

[118] Hayat S, Hayat Q, Alyemeni MN, *et al.* Role of proline under changing environments. Plant Signal Behav 2012; 7: 1456-66.
[http://dx.doi.org/10.4161/psb.21949] [PMID: 22951402]

[119] Wahid A, Close TJ. Expression of dehydrins under heat stress and their relationship with water relations of sugarcane leaves. Biol Plant 2007; 51: 104-9.
[http://dx.doi.org/10.1007/s10535-007-0021-0]

[120] Mansour MMF. Nitrogen containing compounds and adaptation of plants to salinity stress. Biol Plant 2000; 43: 491-500.
[http://dx.doi.org/10.1023/A:1002873531707]

[121] Zhong C, Cao X, Hu J, *et al.* Nitrogen metabolism in adaptation of photosynthesis to water stress in rice grown under different nitrogen levels. Front Plant Sci 2017; 8: 1079.
[http://dx.doi.org/10.3389/fpls.2017.01079] [PMID: 28690622]

[122] Soleimani Z, Afshar AS, Nematpour FS. Responses of antioxidant gene and enzymes to salinity stress in the *Cuminum cyminum* L. Russ J Plant Physiol 2017; 64: 361-7.

[http://dx.doi.org/10.1134/S1021443717030177]

[123] Abid M, Zhang YJ, Li Z, et al. Effect of Salt stress on growth, physiological and biochemical characters of four kiwifruit genotypes. Sci Hortic (Amsterdam) 2020; 271: : e109473.
[http://dx.doi.org/10.1016/j.scienta.2020.109473]

[124] Hasanuzzaman M, Bhuyan MHM, Zulfiqar F, et al. Reactive oxygen species and antioxidant defense in plants under abiotic stress: Revisiting the crucial role of a universal defense regulator. Antioxidants 2020; 9: 681.
[http://dx.doi.org/10.3390/antiox9080681] [PMID: 32751256]

[125] Khan A, Numan M, Khan AL, et al. Melatonin: Awakening the defense mechanisms during plant oxidative stress. Plants 2020; 9: 407.
[http://dx.doi.org/10.3390/plants9040407] [PMID: 32218185]

[126] Bianco C, Defez R. *Medicago truncatula* improves salt tolerance when nodulated by an indole-3-acetic acid-overproducing *Sinorhizobium meliloti* strain. J Exp Bot 2009; 60: 3097-107.
[http://dx.doi.org/10.1093/jxb/erp140] [PMID: 19436044]

[127] Arora M, Kaushik A, Rani N, et al. Effect of cyanobacterial exopolysaccharides on salt stress alleviation and seed germination. J Environ Biol 2010; 31: 701-4.
[http://www.jeb.co.in./journal_issues/201009_sep10/paper_25.pdf]
[PMID: 21387925]

[128] Nguyen PT, Nguyen TT, Bui DC, et al. Exopolysaccharide production by lactic acid bacteria: The manipulation of environmental stresses for industrial applications. AIMS Microbiol 2020; 6: 451-69.
[http://dx.doi.org/10.3934/microbiol.2020027] [PMID: 33364538]

[129] Osińska-Jaroszuk M, Jaszek M, Starosielec M, et al. Bacterial exopolysaccharides as a modern biotechnological tool for modification of fungal laccase properties and metal ion binding. Bioprocess Biosyst Eng 2018; 41: 973-89.
[http://dx.doi.org/10.1007/s00449-018-1928-x] [PMID: 29582151]

[130] Upadhyay SK, Singh JS, Singh DP. Exopolysaccharide plant growth promoting rhizobacteria under salinity condition. Pedosphere 2011; 21: 214-22.
[http://dx.doi.org/10.1016/S1002-0160(11)60120-3]

[131] Sharma A, Shahzad B, Rehman A, et al. Response of phenylpropanoid pathway and the role of polyphenols in plants under abiotic stress. Molecules 2019; 24: 2452.
[http://dx.doi.org/10.3390/molecules24132452] [PMID: 31277395]

[132] Zuccarini P, Okurowska P. Effects of mycorrhizal colonization and fertilization on growth and photosynthesis of sweet basil under salt stress. J Plant Nutr 2008; 31: 497-513.
[http://dx.doi.org/10.1080/01904160801895027]

[133] Sannazzaro AI, Ruiz OA, Albertó EO, et al. Alleviation of salt stress in *Lotus glaber* by *Glomus intraradices*. Plant Soil 2006; 285: 279-87.
[http://dx.doi.org/10.1007/s11104-006-9015-5]

[134] Bromham L, Hua X, Cardillo M. Macroevolutionary and macroecological approaches to understanding the evolution of stress tolerance in plants. Plant Cell Environ 2020; 43: 2832-46.
[http://dx.doi.org/10.1111/pce.13857] [PMID: 32705700]

[135] Hammer EC, Nasr H, Pallon J, et al. Elemental composition of arbuscular mycorrhizal fungi at high salinity. Mycorrhiza 2011; 21: 117-29.
[http://dx.doi.org/10.1007/s00572-010-0316-4] [PMID: 20499112]

[136] Giri B, Kapoor R, Mukerji KG. Improved tolerance of *Acacia nilotica* to salt stress by arbuscular mycorrhiza, *Glomus fasciculatum* may be partly related to elevated K/Na ratios in root and shoot tissues. Microb Ecol 2007; 54: 753-60.
[http://dx.doi.org/10.1007/s00248-007-9239-9] [PMID: 17372663]

[137] Rodríguez-Rosales MP, Jiang X, Gálvez FJ, et al. Overexpression of the tomato K^+/H^+ antiporter

LeNHX2 confers salt tolerance by improving potassium compartmentalization. New Phytol 2008; 179: 366-77.
[http://dx.doi.org/10.1111/j.1469-8137.2008.02461.x] [PMID: 19086176]

[138] Ouziad F, Wilde P, Schmelzer E, *et al.* Analysis of expression of aquaporins and Na^+/H^+ transporters in tomato colonized by arbuscular mycorrhizal fungi and affected by salt stress. Environ Exp Bot 2006; 57: 177-86.
[http://dx.doi.org/10.1016/j.envexpbot.2005.05.011]

[139] Santos J, Al-Azzawi M, Aronson J, *et al.* eHALOPH a database of salt-tolerant plants: Helping put halophytes to work. Plant Cell Physiol 2016; 57: e10.
[http://dx.doi.org/10.1093/pcp/pcv155] [PMID: 26519912]

[140] del Amor FM, Cuadra-Crespo P. Plant growth-promoting bacteria as a tool to improve salinity tolerance in sweet pepper. Funct Plant Biol 2012; 39: 82-90.
[http://dx.doi.org/10.1071/FP11173] [PMID: 32480762]

[141] Pérez-Alfocea F, Albacete A, Ghanem ME, *et al.* Hormonal regulation of source - sink relations to maintain crop productivity under salinity: A case study of root-to-shoot signalling in tomato. Funct Plant Biol 2010; 37: 592-603.
[http://dx.doi.org/10.1071/FP10012]

[142] Mulet JM, Campos F, Yenush L. Editorial: Ion homeostasis in plant stress and development. Front Plant Sci 2020; 11: : 618273.
[http://dx.doi.org/10.3389/fpls.2020.618273] [PMID: 33329682]

[143] Su N, Wu Q, Chen J, *et al.* GABA operates upstream of H^+-ATPase and improves salinity tolerance in *Arabidopsis* by enabling cytosolic K^+ retention and Na^+ exclusion. J Exp Bot 2019; 70: 6349-61.
[http://dx.doi.org/10.1093/jxb/erz367] [PMID: 31420662]

[144] Sarwar M, Saleem MF, Ullah N, *et al.* Role of mineral nutrition in alleviation of heat stress in cotton plants grown in glasshouse and field conditions. Sci Rep 2019; 9: 13022.
[http://dx.doi.org/10.1038/s41598-019-49404-6] [PMID: 31506449]

[145] Shi XL, Zhou DY, Guo P, *et al.* External potassium mediates the response and tolerance to salt stress in peanut at the flowering and needling stages. Photosynthetica 2020; 58: 1141-9.
[http://dx.doi.org/10.32615/ps.2020.070]

[146] Zhang X, Wu H, Chen J, *et al.* Chloride and amino acids are associated with K^+-alleviated drought stress in tea (*Camellia sinesis*). Funct Plant Biol 2020; 47: 398-408.
[http://dx.doi.org/10.1071/FP19221] [PMID: 32138810]

[147] Giri B, Mukerji KG. Mycorrhizal inoculant alleviates salt stress in *Sesbania aegyptiaca* and *Sesbania grandiflora* under field conditions: Evidence for reduced sodium and improved magnesium uptake. Mycorrhiza 2004; 14: 307-12.
[http://dx.doi.org/10.1007/s00572-003-0274-1] [PMID: 14574620]

[148] Rabie GH. Influence of arbuscular mycorrhizal fungi and kinetin on the response of mungbean plants to irrigation with seawater. Mycorrhiza 2005; 15: 225-30.
[http://dx.doi.org/10.1007/s00572-004-0345-y] [PMID: 15765207]

[149] Kaya C, Ashraf M, Sonmez O, *et al.* The influence of arbuscular mycorrhizal colonisation on key growth parameters and fruit yield of pepper plants grown at high salinity. Sci Hortic (Amsterdam) 2009; 121: 1-6.
[http://dx.doi.org/10.1016/j.scienta.2009.01.001]

[150] Daei G, Ardekani MR, Rejali F, *et al.* Alleviation of salinity stress on wheat yield, yield components, and nutrient uptake using arbuscular mycorrhizal fungi under field conditions. J Plant Physiol 2009; 166: 617-25.
[http://dx.doi.org/10.1016/j.jplph.2008.09.013] [PMID: 19100656]

[151] Borde M, Dudhane M, Kulkarni M. Role of Arbuscular Mycorrhizal Fungi (AMF) in Salinity

[152] Ahanger MA, Qin C, Begum N, *et al.* Nitrogen availability prevents oxidative effects of salinity on wheat growth and photosynthesis by up-regulating the antioxidants and osmolytes metabolism, and secondary metabolite accumulation. BMC Plant Biol 2019; 19: 479.
[http://dx.doi.org/10.1186/s12870-019-2085-3] [PMID: 31703619]

[153] Lewis OAM, Leidi EO, Lips SH. Effect of nitrogen source on growth response to salinity stress in maize and wheat. New Phytol 1989; 111: 155-60.
[http://dx.doi.org/10.1111/j.1469-8137.1989.tb00676.x] [PMID: 33874262]

[154] Masclaux-Daubresse C, Daniel-Vedele F, Dechorgnat J, *et al.* Nitrogen uptake, assimilation and remobilization in plants: Challenges for sustainable and productive agriculture. Ann Bot (Lond) 2010; 105: 1141-57.
[http://dx.doi.org/10.1093/aob/mcq028] [PMID: 20299346]

[155] Acosta-Motos J, Ortuño M, Bernal-Vicente A, *et al.* Plant responses to salt stress: Adaptive mechanisms. Agronomy (Basel) 2017; 7: 18.
[http://dx.doi.org/10.3390/agronomy7010018]

[156] Munns R, Tester M. Mechanisms of salinity tolerance. Annu Rev Plant Biol 2008; 59: 651-81.
[http://dx.doi.org/10.1146/annurev.arplant.59.032607.092911] [PMID: 18444910]

[157] Yokoi S, Bressan RA, Hasegawa PM. Salt stress tolerance of plants. JIRCAS working report 2002; 25-33.

[158] Tang H, Niu L, Wei J, *et al.* Phosphorus limitation improved salt tolerance in maize through tissue mass density increase, osmolytes accumulation, and Na^+ uptake inhibition. Front Plant Sci 2019; 10: 856.
[http://dx.doi.org/10.3389/fpls.2019.00856] [PMID: 31333699]

[159] Vance CP, Uhde-Stone C, Allan DL. Phosphorus acquisition and use: Critical adaptations by plants for securing a nonrenewable resource. New Phytol 2003; 157: 423-47.
[http://dx.doi.org/10.1046/j.1469-8137.2003.00695.x] [PMID: 33873400]

[160] Lambers H, Shane MW, Cramer MD, *et al.* Root structure and functioning for efficient acquisition of phosphorus: Matching morphological and physiological traits. Ann Bot (Lond) 2006; 98: 693-713.
[http://dx.doi.org/10.1093/aob/mcl114] [PMID: 16769731]

[161] Pessarakli M, Haghighi M, Sheibanirad A. Plant responses under environmental stress conditions. Adv Plants Agric Res 2015; 2: 276-86.
[http://dx.doi.org/10.15406/apar.2015.02.00073]

[162] Parida AK, Das AB. Effects of NaCl stress on nitrogen and phosphorous metabolism in a true mangrove *Bruguiera parviflora* grown under hydroponic culture. J Plant Physiol 2004; 161: 921-8.
[http://dx.doi.org/10.1016/j.jplph.2003.11.006] [PMID: 15384403]

[163] Shilev S. Plant-growth-promoting bacteria mitigating soil salinity stress in plants. Appl Sci (Basel) 2020; 10: 7326.
[http://dx.doi.org/10.3390/app10207326]

[164] Hayat R, Ali S, Amara U, *et al.* Soil beneficial bacteria and their role in plant growth promotion: A review. Ann Microbiol 2010; 60: 579-98.
[http://dx.doi.org/10.1007/s13213-010-0117-1]

[165] Shokri S, Maadi B. Effects of arbuscular mycorrhizal fungus on the mineral nutrition and yield of *Trifolium alexandrinum* plants under salinity stress. J Agron 2009; 8: 79-83.
[http://dx.doi.org/10.3923/ja.2009.79.83]

[166] Garg N, Manchanda G. Effect of arbuscular mycorrhizal inoculation of salt-induced nodule senescence

in *Cajanus cajan* (pigeonpea). J Plant Growth Regul 2008; 27: 115-24.
[http://dx.doi.org/10.1007/s00344-007-9038-z]

[167] Evelin H, Devi TS, Gupta S, *et al.* Mitigation of salinity stress in plants by arbuscular mycorrhizal symbiosis: Current understanding and new challenges. Front Plant Sci 2019; 10: 470.
[http://dx.doi.org/10.3389/fpls.2019.00470] [PMID: 31031793]

[168] Casares D, Escribá PV, Rselló CA. Membrane lipid composition: Effect on membrane and organelle structure, function and compartmentalization and therapeutic avenues. Int J Mol Sci 2019; 20: 2167.
[http://dx.doi.org/10.3390/ijms20092167] [PMID: 31052427]

[169] Calvo-Polanco M, Sánchez-Romera B, Aroca R. Mild salt stress conditions induce different responses in root hydraulic conductivity of *phaseolus vulgaris* over-time. PLoS One 2014; 9: : e90631.
[http://dx.doi.org/10.1371/journal.pone.0090631] [PMID: 24595059]

[170] Groppa MD, Benavides MP, Zawoznik MS. Root hydraulic conductance, aquaporins and plant growth promoting microorganisms: A revision. Appl Soil Ecol 2012; 61: 247-54.
[http://dx.doi.org/10.1016/j.apsoil.2011.11.013]

[171] Aroca R, Porcel R, Ruiz-Lozano JM. How does arbuscular mycorrhizal symbiosis regulate root hydraulic properties and plasma membrane aquaporins in *Phaseolus vulgaris* under drought, cold or salinity stresses? New Phytol 2007; 173: 808-16.
[http://dx.doi.org/10.1111/j.1469-8137.2006.01961.x] [PMID: 17286829]

CHAPTER 10

Lignocellulose Degrading Bacteria in Soil

Archana Rawat[1], Parul Bhatt Kotiyal[1,*], Soni Singh[1] and Neeraj Verma[2]

[1] *Forest Ecology and Climate Change Division, Forest Research Institute, Dehradun, India*
[2] *Department of Agriculture Science, AKS University, Satna, MP, India*

Abstract: The degradation of wood is a highly complex process involving the activities of several different microbes. It has been explored through research that microorganisms have developed various strategies (enzymatic and nonenzymatic) to utilize wood. In the present article, we are presenting the enzymes that originated from fungi and bacteria and their reactions to decomposing wood. Analysis of enzymes involved in wood degradation will not only be helpful in the study of the wood degradation process but also provide information about various ecological niches of the microorganisms. Genomic and secretome data have revealed the importance of the enzymes secreted by microorganisms such as fungi and bacteria in wood degradation in ecological niches.

Keywords: Lignocellulose, Biodegradation, Enzymes, Lignin-modifying enzymes.

INTRODUCTION

Organic matter is destroyed by microorganisms, such as bacteria and fungi, through biodegradation. There are three stages in the biodegradation process: biodeterioration, biofragmentation, and assimilation [1]. The process of biodeterioration can sometimes be described as a surface-level degradation of materials resulting in changes in mechanical, physical, and chemical properties [2]. Abiotic factors in the outside environment deteriorate the material's structure, allowing for further degradation. Abiotic factors such as mechanical pressure, temperature, light, chemicals, *etc.*, can influence these initial changes. Biodeterioration usually occurs at the beginning of biodegradation, and sometimes it can occur in parallel to biofragmentation. Hueck defined biodeterioration as the unnecessary alteration to the structure of an object caused by the growth of living organisms, involving the breakdown of stone facade,

---

* Corresponding author Parul Bhatt Kotiyal: Forest Ecology and Climate Change Division, Forest Research Institute, Dehradun, India; E-mail: parulbhatt29@gmail.com

Shampi Jain, Ashutosh Gupta and Neeraj Verma
All rights reserved-© 2023 Bentham Science Publishers

corrosion of metals by microorganisms, or simply the aesthetic changes caused by the growth of living organisms on a structure [3].

Biofragmentation of a polymer is a lytic process in which bonds within a polymer are cleaved, resulting in oligomers and monomers. The steps taken to defragment these materials also differ based on the presence of oxygen in the system. Aerobic digestion refers to the breakdown of materials by microorganisms in the presence of oxygen. Anaerobic digestion is the breakdown of materials in the absence of oxygen. The anaerobic reactions produce methane, whereas the aerobic reactions do not (both reactions produce CO_2, H_2O, residual matter, and new biomass). Furthermore, aerobic digestion typically occurs faster than anaerobic digestion, whereas anaerobic digestion reduces the volume and mass of the material. Due to this, anaerobic digestion produces natural gas, while anaerobic digestion is widely used for waste management systems and as a source of local, renewable energy [4].

Biofragmentation products are later integrated into microbial cells, which is the assimilation stage. Some products of fragmentation are easily transported within the cell by membrane carriers. Others, on the other hand, must undergo biotransformation reactions in order to produce products that can then be transported inside the cell [5]. Once inside the cell, the products enter catabolic pathways that either produce adenosine triphosphate (ATP) or elements that contribute to the cell's structure.

FACTORS AFFECTING THE BIODEGRADATION PROCESS

Practically every chemical compound and material is subjected to biodegradation. But the significance lies in the relative rate at which these processes occur, such as days, weeks, years, or centuries. Several factors, such as light, water, oxygen, and temperature, determine the rate of degradation of organic compounds. Several methods are available to measure the rate of biodegradation. Respirometry tests are one of the methods that are used for aerobic microbes [6] in which the resulting amount of CO_2 serves as an indicator of degradation. Biodegradability can also be measured by anaerobic microbes and the amount of methane or alloy that they can produce [7].

It is important to note that factors that affect biodegradation rates during product testing ensure that the results produced are accurate and reliable. Several materials are tested as being biodegradable under optimal conditions in a lab for approval, but these results may not reflect real-world outcomes, where factors are more variable. For example, a material that may have been tested as biodegrading at a high rate in the lab may not degrade at a high rate in a landfill because landfills often lack light, water, and microbial activity that are necessary for degradation to occur. As a result, there must be standards for biodegradable plastic products, which have a significant environmental impact. In order to ensure that all plastics produced and commercialized will biodegrade naturally, it is important to develop and use accurate standard testing methods. The DINV54900 test has been developed for this purpose [8].

Due to the reaction with oxygen in the air, very large polymer molecules, which contain only carbon and hydrogen, enable plastic products to decompose in a week to one to two years. This reaction happens at a very slow pace without prodegradants. Therefore, conventional plastics persist in the environment for a long time. Formulations containing oxy-biodegradable ingredients catalyze and accelerate the biodegradation process. However, it takes an extensive amount of experience and skill to balance the ingredients to give the formulations a useful life, followed by degradation and biodegradation. Polyester is one of the known examples of a biodegradable product. In the bio-medical field, biodegradable technology is especially useful.

Biodegradable polymers can be classified into three groups, *viz.*, (i) Medical, (ii) Ecological, and (iii) Dual application. In terms of origin, they can be divided into two groups, *viz.*, (i) Natural and (ii) Synthetic. The Clean Technology Group (CTG) is exploiting the use of supercritical carbon dioxide, which under high pressure at room temperature is a solvent that can be used to make biodegradable plastics into polymer drug coatings. The polymer is used to encapsulate a drug before injecting it into the body and is based on lactic acid, a compound normally produced in the body, and is thus able to be excreted naturally. As a result, the coating is designed to release controlled levels of medicine over time, reducing the need for injections and maximizing the therapeutic benefit. Professor Howdle suggested that biodegradable polymers are particularly suitable for drug delivery since, once inside the body, they do not require retrieval or further manipulation and degrade into soluble, non-toxic byproducts. The degradation of different polymers may differ in the body, and therefore, the choice of polymer can be made in the right way to achieve the desired release levels [9].

BIODEGRADATION VS. COMPOSTING

There are several definitions for biodegradation and composting, leading to confusion between the terms. These terms are often combined, though they do not have the same meaning. Biodegradation occurs when materials are naturally broken down by microorganisms like bacteria and fungi or by other biological processes. A composting process involves human intervention and a particular set of circumstances leading to biodegradation [10].

Biodegradable materials can be decomposed without an oxygen source (anaerobically) into CO_2, H_2O, and biomass, but the timeline is not very specifically defined. Similarly, compostable material can be broken down into CO_2, H_2O, and biomass [11]. However, compostable materials can also be broken down into inorganic compounds. Essentially, composting is an accelerated biodegradation process due to optimized conditions controlled by human beings. Furthermore, the composted material not only returns to its original state but also boosts the soil with beneficial microorganisms, called humus [12]. By using this organic matter in gardens and farms, plants can grow healthier [13].

At-home and commercial composting are the two main types of composting. Although both methods produce healthy soil for reuse, the main difference is what materials can go into the processes. Usually, at-home composting is done with food scraps and excess garden materials like weeds. Composting for commercial purposes can break down more complex plant-based products like corn-based plastics and larger tree branches. For commercial composting, a grinder or a machine is used to break down the materials manually. At-home composting generally involves smaller scale equipment and is not as efficient as large-scale composting, since the material may not fully decompose. So, there are advantages and disadvantages to both commercial and at-home composting.

The use of biodegradation technology has made it possible to dispose of waste much more efficiently. Now there are compost, recycling, and trash bins for optimal waste disposal. The disposal process will not be optimized if these waste streams are regularly and frequently mixed up. These inventions and efforts are wasted when compostable products are thrown out instead of being composted and sent to landfills. Citizens, therefore, require a better understanding of the differences between these terms to properly and efficiently dispose of materials.

Wildlife is at risk from plastic pollution due to illegal dumping. Animals mistake plastics for food, which causes entanglements in their intestines. There are slow-degrading chemicals found in plastics, such as polychlorinated biphenyls (PCBs), nonylphenol (NP), and pesticides, which can later be consumed by wildlife. An environmentalist, Rachel Carson, conducted one of the first key studies on the consequences of chemical ingestion on wildlife, especially birds, in the 1960s.

Human health is also affected by these chemicals. Several diseases and disorders such as cancers, neurological dysfunction, and hormonal changes have been linked to the consumption of tainted food due to the processes called biomagnification and bioaccumulation. An increase in dangerously high levels of mercury in fish, affecting sex hormones in humans, is an example of biomagnification.

Plastics, detergents, metals, and other pollution created by humans that takes a long time to degrade have become a major concern due to the economic cost of remediation. Undegraded materials can also become home to invasive species like barnacles and tubeworms. A change in ecosystems caused by invasive species can alter resident species and the natural balance of resources, resulting in changes in genetic diversity and species richness [14].

LIGNIN-MODIFYING ENZYMES (LMES)

Lignin-modifying enzymes (LMEs) are various types of enzymes produced by fungi and bacteria that catalyze the breakdown of lignin, a biopolymer commonly found in the cell walls of plants. The terms ligninases and lignases are older names for the same class, but the name "lignin-modifying enzymes" is now preferred, given that these enzymes are not hydrolytic but rather oxidative (electron-withdrawing) by their enzymatic mechanisms. LMEs include peroxidases, such as lignin peroxidase (EC 1.11.1.14), manganese peroxidase (EC 1.11.1.13), versatile peroxidase (EC 1.11.1.16), and many phenoloxidases of the laccase type [15, 16].

LMEs have been known to be produced by many species of white rot basidiomycetous fungi, including *Phanerochaete chrysosporium*, *Ceriporiopsis subvermispora*, *Trametes Versicolor*, *Phlebia radiata*, *Pleurotus ostreatus*, and *Pleurotus eryngii* [17].

LMEs are also produced by litter-decomposing basidiomycetous fungi such as *Agaricus bisporus* (button mushroom), and many *Coprinus* and *Agrocybe* species. The brown-rot fungi, which can colonize wood by degrading cellulose, are only able to partially degrade lignin [18].

Some bacteria also produce LMEs, although fungal LMEs are more efficient in lignin degradation. Fungi are thought to be the most substantial contributors to lignin degradation in natural systems [19].

LMEs and cellulases are crucial to ecological cycles (for example, growth/death/decay/regrowth, the carbon cycle, and soil health), because they allow plant tissue to be decomposed quickly, releasing the matter therein for reuse by new generations of life. LMEs are also crucial to several different industries.

Bacterial Lignin Modifying Enzymes

Although much research has been done to understand fungal LMEs, recently more focus has been placed on characterizing these enzymes in bacteria. The main LMEs in both fungi and bacteria are peroxidases and laccases. Although bacteria lack homologs to the most common fungal peroxidases (lignin peroxidase, manganese peroxidase, and versatile peroxidase), many produce dye decolourizing peroxidases (DyP-type peroxidases) [20]. Bacteria from a variety of classes express DyP peroxidases, including gammaproteobacteria, firmicutes, and actinobacteria. Peroxidases depolymerize lignin by oxidation using hydrogen peroxide. Fungal peroxidases have higher oxidizing power than bacterial DyP-type peroxidases studied so far and can degrade more complex lignin structures. DyP-type peroxidases have been found to work on a large range of substrates, including synthetic dyes, monophenolic compounds, lignin-derived compounds, and alcohols.

Laccases, which are multicopper oxidases, are another class of enzymes found in both bacteria and fungi that have significant lignin-degrading properties. Laccases degrade lignin by oxidation using oxygen [21]. Laccases are also widely distributed among bacterial species, including *Bacillus subtilis*, *Caulobacter crescentus*, *Escherichia coli*, and *Mycobacterium tuberculosis*. Like DyP-type peroxidases, bacterial laccases have a wide substrate range [22].

There is interest in using bacterial laccases and DyP peroxidases for industrial applications, biotechnology, and bioremediation because of the greater ease of manipulation of bacterial genomes and gene expression compared to fungi. The wide range of substrates for these types of enzymes also increases the range of processes in which they may be used. These processes include pulp processing, textile dye modification, decontamination of wastewater, and the production of pharmaceutical building blocks. Furthermore, bacterial laccases function at higher temperatures, alkalinity, and salt concentrations than fungal laccases, making them more suitable for industrial use [23].

Both intracellular and extracellular bacterial DyP-type peroxidases and laccases have been identified, suggesting that some are used as intracellular enzymes, while others are secreted to degrade compounds in the environment. However, their roles in bacterial physiology and their natural physiological substrates have yet to be detailed.

Lignocellulosic Biomass

Lignocellulose refers to plant dry matter (biomass), also called lignocellulosic biomass. It is the most abundantly available raw material on Earth for the production of biofuels, mainly bio-ethanol. It is composed of carbohydrate polymers (cellulose, hemicellulose) and an aromatic polymer (lignin). These carbohydrate polymers contain different sugar monomers (six and five-carbon sugars) and they are tightly bound to lignin. Lignocellulosic biomass can be broadly classified into virgin biomass, waste biomass, and energy crops. Virgin biomass includes all naturally occurring terrestrial plants such as trees, bushes, and grass. Waste biomass is produced as a low-value byproduct of various industrial sectors, such as agriculture (corn stover, sugarcane bagasse, straw, *etc.*) and forestry (sawmill and paper mill discards). Energy crops are crops with a high yield of lignocellulosic biomass produced to serve as a raw material for the production of second-generation biofuels, *e.g.*, switchgrass (*Panicum virgatum*) and elephant grass [24].

Lignocellulose has evolved from ancient pteridophytes and gymnosperms to the most evolved grasses, becoming one of the most important complex components of the cell wall of woody plants, with mechanical function to land plants and protection against pests and grazing animals. The lignin biosynthetic pathway is widely believed to be highly conserved throughout evolution. Analyses based on a molecular clock indicate land plants are a monophyletic group that diverged from green algae about 700 million years ago (Ma). Despite the requirement for rigid tubular structures in higher plants, which transport water and nutrients, the presence of cellulose microfibrils, lignin, and hemicellulose creates a natural complex system that can support their huge weight. Due to this, trees are able to face harsh weather conditions such as very strong winds and can also live for thousands of years. An example of this is the North American sequoias, which are the largest single living trees and some of the oldest living trees on the planet [25]. Lignin strengthens plant cell walls by adhering to cellulose microfibrils, letting plants grow significantly in size, enhancing water transport and keeping them resistant to diseases [26].

Lignin degradation is caused by certain fungi as well as several bacterial species. Fungi are more efficient in the breakdown of lignin than bacteria, in which delignification is slower and more limited. Microbial degradation of lignin other than fungi can also be carried out by bacteria. Such lignin-degrading bacteria mainly belong to three classes: Actinomycetes, α-Proteobacteria, and γ-Proteobacteria. These microbes are widely distributed worldwide and occur both in terrestrial and aquatic habitats. They resemble filamentous fungi in their apical and branching growth and morphogenetic development [27]. These bacteria are considered major decomposers of lignocellulose in soils as they produce secondary metabolites and use extracellular enzymes.

Laccase, a ligninolytic enzyme, is one of the most frequently identified and studied in prokaryotes. Phylogenic analysis has revealed that members of four phyla belonging to Archaea have laccase genes. The greatest members of Proteobacteria, Actinobacteria, and Firmicutes have laccase genes, while cyanobacteria and Bacteroidetes have the lowest members [28]. Recent research showed that *Sphingobacterium* (phylum Bacteroides) produces manganese superoxidedismutase, which can oxidize lignin *via* a hydroxyl radical mechanism.

Many strains from the *Brucella*, *Ochrobactrum*, *Sphingobium*, and *Sphingomonas* genera, belonging to the class α-Proteobacteria, decompose lignin. Within another group of ligninolytic bacteria (γ-Proteobacteria), the best known for their degrading abilities are *Pseudomonas fluorescens* (producing lignin peroxidase-LiP), *P. putida* (dye-decolorizing peroxidase-DyP and manganese peroxidase-MnP), *Enterobacter lignolyticus* (catalase/peroxidase HPI and DyP), and *E. coli* (laccase) [29].

Streptomyces viridosporus, *S. paucimobilis*, and *Rhodococcus jostii* belong to the Actinomycetes and can produce catabolic enzymes. The genus *Streptomyces* contains several filamentous species that can degrade lignin. Enzymes from some of these species for lignin degradation have been characterized. Several actinomycete species are known to produce laccase, LiP, and MnP enzymes associated with lignin biodegradation and biochemistry, *e.g.*, *S. coelicolor*, *S. griseus*, and *S. psammoticus* [30]. Laccase and LiP activity have been observed in *S. cinnamomeus*. The best-studied *S. viridosporus* has been reported to produce several extracellular peroxidases and laccase. *S. ipomoea* also has laccase activity. These studies have implicated bacterial degradation of lignin as being more important, particularly in soil [31].

The exact number of fungal species capable of degrading lignin is not known. However, Gilbertson reported approximately 1600 to 1700 species of wood-degrading fungi in North America. Generally, wood-degrading fungi live as saprophytes or weak parasites in natural and man-made forest ecosystems. Saprotrophic fungi live on dead and decaying organic matter and thus prevent the accumulation of organic matter. A large proportion of the removable organic material in the biosphere is composed of lignin, which is degraded by wood-rotting fungi, a group of microorganisms that are only capable of mineralizing this abundant plant substance [32]. Based on wood decomposition, the saprophytic fungi have been classified into three main groups: (i) White rot, (ii) Brown rot, and (iii) Soft rot fungi. These groups can be further subdivided into (iv) Litter-decomposing and (v) Dung-dwelling (coprophilic) fungi that can also decompose lignin [33]. Of these fungi, only white rot is able to decompose wood completely into CO_2 and H_2O [34].

Although over 90% of all known wood-decaying species belong to the white rot type, the members of Basidiomycota are white and brown rot, litter-decomposing, plant-pathogenic, and ectomycorrhizal (ECM) fungi. The fungi belonging to Basidiomycetes can grow on various substrates, *e.g.*, deciduous forest trees, conifers, grassland soils, crops, forest litter, roots, *etc* [35]. The fungi belonging to Polyporales and Agaricales, *e.g.*, *Ganoderma* spp., *Phlebia radiata*, *Lentinula edodes*, *Pleurotus* spp., *etc.*, are white rot fungi. *Gloeophyllum trabeum* is one of the most studied brown rot fungi, along with *Serpula lacrymans* and *Coniophora puteana* (Boletales), that can cause problems in wood used for construction [36]. While white rot Basidiomycetes are found more often on hardwoods of angiospermic plants than on softwoods of gymnosperms, brown rot fungi more commonly attack softwoods and are more often found in coniferous environments.

Lignin degradation mechanisms have been studied in detail in the white rot fungus *Phanerochaete chrysosporium* (currently *Phanerodontia chrysosporium*), producing several isoenzymes of lignin and manganese peroxidases instead of producing laccase. The fungus appears later in wood degradation when available nitrogen is used up by other microorganisms [37].

Ligninolytic enzymes have been isolated from litter-decomposing Basidiomycota. *Agaricus bisporus* has been found to produce both laccase and MnP, while *Marasmius quercophilus* produces only laccase. ECM fungi also have genes to produce enzymes for lignocellulose degradation, but they can degrade lignocellulose slower than other fungi. By assimilating reduced carbon compounds, they are able to survive in the absence of host plants [38].

These fungi can degrade lignin to a limited extent. Anamorphic fungi, such as the commonly occurring *Alternaria alternata*, also cause soft rot decay. A few *Eutypella* species produce white rot decay in the late stages of wood decay, but soft rot in the early stages. Other studied microfungi (*e.g.*, *Penicillium chrysogenum*, *Fusarium solani*, and *F. oxysporum*) have shown lower degrees of lignin decomposition when compared to white rot fungi. Production of extracellular laccases has been observed in the plant pathogenic, anamorphic fungus *Botrytis cinerea*, which causes soft rot-like decay in several horticultural crops (*e.g.*, *Daucus* and *Cucumis*) and noble rot and grey rot diseases in *Vitis vinifera* [39]. Very little research data indicates that lower fungi (Zygomycota and Chytridiomycota) are also involved in wood decomposition. Cellulose and lignin are rarely used by the members of Zygomycota. However, some members of this group can decompose lignin and are confined to the outer layers of decaying plant tissue.

ENZYMES RESPONSIBLE FOR LIGNIN DEGRADATION

Enzymes responsible for lignin degradation can be broadly divided into two main groups, *i.e.*, (1) Lignin-modifying (LMEs) and (2) Lignin-degrading auxiliary (LDAEs) enzymes. LDA enzymes can not degrade lignin alone, but they are necessary to complete the degradation process. LMEs produced by different microorganisms are classified as phenol oxidases (laccases) and heme-containing peroxidases (POD), namely lignin, manganese, and multifunctional (versatile) peroxidase.

Recently, a new group of enzymes, heme-thiolate haloperoxidases (HTPs), has been suggested to have a role in lignin degradation. This class of enzymes has yet to be classified into either the LME or LDA group. These enzymes are supposed to be the most versatile biocatalysts in the hemeprotein family. HTPs share catalytic properties with at least three other classes of heme-containing oxidoreductases, *viz.*, classic plant and fungal peroxidases, cytochrome P450 monooxygenases, and catalases [40]. Since there was no experimental proof that these enzymes alone could degrade lignin, it was concluded to consider them as LDA enzymes [41].

Auxiliary enzymes facilitate lignin breakdown by coordinating the actions of multiple proteins, which may include the production of H_2O_2. They include glyoxal oxidase (GLOX; EC 1.2.3.5), aryl alcohol oxidases (AAO; EC 1.1.3.7), pyranose 2-oxidase (POX; EC 1.1.3.10), cellobiose dehydrogenase (CDH; EC 1.1.99.18) and glucose oxidase (EC 1.1.3.4). Such enzymes are frequently found in the secretomes of white rot fungi [42].

In nature, aerobic white rot fungi of Basidiomycota can completely degrade lignin. Due to a lack of enzymatic machinery, anaerobic fungi are unable to degrade lignin. It is necessary to cleave the aromatic ring, which requires oxygen or its partially reduced form (reactive oxygen species-ROS), and therefore, cannot take place under anaerobic conditions [43].

Important Groups of LMEs

Lignin-modifying peroxidases (LMPs: LiP, MnP, and VP), also called POD, belong to class II peroxidases within the superfamily of catalase-peroxidases (formally called plant and fungal peroxidases). All members of this group contain protoporphyrin IX as a prosthetic group. Based on their amino acid sequences, similarities and catalytic properties, this superfamily can be divided into three subclasses *viz.*, (Class I) including catalase-peroxidases, prokaryotic and organelle-localized eukaryotic heme peroxidases [44], (Class II) including all extracellular fungal heme peroxidases and is classified as the auxiliary activities (AA2) family in carbohydrate-active enzyme (CAZy), and (Class III) containing all secreted plant heme peroxidases. Very recently, a new superfamily of heme peroxidases has been identified in fungi and later in bacteria and is called dye-peroxidases (DyP), originally named 'dye-decolorizing peroxidases'. The DyP enzyme from the Basidiomycete, *Auricularia auricula-judae*, has the potential to cleave lignin substructure linkages [45].

Lignin Peroxidases (LiP)

Lignin peroxidases were first discovered in *Phanerochaete chrysosporium* and later in *Trametes versicolor*, *Bjerkandera* sp., and *Phlebia tremellosa*, which are well known white rot fungi [46]. The activity of LiP has previously been detected in some bacteria, such as *Acinetobacter calcoaceticus* and *Streptomyces viridosporus*. LiP, which can attack lignin polymers and is generally non-specific to the substrate. LiPs have been shown to oxidize several aromatic phenolic compounds, non-phenolic lignin compounds, and various other organic molecules. Certain microorganisms secrete these enzymes in the form of isozymes with different compositions and isoelectric points (pI), often depending on the growth medium and nutrient conditions [47]. The molecular mass of LiPs ranges from 35 to 48 kDa and it has a pI between 3.1 and 4.7. These enzymes can oxidize substrates that are not oxidized by other peroxidases due to their high redox potential (approximately 1.2 V, pH 3.0).

Manganese Dependent Peroxidase (MnP)

One of the most important LMEs, manganese peroxidase, is also found in several isoforms and was first discovered more than 30 years ago in *P. chrysosporium* [48]. They have also been discovered in other Basidiomycetes, including *Panus tigrinus*, *Lenzites betulina*, *Agaricus bisporus*, and *Nematoloma frowardii*. The presence of MnP in bacteria, yeast, and mould has also been reported. Their activity has been observed in *Bacillus pumilus*, *Paenibacillus* sp., and *Azospirillum brasilense* in the absence and presence of various inducers [49], as well as in the actinobacterium, *S. psammoticus* [50]. MnPs contain 4%–18% glycans. The molecular mass and pI range from 38 to 62.5 kDa and 2.9 to 7.1, respectively.

Laccases

Among the most important components of the ligninolytic complex of wood-degrading microorganisms are laccases. Using molecular oxygen as an electron acceptor, all laccases can oxidize several aromatic compounds, including phenolic moieties found in lignin, aromatic amines, benzenothiols, and hydroxyindols. Laccases can also use inorganic or organic metal compounds as substrates, *e.g.*, Mn (II) can be oxidized to Mn (III) and $[Fe(CN)_6]^{2-}$ to $[Fe(CN)_6]^{3-}$. Laccases have been reported from plants, fungi, and certain bacteria (*A. lipoferum*, *B. subtilis*, *S. lavendulae*, and *Sinorhizobium meliloti*) [51] and also in insects. The laccase is localised based on the organism and various physiological functions.

These are extracellular, periplasmic, and intracellular proteins. Intracellular laccases take part in the synthesis of lignin in plants. Intracellular and extracellular laccases are found in most fungi. The fungal laccases take part in various reactions, *e.g.*, pathogenesis, detoxification, lignin degradation, developmental processes, and even the morphogenesis of higher fungi. Laccases are also known as blue multicopper oxidases since they are metalloproteins belonging to the group of polyphenol oxidases that contain copper atoms in the catalytic site [52].

Due to the random polymer nature of lignin and the lower redox potential of laccases, they react with free phenolic fragments of lignin. However, laccase can also react with non-phenolic aromatic compounds with high redox potentials in the presence of some low molecular weight compounds. Substrate oxidation by laccase is a single-electron reaction that produces a free radical [53]. Several isozymes of laccase are produced by different fungi and are guided through the same or different genes. Their number depends on fungi, growth conditions, and the inducer.

Glyoxal Oxidase (GLOX)

It generates extracellular H_2O_2 by the oxidation of various simple dicarbonyl and hydroxycarbonyl compounds, mainly glyoxal and methylglyoxal compounds. There are some reports that, apart from these substrates, glycol aldehyde (derived from lignin) may also be a substrate for GLOX. GLOX, a copper-containing enzyme with an active site, belongs to the CAZy family [54]. It regulates peroxidase activity and is activated *in vitro* in the presence of LiP. GLOX (MW = 68 kDa) has been well characterized in *Phanerochaete chrysosporium*, but GLOX genes have also been reported in other white rot fungi [55].

Aryl Alcohol Oxidase (AAO)

They belong to the glucose-methanol-choline oxidase/dehydrogenase family (GMC). These were initially reported in several agaricales and later from *Aspergillus*, *Fusarium* and in some bacteria also. AAO is a monomeric enzyme composed of two domains with a non-covalently bound FAD cofactor. The molecular weight ranged from 66.7 to 334 kDa. This enzyme produces H_2O_2 by oxidative dehydrogenation of phenolic and non-phenolic arylalcohols, polyunsaturated (aliphatic) primary alcohols or aromatic secondary alcohols [56].

Heme-Thiolate Haloperoxidases

These are very recently discovered enzymes and are of two types: chloroperoxidases (CPO, EC 1.11.1.10) and aromatic peroxygenases (APO, unspecific peroxygenases, EC 1.11.2.1). CPO was first isolated from *Caldariomyces fumago*. There are reports that CPO not only chlorinates structures in lignin but can also break them.

APO (MW 45 kDa) was first isolated from cultures of *Agrocybe aegerita*. It can transfer oxygen to aromatic and aliphatic substrates. APO utilizes H_2O_2 instead of directly transferring molecular O_2. This enzyme has a very high reactivity towards benzylic C–H and phenolic substrates as compared to other peroxidases due to Fe at their active site.

Flavin adenine dinucleotide (FAD)-dependent glucose dehydrogenase (GDH)

GDH, an extracellularly secreted enzyme by several fungi, plays an important role in detoxification by reducing quinones produced by plants [57]. It can also protect the fungal cells from phenoxy radicals of the same origin [58]. During the catalyzation of the anomeric hydroxy group of glucose by GDH, FAD acts as the primary electron acceptor. A detailed study of GDH has been carried out in *Aspergillus*. Recently, studies revealed that GDH also plays an important role in the metabolism of *P. cinnabarinus* [59].

Pyranose 2-Oxidase (POX)

POX [pyranose oxidase (pyranose: oxygen 2-oxidoreductase)](mw up to 300 kDa) has been reported from members of the Basidiomycetes, mainly in the order Polyporales. It catalyzes the C-2 oxidation of aldopyranoses related to the reduction of O_2 to H_2O_2.

Cellobiose Dehydrogenase (CDH)

Cellobiose dehydrogenase (cellobiose: [acceptor] 1-oxi-doreductase) is secreted by white rot fungi, *e.g.*, *P. chrysosporium*, *Trametes versicolor*, *P. sanguineus*, *I. lacteus*, etc. Except for *Coniophora puteana*, the brown rot fungi do not produce CDH. The CDH has also been isolated from the *Sclerotium rolfsii* (soil-plant pathogen) and has been found in *Pisolithus tinctorius*, *Suillus variegatus*, and *Cortinarius* sp. CDH has not been isolated from bacteria [60]. Research has revealed its association with the cellulose degradation system and can reduce Fe^{3+} to Fe^{2+} and O_2 to H_2O_2 [61]. It is the first non-hydrolytic enzyme involved in the breakdown of cellulose.

CONCLUDING REMARKS

Though fungi are known as one of the first eukaryotes, their evolution is not yet finished. Their evolution is still going on to grow faster, to become more cosmic and utilize minimum energy. Some microorganisms have modified themselves to live in an environment that may not be suitable for others. Previously, white rot fungi had been proposed to be the most effective wood decomposers. However, it has been established that brown rot fungi can also carry out the depolymerization of holocellulose and modify lignin efficiently. These are suggested to have evolved from saprotrophic white rot fungal ancestors, having the whole lignocellulose-degrading system.

Due to this being a very complex process, there is a need for multidimensional research to find out the complete mechanism and evolution, including the involvement of various microorganisms in lignin degradation. The fungi can successfully propagate in different ecosystems, from rotting plants to forests and even to wood used in houses. It is believed that fungi can make themselves well adjusted and prepared as per the environmental conditions. However, the rapidly changing climate and global warming may influence the evolution of fungi further.

Extensive research has been carried out to describe the complex process of degradation of wood. Microorganisms have developed various enzymatic and non-enzymatic tactics to utilize wood.

REFERENCES

[1]   Allsopp D, Seal KJ, Gaylarde CC, Eds. Introduction to Biodeterioration. Cambridge: Cambridge University Press 2004; p. 256.
[http://dx.doi.org/10.1017/CBO9780511617065]

[2]   Aarti C, Arasu MV, Agastian P. Lignin degradation: A microbial approach. South Indian J Biol Sci 2015; 1: 119-27.
[http://dx.doi.org/10.22205/sijbs/2015/v1/i3/100405]

[3]   Adamcová D, Fronczyk J, Radziemska M, *et al.* Research of the biodegradability of degradable/biodegradable plastic material in various types of environments. Prz Nauk Inz Ksztalt Sr 2017; 26: 3-14.
[http://dx.doi.org/10.22630/PNIKS.2017.26.1.01]

[4]   Gómez EF, Michel FC Jr. Biodegradability of conventional and bio-based plastics and natural fiber composites during composting, anaerobic digestion and long-term soil incubation. Polym Degrad Stabil 2013; 98: 2583-91.
[http://dx.doi.org/10.1016/j.polymdegradstab.2013.09.018]

[5]   Levasseur A, Piumi F, Coutinho PM, *et al.* FOLy: An integrated database for the classification and functional annotation of fungal oxidoreductases potentially involved in the degradation of lignin and related aromatic compounds. Fungal Genet Biol 2008; 45: 638-45.
[http://dx.doi.org/10.1016/j.fgb.2008.01.004] [PMID: 18308593]

[6]   Hatakka A. Biodegradation of Lignin. In: Steinbüchel A, Ed. Biopolymers Online. Wiley-VCH Verlag GmbH & Co, KGaA 2005; pp. 129-80.

[7]   Müller RJ. Biodegradability of Polymers: Regulations and Methods for Testing. In: Steinbüchel A, Ed Biopolymers Online. Wiley-VCH, Verlag GmbH & Co, KGaA 2005; pp. 365-74.

[8]   Akullian A, Karp C, Austin K, *et al.* Plastic bag externalities and policy in Rhode Island. Brown Policy Review 2006; p. 11. [http://seattlebagtax.org/referencedpdfs/en-akullianetal.pdf]

[9] Haider TP, Völker C, Kramm J, *et al.* Plastics of the future? The impact of biodegradable polymers on the environment and on society. Angew Chem Int Ed 2019; 58: 50-62.
[http://dx.doi.org/10.1002/anie.201805766] [PMID: 29972726]

[10] Kranert M, Behnsen A, Schultheis A, *et al.* Composting in the Framework of the EU Landfill Directive. In: Insam H, Riddech N, Klammer S, Eds. Microbiology of Composting. Berlin, Heidelberg: Springer 2002; pp. 473-86.
[http://dx.doi.org/10.1007/978-3-662-08724-4_39]

[11] Żenkiewicz M, Malinowski R, Rytlewski P, *et al.* Some composting and biodegradation effects of physically or chemically crosslinked poly(lactic acid). Polym Test 2012; 31: 83-92.
[http://dx.doi.org/10.1016/j.polymertesting.2011.09.012]

[12] Berg B, McClaugherty CA. Decomposer Organisms. In: Berg B, McClaugherty CA, Eds. Plant Litter: Decomposition, Humus Formation, Carbon Sequestration. Verlag Berlin and Heidelberg GmbH & Co KG: Springer 2014; pp. 31-48.
[http://dx.doi.org/10.1007/978-3-642-38821-7_3]

[13] Martínez-Blanco J, Colón J, Gabarrell X, *et al.* The use of life cycle assessment for the comparison of biowaste composting at home and full scale. Waste Manag 2010; 30: 983-94.
[http://dx.doi.org/10.1016/j.wasman.2010.02.023] [PMID: 20211555]

[14] Jing D, Wang J. Controlling the simultaneous production of laccase and lignin peroxidase from *Streptomyces cinnamomensis* by medium formulation. Biotechnol Biofuels 2012; 5: 15.
[http://dx.doi.org/10.1186/1754-6834-5-15] [PMID: 22429569]

[15] Choinowski T, Blodig W, Winterhalter KH, *et al.* The crystal structure of lignin peroxidase at 1.70 Å resolution reveals a hydroxy group on the C^β of tryptophan 171: A novel radical site formed during the redox cycle. J Mol Biol 1999; 286: 809-27.
[http://dx.doi.org/10.1006/jmbi.1998.2507] [PMID: 10024453]

[16] Hoshino F, Kajino T, Sugiyama H, *et al.* Thermally stable and hydrogen peroxide tolerant manganese peroxidase (MnP) from *Lenzites betulinus*. FEBS Lett 2002; 530: 249-52.
[http://dx.doi.org/10.1016/S0014-5793(02)03454-3] [PMID: 12387901]

[17] Fernández-Fueyo E, Ruiz-Dueñas FJ, Miki Y, *et al.* Lignin-degrading peroxidases from genome of selective ligninolytic fungus *Ceriporiopsis subvermispora*. J Biol Chem 2012; 287: 16903-16.
[http://dx.doi.org/10.1074/jbc.M112.356378] [PMID: 22437835]

[18] Baldrian P, Valášková V. Degradation of cellulose by basidiomycetous fungi. FEMS Microbiol Rev 2008; 32: 501-21.
[http://dx.doi.org/10.1111/j.1574-6976.2008.00106.x] [PMID: 18371173]

[19] Liers C, Arnstadt T, Ullrich R, *et al.* Patterns of lignin degradation and oxidative enzyme secretion by different wood- and litter-colonizing basidiomycetes and ascomycetes grown on beech-wood. FEMS Microbiol Ecol 2011; 78: 91-102.
[http://dx.doi.org/10.1111/j.1574-6941.2011.01144.x] [PMID: 21631549]

[20] Goszczynski S, Paszczynski A, Pasti-Grigsby MB, *et al.* New pathway for degradation of sulfonated azo dyes by microbial peroxidases of *Phanerochaete chrysosporium* and *Streptomyces chromofuscus*. J Bacteriol 1994; 176: 1339-47.
[http://dx.doi.org/10.1128/jb.176.5.1339-1347.1994] [PMID: 8113173]

[21] Brückmann M, Termonia A, Pasteels JM, *et al.* Characterization of an extracellular salicyl alcohol oxidase from larval defensive secretions of *Chrysomela populi* and *Phratora vitellinae* (Chrysomelina). Insect Biochem Mol Biol 2002; 32: 1517-23.
[http://dx.doi.org/10.1016/S0965-1748(02)00072-3] [PMID: 12530219]

[22] Zámocký M, Gasselhuber B, Furtmüller PG, *et al.* Turning points in the evolution of peroxidase – catalase superfamily: Molecular phylogeny of hybrid heme peroxidases. Cell Mol Life Sci 2014; 71: 4681-96.

[http://dx.doi.org/10.1007/s00018-014-1643-y] [PMID: 24846396]

[23] Ek M, Gellerstedt G, Henriksson G, Eds. Wood Chemistry and Wood Biotechnology. Walter De Gruyter 2009; p. 308.
[http://dx.doi.org/10.1515/9783110213409]

[24] Couturier M, Navarro D, Olivé C, *et al.* Post-genomic analyses of fungal lignocellulosic biomass degradation reveal the unexpected potential of the plant pathogen *Ustilago maydis*. BMC Genomics 2012; 13: 57.
[http://dx.doi.org/10.1186/1471-2164-13-57] [PMID: 22300648]

[25] Manavalan T, Manavalan A, Heese K. Characterization of lignocellulolytic enzymes from white-rot fungi. Curr Microbiol 2015; 70: 485-98.
[http://dx.doi.org/10.1007/s00284-014-0743-0] [PMID: 25487116]

[26] Kuhad RC, Singh A, Eriksson KEL. Microorganisms and enzymes involved in the degradation of plant fiber cell walls. Adv Biochem Eng Biotechnol 1997; 57: 45-125.
[http://dx.doi.org/10.1007/BFb0102072] [PMID: 9204751]

[27] Crawford RH, Carpenter SE, Harmon ME. Communities of filamentous fungi and yeast in decomposing logs of *Pseudotsuga menziesii*. Mycologia 1990; 82: 759-65.
[http://dx.doi.org/10.1080/00275514.1990.12025957]

[28] Fernandes TAR, Silveira WB, Passos FML, *et al.* Laccases from *Actinobacteria*. What we have and what to expect. Adv Microbiol 2014; 4: 285-96.
[http://dx.doi.org/10.4236/aim.2014.46035]

[29] Lisov AV, Leontievsky AA, Golovleva LA. Hybrid Mn-peroxidase from the ligninolytic fungus *Panus tigrinus* 8/18. Isolation, substrate specificity, and catalytic cycle. Biochemistry (Mosc) 2003; 68: 1027-35.
[http://dx.doi.org/10.1023/A:1026072815106] [PMID: 14606947]

[30] Dashtban M, Schraft H, Syed TA, *et al.* Fungal biodegradation and enzymatic modification of lignin. Int J Biochem Mol Biol 2010; 1: 36-50.
[PMID: 21968746]

[31] Lombard V, Golaconda Ramulu H, Drula E, *et al.* The carbohydrate-active enzymes database (CAZy) in 2013. Nucleic Acids Res 2014; 42: D490-5.
[http://dx.doi.org/10.1093/nar/gkt1178] [PMID: 24270786]

[32] Gilbertson RL. Wood-rotting fungi of North-America. Mycologia 1980; 72: 1-49.
[http://dx.doi.org/10.1080/00275514.1980.12021153]

[33] Heinzkill M, Bech L, Halkier T, *et al.* Characterization of laccases and peroxidases from wood-rotting fungi (family Coprinaceae). Appl Environ Microbiol 1998; 64: 1601-6.
[http://dx.doi.org/10.1128/AEM.64.5.1601-1606.1998] [PMID: 9572923]

[34] Zimmerman AR, Kang DH, Ahn MY, *et al.* Influence of a soil enzyme on iron-cyanide complex speciation and mineral adsorption. Chemosphere 2008; 70: 1044-51.
[http://dx.doi.org/10.1016/j.chemosphere.2007.07.075] [PMID: 17845813]

[35] Lindahl BD, Tunlid A. Ectomycorrhizal fungi – potential organic matter decomposers, yet not saprotrophs. New Phytol 2015; 205: 1443-7.
[http://dx.doi.org/10.1111/nph.13201] [PMID: 25524234]

[36] Boer W, Folman LB, Summerbell RC, *et al.* Living in a fungal world: Impact of fungi on soil bacterial niche development. FEMS Microbiol Rev 2005; 29: 795-811.
[http://dx.doi.org/10.1016/j.femsre.2004.11.005] [PMID: 16102603]

[37] Cragg SM, Beckham GT, Bruce NC, *et al.* Lignocellulose degradation mechanisms across the Tree of Life. Curr Opin Chem Biol 2015; 29: 108-19.
[http://dx.doi.org/10.1016/j.cbpa.2015.10.018] [PMID: 26583519]

[38] Blanchette RA. Degradation of the lignocellulose complex in wood. Can J Bot 1995; 73: 999-1010.
[http://dx.doi.org/10.1139/b95-350]

[39] Pildain MB, Novas MV, Carmarán CC. Evaluation of anamorphic state, wood decay and production of lignin-modifying enzymes for diatrypaceous fungi from Argentina. J Agric Technol 2005; 1: 81-96.

[40] Beeson WT, Vu VV, Span EA, *et al.* Cellulose degradation by polysaccharide monooxygenases. Annu Rev Biochem 2015; 84: 923-46.
[http://dx.doi.org/10.1146/annurev-biochem-060614-034439] [PMID: 25784051]

[41] Fernández IS, Ruíz-Dueñas FJ, Santillana E, *et al.* Novel structural features in the GMC family of oxidoreductases revealed by the crystal structure of fungal aryl-alcohol oxidase. Acta Crystallogr D Biol Crystallogr 2009; 65: 1196-205.
[http://dx.doi.org/10.1107/S0907444909035860] [PMID: 19923715]

[42] de Koker TH, Mozuch MD, Cullen D, *et al.* Isolation and purification of pyranose 2-oxidase from *Phanerochaete chrysosporium* and characterization of gene structure and regulation. Appl Environ Microbiol 2004; 70: 5794-800.
[http://dx.doi.org/10.1128/AEM.70.10.5794-5800.2004] [PMID: 15466516]

[43] Martinez D, Challacombe J, Morgenstern I, *et al.* Genome, transcriptome, and secretome analysis of wood decay fungus *Postia placenta* supports unique mechanisms of lignocellulose conversion. Proc Natl Acad Sci USA 2009; 106: 1954-9.
[http://dx.doi.org/10.1073/pnas.0809575106] [PMID: 19193860]

[44] Zámocký M, Furtmüller PG, Obinger C. Two distinct groups of fungal catalase/peroxidases. Biochem Soc Trans 2009; 37: 772-7.
[http://dx.doi.org/10.1042/BST0370772] [PMID: 19614592]

[45] Maciel GM, Castoldi R, Mariano SdaS, *et al.* Involvement of lignin-modifying enzymes in the degradation of herbicides. Herbicides-Advances in Research. Price AJ, Kelton JA, Eds. IntechOpen 2013; pp. 165-87.

[46] Paszczyński A, Huynh VB, Crawford R. Comparison of ligninase-I and peroxidase-M2 from the white-rot fungus *Phanerochaete chrysosporium*. Arch Biochem Biophys 1986; 244: 750-65.
[http://dx.doi.org/10.1016/0003-9861(86)90644-2] [PMID: 3080953]

[47] Palma C, Martínez AT, Lema JM, *et al.* Different fungal manganese-oxidizing peroxidases: A comparison between *Bjerkandera* sp. and *Phanerochaete chrysosporium*. J Biotechnol 2000; 77: 235-45.
[http://dx.doi.org/10.1016/S0168-1656(99)00218-7] [PMID: 10682282]

[48] Giatti Marques De Souza C, Kirst Tychanowicz G, Farani De Souza D, *et al.* Production of laccase isoforms by *Pleurotus pulmonarius* in response to presence of phenolic and aromatic compounds. J Basic Microbiol 2004; 44: 129-36.
[http://dx.doi.org/10.1002/jobm.200310365] [PMID: 15069672]

[49] Diamantidis G, Effosse A, Potier P, *et al.* Purification and characterization of the first bacterial laccase in the rhizospheric bacterium *Azospirillum lipoferum*. Soil Biol Biochem 2000; 32: 919-27.
[http://dx.doi.org/10.1016/S0038-0717(99)00221-7]

[50] Niladevi KN, Jacob N, Prema P. Evidence for a halotolerant-alkaline laccase in *Streptomyces psammoticus*: Purification and characterization. Process Biochem 2008; 43: 654-60.
[http://dx.doi.org/10.1016/j.procbio.2008.02.002]

[51] Pawlik A, Wójcik M, Rułka K, *et al.* Purification and characterization of laccase from *Sinorhizobium meliloti* and analysis of the lacc gene. Int J Biol Macromol 2016; 92: 138-47.
[http://dx.doi.org/10.1016/j.ijbiomac.2016.07.012] [PMID: 27392777]

[52] Baldrian P. Fungal laccases – occurrence and properties. FEMS Microbiol Rev 2006; 30: 215-42.
[http://dx.doi.org/10.1111/j.1574-4976.2005.00010.x] [PMID: 16472305]

[53] Leonowicz A, Cho N, Luterek J, *et al.* Fungal laccase: Properties and activity on lignin. J Basic Microbiol 2001; 41: 185-227.
[http://dx.doi.org/10.1002/1521-4028(200107)41:3/4<185::AID-JOBM185>3.0.CO;2-T] [PMID: 11512451]

[54] Kersten P, Cullen D. Copper radical oxidases and related extracellular oxidoreductases of wood-decay Agaricomycetes. Fungal Genet Biol 2014; 72: 124-30.
[http://dx.doi.org/10.1016/j.fgb.2014.05.011] [PMID: 24915038]

[55] Ferreira P, Hernández-Ortega A, Herguedas B, *et al.* Kinetic and chemical characterization of aldehyde oxidation by fungal aryl-alcohol oxidase. Biochem J 2010; 425: 585-93.
[http://dx.doi.org/10.1042/BJ20091499] [PMID: 19891608]

[56] Hernández-Ortega A, Ferreira P, Martínez AT. Fungal aryl-alcohol oxidase: A peroxide-producing flavoenzyme involved in lignin degradation. Appl Microbiol Biotechnol 2012; 93: 1395-410.
[http://dx.doi.org/10.1007/s00253-011-3836-8] [PMID: 22249717]

[57] Pitsawong W, Sucharitakul J, Prongjit M, *et al.* A conserved active-site threonine is important for both sugar and flavin oxidations of pyranose 2-oxidase. J Biol Chem 2010; 285: 9697-705.
[http://dx.doi.org/10.1074/jbc.M109.073247] [PMID: 20089849]

[58] Piumi F, Levasseur A, Navarro D, *et al.* A novel glucose dehydrogenase from the white-rot fungus *Pycnoporus cinnabarinus*: production in *Aspergillus niger* and physicochemical characterization of the recombinant enzyme. Appl Microbiol Biotechnol 2014; 98: 10105-18.
[http://dx.doi.org/10.1007/s00253-014-5891-4] [PMID: 24965558]

[59] Kumar AK, Goswami P. Functional characterization of alcohol oxidases from *Aspergillus terreus* MTCC 6324. Appl Microbiol Biotechnol 2006; 72: 906-11.
[http://dx.doi.org/10.1007/s00253-006-0381-y] [PMID: 16547701]

[60] Kajisa T, Yoshida M, Igarashi K, *et al.* Characterization and molecular cloning of cellobiose dehydrogenase from the brown-rot fungus *Coniophora puteana*. J Biosci Bioeng 2004; 98: 57-63.
[http://dx.doi.org/10.1016/S1389-1723(04)70242-X] [PMID: 16233666]

[61] Henriksson G, Johansson G, Pettersson G. A critical review of cellobiose dehydrogenases. J Biotechnol 2000; 78: 93-113.
[http://dx.doi.org/10.1016/S0168-1656(00)00206-6] [PMID: 10725534]

CHAPTER 11

An Overview of Diverse Yeast Species Performing the Biocontrol Function in Agriculture

Abhishek Sinha[1,*]

[1] *Department of Microbiology, Swami Vivekanand University, N.H. 26 Narsinghpur road, Sagar, Madhya Pradesh, India*

Abstract: The agricultural economy has been suffering from various pathogenic diseases of crops, fruits, grains, and vegetables for a long time. Controlling these diseases is pivotal to the growth of agricultural production and the availability of harvests. Compared with relatively harmful chemical agents, biocontrol agents are now being used as safer and less toxic alternatives to control crop losses. Yeasts survive in all environmental conditions and have been described as potent antagonists to various plant pathogens. Due to their antagonistic activity towards pathogens, relatively simple cultivation requirements, and very limited biosafety concerns, many of these unicellular fungi are being considered for biocontrol applications. In this chapter, we have discussed the pros and cons of yeasts as biocontrol agents, various yeast species, and their modes of antagonistic action. To survive in the environment, yeasts need to tolerate various biotic and abiotic stresses. Those stresses have been discussed here, and how different yeast strains overcome these harsh conditions and carry out antagonistic activities have also been highlighted. Yeast biocontrol activities to date represent a largely unexplored field of research and plenty of opportunities remain for the development of commercial, yeast-based applications for plant protection.

Keywords: Anti-fungal, Biocontrol, Biofilm, Plant protection, Yeast.

INTRODUCTION

Various plant diseases caused by plant pathogens account for major crop losses throughout the year. It has huge socio-economic and food safety implications. Only plant diseases cause a worldwide estimated loss of US $40 billion each year. Reports confirmed that, since the year 2000, the total varieties of new fungal plant pathogens have increased seven-fold. Among crop diseases, the share of fungal-generated diseases constitutes 64–67% and a 20% loss of production. The application of chemical fungicides is one method that is used to prevent plant diseases and protect crops from pests and pathogens. But the redundant use of

---

* **Corresponding author Abhishek Sinha:** Department of Microbiology, Swami Vivekanand University, N.H. 26, Narsinghpur road, Sagar, Madhya Pradesh, India; E-mail: asinha1984@gmail.com

Shampi Jain, Ashutosh Gupta and Neeraj Verma
All rights reserved-© 2023 Bentham Science Publishers

chemical fungicides confers the development of fungicide resistance in the fungal pathogens and, in the absence of other control measures, leads to the resurgence of the disease. Therefore, although the use of pesticides has brought clear improvements in crop production since their induction, their potency to treat some of the most harmful plant diseases is in decline. The other downside of chemical fungicides is their heavy impact on the microflora of agrarian soil and their indiscriminate destruction of beneficial microbes, *e.g.*, endophytic bacteria and fungi, as well as soil-inhabitant helminths, arthropods, and other animals.

One reliable approach to controlling fungal diseases of plants might be the use of biopesticides or biological control agents. In phytopathology, this term is used to define the use of introduced or resident living organisms to eliminate or suppress populations of pathogens. There are a few of these agents on the market, mostly used in the control of pests on a small scale. They comprise dry biomass of bacterial or fungal strains isolated from the endosphere or the rhizosphere of plants. These strains are cultured and live reproductive microorganisms are used as biocontrol agents. These include a diverse group of endophytic fungi that are characterized by their ability to colonize the plant host tissues without causing any external disease symptoms. These endophytes can prevent pathogen infection and propagation directly by competition, mycoparasitism, or antibiosis, or indirectly by inducing resistance responses in the host. Despite this, the biocontrol agents that could be applied in phytopathology are not restricted to these two groups of organisms. Endophytic microorganisms also include Ascomycota and Basidiomycota yeasts, found in many species of trees from very diverse climates, but also in agricultural species. The present review tabulates different kinds of yeast that are beneficial and can be used as biocontrol agents in agricultural ecosystems.

BENEFITS AND DRAWBACKS OF USING YEASTS AS BIOCONTROL AGENTS

The most desired characteristic of the organism to be used as an active biocontrol agent must be its effectiveness against a target disease. Other properties such as biosafety and registration issues, easy production and handling conditions, and the required application equipment are just as important as their effectiveness. Although in a snapshot, the lack of invasive, filamentous growth of most yeasts may seem a disadvantage, a closer analysis proves that the yeast-like morphology makes their culturability in fermenters easy and increases strategic options to prepare formulations, hence ample application options prevail [1]. Like bacteria, unicellular yeasts also favour attachment and biofilm formation, which influences their environmental persistence and survival, thereby improving biocontrol activity. Another important benefit of yeasts is their easy culturing conditions.

Most of the yeasts can be modified or improved using various biotechnological and molecular biology methods in the laboratory. Hence, direct intervention with recombinant DNA (rDNA) technology can improve their biocontrol potential. Budding yeast like *Saccharomyces cerevisiae* strains contains plasmids, whereas most non-conventional yeasts lack plasmids. But these yeasts can be engineered to maintain extra-chromosomal DNA by designing a plasmid vector from their sister species, which contain autonomously replicating and centromere sequences. Yeasts, thus, share growth characteristics and biocontrol activities like bacteria but are safer compared to antibiotic resistance, horizontal gene transfer, and toxin synthesis properties of bacteria.

Yeasts have been used for food and beverage production for generations. They are consumed directly as food supplements and are widely employed in the food processing and bakery industries. In some cases, yeasts used for consumption and biocontrol belong to the same genus or even species (*e.g.*, *Saccharomyces cerevisiae, Candida sake*, and *Metschnikowia pulcherrima*) [1 - 3]. Because yeasts are considered safe, their use in crops and food products causes less concern than introducing bacteria or filamentous fungi. Although some yeasts, such as certain *Candida* or *Cryptococcus* species, are also human pathogens. Among the drawbacks of yeast is dimorphism, *i.e.*, the switch to an invasive growth form needs to be kept in mind before considering yeasts for biocontrol. On the other hand, yeasts that show resistance to fungicides or common antifungals also need to be studied and carefully considered before making their application ready.

DIFFERENT YEASTS FOR BIOCONTROL AND THEIR MODES OF ACTION

Different microorganisms compete with each other and host species for nutrients, minerals, and space, and the winner generally predominates [4]. This competition is difficult to predict but is of immense importance in natural environments where resources are limited and highly competitive. In a natural environment, the limitation of space plays a minor role in species diversity and selection. Yeasts generally grow well on agar plates, but large differences in their antifungal activities are reported when compared between laboratory conditions and field trials. Most of the antifungal effects of yeasts are not species-specific. A particular yeast is either strongly or weakly antagonistic against most fungi in laboratory conditions. However, growing under field conditions activates diverse survival mechanisms, and the competition for the physical niche might gain importance in such circumstances.

For biocontrol yeasts, iron is possibly the most sought-after nutrient, and iron competition is recognized as an important mode of action. For example, *Aureobasidium pullulans* exerts its antibacterial activity through a siderophore named fusarinine C (fusigen). The red colour pigment of *M. pulcherrima* colonies is due to the formation of a cyclic dipeptide, pulcherriminic acid, which forms a complex with iron. That pigment component is crucial for antifungal activity, as pigment-less mutants were shown to exhibit much lower antifungal activity. Sequestering of iron in different complexes in biocontrol yeasts also creates iron deficiency in the fungal pathogen [5]. However, it needs to be made clear that mutants of *M. pulcherrima* incapable of synthesizing pulcherriminic acid still inhibit filamentous fungi quite effectively, suggesting the presence of an alternate mechanism of inhibition.

Another efficient mechanism by which yeast performs its biocontrol activity is by competing for sulfur-containing molecules with target pathogens. Yeast (*Saccharomycopsis schoenii*) lacks several enzymes of the sulfur assimilation pathway and acquires methionine from its target. Similar sulfur-containing compound assimilation defects exist among fungal pathogens. Methionine is one such amino acid that contains sulfur. Competition for methionine sequestration thus becomes one of the major modes of action of the antifungal activity of *S. schoenii*. *Trichoderma* species also show a similar mode of killing.

In the past decade, biofilm formation has emerged as a successful strategy used by microbes to effectively use space to achieve growth needs and effectively increase survival opportunities by presenting themselves as a population. The communities of microbes in biofilms that live and grow on surfaces can encompass a single species or multi-species consortia. Biofilms may exhibit vastly different characteristics as compared to individual cells and have been shown to be involved in virulence during a lot of microbial diseases. Some of the yeasts use the same mechanism to perform their biocontrol activity. Biofilm formation starts with the attachment of an individual cell to a surface through various hydrophobic molecules present on the outer surface or attached to the cell wall. The next steps consist of the secretion of an extracellular matrix and often the formation of hyphae or pseudohyphae in some yeasts.

Biocontrol yeasts mainly form biofilms in the wounds of plants, which are vulnerable to infection from pathogenic fungi or bacteria. Although this mechanism of action is widely accepted in the field of mycology, the molecular mechanisms of the process and the composition of different biofilms have not been well characterized. *Pichia fermentans* is a rare species of yeast where the above-mentioned puzzle is somewhat better understood. This species forms a biofilm on apple wounds and protects them against postharvest diseases. In peaches, the same yeast switches from the yeast like to the hyphal phenotype and causes a rapid decay of inoculated fruits in the absence of a plant pathogen. Due to their activity as biocontrol agents as well as pathogenesis, this organism is proposed as a potential biohazard factor for biocontrol yeasts. Other organisms that perform their biocontrol activities through the formation of biofilms are *S. cerev*isiae, *A. pullulans, Pichia kudriavzevii, Wickerhamomyces anomalus, Kloeckera apiculata,* and *M. pulcherrima* [6]. In *S. cerevisiae* biocontrol strains, biofilm-forming cells were far more efficient than planktonic cells in colonizing the inner surface of apple wounds and controlling the development of blue mould caused by *P. expansum* [7]. Table **1** provides a detailed list of major yeasts that are being used as biocontrol agents. Although the yeasts were found to be efficient in dealing with pathogens in pre and post-harvest plants, fruits, and vegetables, the on-field application has to counter various environmental stresses that are prevalent in on-field applications.

Table 1. List of major yeast strains functioning as biocontrol agents.

Name of Yeast	Host	Pathogen
Aureobasidium pullulans	Pear and apple	*Penicillium expansum, Botrytis cinerea,* and *Penicillium expansum*
Candida pelliculosa	Tomato	*Botrytis cinerea*
Candida sake	Apple	*Penicillium expansum*
Candida (Pichia) guilliermondii	Cherry tomato, chili, tomato, kiwifruit, and papaya	Fruit decay agents, *Colletotrichum capsici, Rhizopus stolonifer, Botrytis cinerea,* and *Colletotrichum gloeosporioides*
Candida oleophila	Banana	*Colletotrichum musae, Fusarium moniliforme,* and *Cephalosporium* sp.
Cryptococcus infirmominiatus	Sweet cherry	*Monilinia fructicola*
Metschnikowia fructicola	Apple and grapefruit	*Penicillium expansum* and *P. digitatum*
Aureobasidium pullulans	Pome	*Penicillium, Botrytis,* and *Monilinia*
Rhodosporidium paludigenum	Cherry tomato	*Botrytis cinerea*

Note: The list consists of a few major yeast strains that are used as biocontrol. Numerous other yeast species are being used as antagonists, which have not been included here.

ENVIRONMENTAL CONDITIONS AFFECTING BIOCONTROL BY YEASTS

The biocontrol zone comprises a tripartite interaction between a plant host, a pathogen, and a yeast species. All of these biotic components are affected by environmental factors such as pH, temperature, and UV light as well as osmotic and oxidative stresses. During the production process, biocontrol agents encounter various abiotic stresses that can also impact their viability. Therefore, understanding the ecological fitness of the potential yeast biocontrol agents and their stress tolerance ability are essential to their success in a commercial application.

Heat Stress

This is the earliest stress that the biocontrol yeasts encounter upon their application to the plant surface. Elevated temperatures can severely impact the viability of the yeast. The viability of *Candida oleophila* was shown to decrease with increasing temperature within 30 minutes (39 to 41°C). In contrast, *M. fructicola* exhibits a high degree of thermotolerance, with viability that remains stable at temperatures above 43°C. More specifically, after a 30-min exposure at 41°C, the viability of *C. oleophila* cells dropped to 22%, while the viability of *M. fructicola* cells was 87% at 43°C. Reports also suggest that *Pichia anomala* J121, the yeast used to preserve grains in storage, could grow at a wide range of temperatures, ranging from 3 to 37°C. Thus, diverse species and strains of yeast antagonists have varying degrees of heat tolerance, which makes them more or less capable of surviving under different agricultural fields or packing-house conditions [8].

Desiccation Shock

The ability to withstand osmotic shock makes yeasts well suited as biocontrol agents. Many antagonistic yeast species can grow under conditions of low water activity. The environmental conditions within plant hosts, fruit, or food grains on which yeast species serve as biocontrol agents, contain varying amounts of water activity. Both *P. anomala* and *C. sake* could grow at a lower level of water activity. Yeasts isolated from high-osmosis niches, *e.g.*, raw fruit juices, soy sauces, or honey, may be very useful as antagonists. On the other hand, the ability of some yeasts to grow in halophilic conditions in various salt solutions also introduced the idea of combining yeast antagonists with various salts, such as $CaCl_2$, $MgCl_2$, and $NaHCO_3$, which is now used as a routine approach to enhance biocontrol efficacy. Combining both characteristics of yeasts (tolerance of low water activity and the component salt ions) makes them an attractive agent for biocontrol purposes.

The effect of different solutes (NaCl, glycerol, and glucose) on the growth of *Rhodosporidium paludigenum*, an antagonistic yeast, indicated that low water activity inhibited its growth in nutrient yeast dextrose broth (NYDB), but the same yeast grew better in a medium supplemented with NaCl solute than with nonionic solutes (glycerol and glucose). *R. paludigenum* grown in 6.6% NaCl-enriched NYBD medium showed higher viability (92.1%) under low water activity conditions than the control (81.1%) when incubated for 48 hrs. Studies also indicated that salt-adapted *R. paludigenum* shows higher tolerance of freeze stress and better biocontrol performance on pears and jujubes than the non-adapted culture [9]. Overall, the ability of a yeast antagonist to grow at a low water potential may represent an important attribute regarding its use as a biocontrol agent and so should be considered when selecting and developing a potential biocontrol strain.

Oxidative Stresses

An oxidative stress response is a crucial characteristic of yeast that makes them more suitable as biocontrol agents. Proteomic analysis indicated that an antagonistic yeast, *P. membranaefaciens*, induced the expression of six antioxidant proteins in peach fruit. Out of these six, catalase, glutathione peroxidase, and peroxiredoxin are important proteins contributing to reactive oxygen species (ROS) scavenging. It is known that ROS levels are elevated in fruits as a result of pathogen invasion. Hence, biocontrol agents must confer ROS resistance to host plants as well as possess the characteristics to tolerate ROS-derived oxidative stress, which can affect their viability and efficacy.

Based on transcriptome analysis, it has been reported that the production of peroxidases and superoxide dismutase in grapefruit was induced by the application of the yeast *M. fructicola*. Relative comparison between oxidative stress resistance and the fitness of the postharvest biocontrol yeasts *Cryptococcus laurentii* LS-28 and *Rhodotorula glutinis* LS-11 showed that *C. laurentii* LS-28 possesses higher resistance to ROS-generated oxidative stress than *R. glutinis* LS-11. As a result, *C. laurentii* LS-28 exhibited better colonization and biocontrol efficacy on apple fruit. Examinations of the responses of *M. fructicola*, *C. oleophila*, and *Cystofilobasidium infirmominiatum* PL1 to oxidative stress showed that *C. infirmominiatum* was the most sensitive to exogenous H_2O_2, while *M. fructicola* was the most tolerant. *M. fructicola* survived at 88% even in 200 mM H_2O_2, whereas *C. oleophila* and *C. infirmominiatum* survived at 28% and 23%, respectively, even at much lower H_2O_2 concentrations. *Pichia caribbica*, another biocontrol yeast, showed relatively less oxidative stress tolerance as its viability decreased exponentially when the concentration of H_2O_2 increased from 5 to 20 mM. Almost all of the yeast cells died upon exposure to 20 mM H_2O_2 for 1 hour. Thus, very similar to their thermotolerance, different yeast strains can differ dramatically in their levels of tolerance of oxidative stress.

Other Environmental and Production Condition Stresses

When antagonists occupy a plant host (fruit, vegetable, or grain), the host pH, either in the macroenvironment or in the microenvironment, can have a direct impact on the growth and establishment of plant protection activities. *C. sake* is known to tolerate a wide pH range (3–7) regardless of water activity. It has also been reported that oxidative-stress-adapted *C. oleophila* grew better in liquid culture at pH 4.0 (the same pH that is used to produce Golden Delicious apple flesh commercially) than in non-stress-adapted cultures. Stress-adapted yeast at pH 4.0 showed better *in vivo* growth and resulted in greater efficacy in inhibiting apple rots caused by *P. expansum* than non-stress-adapted yeast. The pH tolerance assay in *R. paludigenum* indicated that pre-exposure of the yeast to pH 4.0 or 5.5 for 36 h by incubating with either malic acid or lactic acid improved the survival of the yeast when it was subsequently exposed to a lethal pH of 1.0 to 3.0 for 1 h. The pre-exposure to pH 4.0 or 5.5 also resulted in better biocontrol efficacy. UV-B radiation (280 to 320 nm) is also a major environmental factor that can affect the viability of biocontrol agents when they are applied under field conditions. Freeze-drying is one of the most common methods used to obtain dry preparations of microorganisms. It was reported that freeze-drying at 20°C using 10% skim milk as a protectant was the best method to preserve the viability of yeast (*C. sake*) cells. Freezing cells in liquid nitrogen results in injury to cells, which reduces their viability significantly. The highest level of biocontrol activity of yeast (*C. sake*) cells was obtained when lactose and skim milk (10% lactose–10% skim milk) were used as protectants during the freeze-drying process and 1% peptone was added when cells were rehydrated before use.

Comparing different formulations (using freeze-drying, vacuum drying, fluidized bed drying, and liquid formulation) of *P. anomala* J121, it was found that the best viability was obtained by freeze-drying the yeast. The initial viability after freeze-drying was as high as 80% when trehalose was used as a protectant during the freeze-drying process. Trehalose was also applied during freeze-drying of *Cryptococcus laurentii* and *Rhodotorula glutinis*, and it was found to play a critical role in enhancing cell viability. Many studies have explored alternative drying methods for producing biocontrol agents, including spray drying, fluidized bed drying, and vacuum drying. Some of the studies indicated that spray drying of *C. sake* resulted in a very high level of cell damage, thus greatly reducing viability. Although a mild heat treatment could induce thermotolerance, it did not improve the survival of yeast (*C. sake*) cells exposed to spray drying enough for the method to be considered a suitable approach to commercial production. Overall, it can be summarized that a variety of environmental factors control the suitability of yeasts as biocontrol agents. Biotechnology or genetic engineering can become useful tools in the future and should be explored to increase the environmental viability of yeast cells as well as increase their biocontrol properties.

CONCLUDING REMARKS

In this chapter, a discussion is made on the benefits and drawbacks of yeast as a biocontrol agent. The benefits include its easy culturability, non-toxicity, and ease of genetically engineering. The downsides include dimorphism, *i.e.*, the ability to switch to invasive growth and resistance to some well-known fungicides. Yeasts exert their biocontrol activity by sequestering iron and creating iron deficiency for surrounding pathogens, acquiring sulphur from target organisms, forming biofilm, *etc.* Hence, it is extremely important to understand the type of target pathogen and the proper selection of yeast to carry out biocontrol activity. Apart from the names provided here, various other yeasts can perform biocontrol activities under suitable conditions. As the application of yeasts for biocontrol functions is performed in field conditions, it is also imperative to consider the abiotic stress that the organisms will face during their application. Here we discussed some stresses like heat, desiccation, pH, oxidative changes, *etc.*, in the chapter. This chapter provides a birds' eye view of different aspects of yeasts that need to be considered for industrial-scale production of yeasts as biocontrol agents and applications in agriculture.

ACKNOWLEDGEMENT

The author would like to thank Dr. Virendra Kumar, for carefully reading the manuscript and suggesting corrections. I would also like to confer my gratitude to Dr. Umesh Mishra, Swami Vivekananda University for English editing assistance.

REFERENCES

[1] Freimoser FM, Rueda-Mejia MP, Tilocca B, *et al*. Biocontrol yeasts: Mechanisms and applications. World J Microbiol Biotechnol 2019; 35: 154.
[http://dx.doi.org/10.1007/s11274-019-2728-4] [PMID: 31576429]

[2] Suzzi G, Romano P, Ponti I, *et al*. Natural wine yeasts as biocontrol agents. J Appl Microbiol 1995; 78: 304-8.
[http://dx.doi.org/10.1111/j.1365-2672.1995.tb05030.x]

[3] Pretscher J, Fischkal T, Branscheidt S, *et al*. Yeasts from different habitats and their potential as biocontrol agents. Fermentation (Basel) 2018; 4: 31.
[http://dx.doi.org/10.3390/fermentation4020031]

[4] El-Tarabily KA, Sivasithamparam K. Potential of yeasts as biocontrol agents of soil-borne fungal plant pathogens and as plant growth promoters. Mycoscience 2006; 47: 25-35.
[http://dx.doi.org/10.1007/S10267-005-0268-2]

[5] Spadaro D, Droby S. Development of biocontrol products for postharvest diseases of fruit: The importance of elucidating the mechanisms of action of yeast antagonists. Trends Food Sci Technol 2016; 47: 39-49.
[http://dx.doi.org/10.1016/j.tifs.2015.11.003]

[6] Perez MF, Contreras L, Garnica NM, *et al.* Native killer yeasts as biocontrol agents of postharvest fungal diseases in lemons. PLoS One 2016; 11: e0165590.
[http://dx.doi.org/10.1371/journal.pone.0165590] [PMID: 27792761]

[7] Ferraz P, Cássio F, Lucas C. Potential of yeasts as biocontrol agents of the phytopathogen causing cacao witches' broom disease: Is microbial warfare a solution? Front Microbiol 2019; 10: 1766.
[http://dx.doi.org/10.3389/fmicb.2019.01766] [PMID: 31417539]

[8] Sui Y, Wisniewski M, Droby S, *et al.* Responses of yeast biocontrol agents to environmental stress. Appl Environ Microbiol 2015; 81: 2968-75.
[http://dx.doi.org/10.1128/AEM.04203-14] [PMID: 25710368]

[9] Robiglio A, Sosa MC, Lutz MC, *et al.* Yeast biocontrol of fungal spoilage of pears stored at low temperature. Int J Food Microbiol 2011; 147: 211-6.
[http://dx.doi.org/10.1016/j.ijfoodmicro.2011.04.007] [PMID: 21546110]

CHAPTER 12

Macrophomina Phaseolina: An Agriculturally Destructive Soil Microbe

Ramesh Nath Gupta[1,*], Kishor Chand Kumhar[2] and J.N. Srivastava[3]

[1] *Department of Plant Pathology, Bihar Agricultural University, Sabour-813210, India*

[2] *Plant Protection, Deen Dayal Upadhyay Centre of Excellence for Organic Farming, CCS HAU, Hisar – 125004, India*

[3] *Department of Plant Pathology, Bihar Agricultural University, Sabour-813210, India*

Abstract: *Macrophomina phaseolina* (Tassi) Goid. is a destructive fungal soil microbe, a cause of charcoal rot disease and causes heavy losses in agricultural production. It is non-specific and appears in moderate to severe form every year worldwide. Due to the seriousness and economic importance of the pathogen as well as disease, it requires multiple approaches like epidemiological study, induction of systemic resistance through non-conventional chemicals, host-pathogen resistance and chemical as well as phytoextract application for its management. Epidemiological studies reveal that the onset of charcoal rot varied in different varieties during different dates of sowing. Timely sowing of crops is an important tool for reducing disease incidence. The intensity of disease in a timely sown crop is less, with higher production and productivity. The non-conventional chemicals like salicylic acid, acetylsalicylic acid, indole acetic acid, indole butyric acid, riboflavin, and thiamine induce systemic acquired resistance (SAR) and effectively inhibit mycelial growth of the pathogen. These non-conventional chemicals showed a reduction of charcoal rot disease under field conditions. It also enhances the yield-attributing traits and yield. It induces total phenol content, peroxidase, polyphenol oxidase, phenylalanine ammonia lyase, and catalase activity by the treatment of these chemicals. These activities showed a differential reaction after inoculation of the pathogen on different varieties. However, resistant varieties showed higher induction of biochemical activities than susceptible ones. Different phytoextracts showed inhibition of mycelial growth and a reduction of disease incidence in different crops. Seed treatment with fungicides is an effective method for controlling the pathogen and ultimately enhances the production of the crop. Genotype evaluation for host resistance is an effective, economical, and continuous way of managing the pathogen and disease.

Keywords: *Macrophomina phaseolina*, Soil microbes, Charcoal rot, Induction, Disease management.

---

[*] ***Corresponding author Ramesh Nath Gupta:** Department of Plant Pathology, Bihar Agricultural University, Sabour-813210, India; E-mail: rameshnathgupta@gmail.com

Shampi Jain, Ashutosh Gupta and Neeraj Verma
All rights reserved-© 2023 Bentham Science Publishers

INTRODUCTION

Macrophomina phaseolina (Tassi) Goidanich is a destructive, omnipresent, and non-specific fungal pathogen with a vast host range. Tassi [1] first identified the pycnidial stage of the fungus as *Macrophomina phaseolina*. It is best known as the cause of a disease, which is aptly called "charcoal rot". Taubenhaus [2] discovered a sclerotial stage, *i.e.*, *Sclerotium bataticola* Taub, and identified the fungus as a causal agent for charcoal rot of sweet potatoes in the USA. Petrak [3] established the genus *Macrophomina*, which was parasitic on sesame. Because of the sclerotial-bearing mycelial stage, the fungus known as *Macrophomina phaseolina* causes characteristic root rot disease. In India, Butler [4] identified a similar sclerotial-bearing fungus and compared it with isolates of Taubenhaus and named it *Rhizoctonia bataticola* (Taub.) Butl, and subsequently, it was transferred to *Macrophomina*. The pathogen attacks several crops in many parts of the world [5]. Generally, the pathogen appears to be non-specific with a wide host range and causes charcoal rot in maize, sesame, sorghum, soybean and other economically important crops and is responsible for huge losses every year in India [6]. *Macrophomina phaseolina* (Tassi) Goidanich belongs to the division Ascomycota, class Dothideomycetes, order Botryosphaeriales, family Botryosphaeriaceae, genus *Macrophomina* and species *phaseolina*.

The hyphae of the pathogen are branched at the right angle with constriction and a septum just after the constriction. Chaudhary and coworkers [7] reported that the pathogen continuously changes its nature, and rapidly resistant cultivars become susceptible. The absence of a known teleomorph stage has stalled its taxonomy for many years [8]. The severity of the disease is directly related to the presence of viable sclerotia in the soil. The hyphae are septate and filiform. Initially, hyphae are hyaline, afterward becoming grey to black and producing jet black oval to round microsclerotia of size 80-90 μm in diameter. Akhtar and coworkers [9] proved the necrotrophic behaviour of pathogens in sesame and found that the seed infection efficiency of *M. phaseolina* was 100% with a significant reduction in the seed index. Infection stages of the charcoal rot fungus *M. phaseolina* in sesame revealed a transition phase from biotrophy *via* BNS (biotrophy-to-necrotrophy switch) to necrotrophy [10]. The pathogen switched its strategy of infection; the host tailored its defence strategy to meet the changing situation. Less reactive oxygen species accumulation, up-regulation of its signalling genes and higher antioxidant enzyme activities post-BNS resulted in resistance. There was a greater accumulation of secondary metabolites and upregulation of secondary metabolite-related genes after BNS. A total of twenty genes functioning in different aspects of plant defence that were monitored over a period of time during the changing infection phases showed a coordinated response.

SURVIVAL OF THE PATHOGEN

M. phaseolina is a diverse, soil-borne pathogen causing stem and root rot on a number of economically important crops like pulses, oilseeds, vegetables, and fruit crops. It causes a monocyclic disease cycle, has a vast host range, and also survives in the seed. The occurrence of sclerotia in plant trash allows the fungus to grow in soil in the absence of the host for two or more years, depending on soil conditions [11]. Meyer and coworkers [12] did not consider the presence of mycelium in the soil as a primary source of inoculum. Reuveni and coworkers [13] observed that the pathogen survives in soil mainly as microsclerotia, which are the primary source of inoculum. Soil, seed, and plant remains are the primary sources of inoculums, and disease severity is directly related to the number of sclerotia present in the soil. These microsclerotia can survive from three months to three years under stressful field conditions [14]. Microsclerotia are formed in vascular tissues, giving a greyish-black appearance in sub-epidermal tissues of the stem. These are vegetative propagules highly resistant to unfavourable climatic conditions, developed on host tissues and then dispersed to the soil after the host plant dies [15]. The fungus survives in soil by multicellular jet black microsclerotia produced enormously during the saprophytic and/or parasitic phases [16]. According to Baggio and coworkers [17], the fungus surviving in the soil through the production of microsclerotia is reduced by pre-plant fumigation of the soil.

THE DISEASE

Charcoal rot is one of the most common, worldwide-dispersed, destructive root and stem rot diseases. In India, charcoal rot of sesame incited by *M. phaseolina* (Tassi) Goidinch was first reported from Uttar Pradesh [18]. Jain and Kulkarni [19] reported this disease from the Jabalpur and Gwalior divisions of Madhya Pradesh, India. It is now widely available throughout the sesame-growing states of Uttar Pradesh, Madhya Pradesh, Gujarat, Maharashtra, Haryana, Punjab, Bihar, West Bengal, Orissa, Tamil Nadu, Karnataka, and Kerala. Generally, the pathogen of charcoal rot appears to be non-specific in nature and causes disease in sesame, maize, sorghum, soybean, sunflower, and other economically important crops, resulting in huge losses every year in India [6]. Generally, about 25% yield losses by charcoal rot in the United States, Uruguay, Spain, and the Soviet Union have been observed, but under favourable weather conditions for the growth and development of pathogens, total crop failure in specific areas has also been recorded [20]. Murugesan and coworkers [21] reported that 1.8 kg/ha of sesame yield was lost for every one percent increase in charcoal rot disease intensity. The disease was reported from North and South America, Asia, Africa and Europe but was more prevalent in subtropical and tropical countries with a semi-arid climate.

Yu and Park [22] found that *M. phaseolina* causes a severe reduction in seed germination and seedling stand. Vyas [23] observed that charcoal rot is a very serious and destructive disease in sesame growing areas and causes 5–100% yield losses. This disease is a serious threat to crops, especially in arid regions of the world. Maiti and coworkers [24] estimated a 57 percent yield loss at 40% disease incidence. Dinakaran and Mohammed [25] found charcoal rot caused by *M. phaseolina* is the most devastating disease among diseases of sesame. Chattopadhyay and Kalpana [26] reported a 50% incidence of charcoal rot resulting in heavy yield losses in India. Deepthi and coworkers [27] estimated yield losses in sesame due to *M. phaseolina* at the capsule formation stage and observed that plants protected with fungicides had a greater number of capsules per plant and test weight of healthy than the infected capsule. Min and Toyoto [28] reported that the incidence of sesame charcoal rot was 10–30%, which caused 10–75% yield losses in Myanmar.

DISEASE SYMPTOMS

The most common symptom of charcoal rot is the sudden wilting of plants from top to bottom throughout the crop growth period. Irregular and deep necrotic lesions tend toward the hypocotyls and root surfaces. Lesions coalesce to form larger patches on branches or other plant parts, which leads to premature senescence and death of the plant. Toxin production and sclerotia formation by the pathogen in the xylem play an important role in the dehydration of adult plants and lead to wilting. Plants infected with *M. phaseolina* dry out and their roots decay, resulting in a shredded appearance. Seedling blight, root rot, and stem rot are the salient symptoms of charcoal rot. Irregular and deep necrotic lesions in hypocotyls and root surfaces can also be observed. Colonization of epidermal and cortical cells is followed by colonization of vascular cambium and phloem cells. Microsclerotia germinate and produce hypha, which penetrates and grows intercellularly. This infection leads to cellular collapse and epidermal and cortical cell necrosis in the roots and hypocotyls [29]. Fungal sclerotia overwinter on weeds and attack plant roots, followed by the decay of fibrous roots and blackening of the stems [30]. The fungus interrupts the function of xylem vessels, causing wilting and premature death of plants [31]. *M. phaseolina* is widely distributed, infects different parts of the plant and causes root rot, stem rot, and seedling blight [32]. Hyphae of the pathogen are septate, filiform, initially hyaline, subsequently changing grey to black and producing jet black oval to round microsclerotia of 80-90 μm in diameter [33]. The pathogen mainly attacks the basal region of the plant and causes lesions on roots, stems, pods, and seeds [34]. According to Khalili and coworkers [35], the plant undergoes several morphological changes after an attack of the pathogen, including irregular lesions having a grey centre with a dark brown border, death of nodes, and finally wilting.

INFLUENCE OF WEATHER CONDITIONS ON DISEASE DEVELOPMENT

Environmental factors play a critical role in the development of charcoal rot of sesame incited by *M. phaseolina*. Proper knowledge of pathogen vulnerable stages and timing of incidence on crop plants with weather factors play an important role in disease management that is effective, efficient, economical, and environmentally friendly. Gemawat and Verma [36] studied the effect of environmental factors (temperature and relative humidity) on the development of charcoal rot of sesame and recorded maximum disease intensity at 35°C temperature and 76% relative humidity. They also observed that charcoal rot was favoured by high temperatures and low relative humidity. Patel and Patel [37] also reported that the maximum disease incidence of sesame charcoal rot was found at a temperature of 35°C and relative humidity of 76%. Ratnoo and coworkers [38] observed that infection and development of ashy grey stem blight of cowpea caused by *M. phaseolina* were favoured at high temperatures between 25°C and 40°C. Singh and coworkers [39] found that the development of charcoal rot on sorghum was favoured at a temperature of 28 to 35°C, and blighting symptoms developed at temperatures equal to or above 30°C. The correlation between disease incidence and temperature showed positive and non-significant, while negative and significant was noticed between disease development and relative humidity. Parmar and coworkers [40] evaluated different temperatures suitable for charcoal rot fungal growth and found that a temperature between 25°C and 35°C was optimum for fungal growth and sclerotial formation in castor. Satpathi and Gohel [41] observed that the incidence of charcoal rot was more progressive during the 37th and 39th standard meteorological weeks along with bright sunshine range (8.70 to 9.10 hrs), maximum temperature (33.30 to 34.50°C) and minimum temperature (24.20 to 24.90°C).

MANAGEMENT OF DISEASE

Management With Non-Conventional Chemicals

Non-conventional chemicals act as elicitors and disease control agents and reduce the negative effects of chemical pesticides. It reduces environmental pollution and the economics of farmers and has a long-lasting effect [42]. The result of plant and pathogen interactions is significantly influenced by endogenous small molecules, which coordinate plant defence responses. Exogenous application of non-conventional chemicals activates plant defences and helps in protection from pathogen infection. Novel non-conventional chemicals induce resistance in plants and are aptly known as plant defence activators or plant activators.

Salicylic Acid (SA)

Salicylic acid ($C_7H_6O_3$, 2-Hydroxybenzoic acid) is an important non-conventional chemical that plays a role in systemic acquired resistance (SAR), signaling, and disease resistance in plants. Exogenous application of SA induces SAR and SAR gene expression. Pieterse and coworkers [43] reported that SA signalling provides resistance against biotrophic pathogens, whereas necrotrophic resistance is controlled by jasmonic acid and ethylene-signaling pathways. Salicylic acid is a key regulator of plant defence that primarily mediates responses to biotrophic pathogens [44]. SA is considered an important signalling molecule and is involved in local and systemic disease resistance to plants against various pathogenic attacks [45]. Salicylic acid acts as a potential non-enzymatic antioxidant and plant growth regulator and plays an important role in many plant physiological processes. It has been identified as a signalling component in numerous plant responses against pathogen attack and induced responses to growth and biochemical constituents.

Acetylsalicylic Acid (ASA)

Acetylsalicylic acid ($C_9H_8O_4$, 2-Acetoxybenzoic acid) is a derivative of salicylic acid. Ojaghian and coworkers [46] evaluated acetylsalicylic acid *in vitro* and *in vivo* conditions, which significantly reduced the mycelial growth, sclerotial formation, and carpogenic germination of the pathogen. They observed the enzymatic activity of peroxidase, polyphenol oxidase, and phenylalanine ammonia lyase were increased in the fungal inoculated carrots three days and four hours after application of ASA. Klessig and coworkers [47] reported that ASA is an important plant hormone involved in many processes, including root initiation, seed germination, stomatal closure, floral induction, thermo-genesis, and response to biotic and abiotic stresses. El-Hai and coworkers [48] reported that the application of ASA (15 mM) reduced the incidence of the root rot pathogen and enhanced the growth parameters, yield parameters, and seed quality.

Indole-3-Acetic Acid (IAA)

Indole acetic acid/indole-3-acetic acid [$C_{10}H_9NO_2$; 2-(1H-indol-3yl) acetic acid] is a basic naturally occurring auxin and has the property of inducing resistance in plants. Hahn and Strittmatter [49] observed the regulation of the enzymatic activity of the glutathione S-transferase in potato by IAA and regulated the induction of plant defence mechanisms. IAA plays an important role in numerous plant-pathogen interactions. Noel and coworkers [50] showed that exogenous application of IAA regulates b-1, 3-glucanase and chitinase accumulation in fungal inoculated potato leaves. IAA acts as an activator of plant resistance and enhances the enzymatic activity of peroxidase and phenylalanine ammonia-lyase in barley. Awasthi [51] evaluated non-conventional chemical IAA at 10^{-4} and 10^{-5} M concentrations for 24 hours to determine its bio-efficacy against stem rot disease in groundnut and found the lowest disease incidence (25.7%) and the highest significant yield (486.8g/m²) by the treatment of IAA at 10^{-5} M concentration.

Indole Butyric Acid (IBA)

Indole butyric acid ($C_{12}H_{13}NO_2$; 1H-indole-3-butanoic acid) is a plant hormone of the auxin family, performing a similar role to IAA and is regarded as synthetic auxin. El-Fiki and coworkers [30] observed that presoaking sesame seed in IBA was found to be an effective treatment for reducing charcoal rot incidence. They also observed significant induction of total phenols, total sugars, and defence enzyme activities like peroxidase, polyphenol oxidase, and catalase as compared to control.

Riboflavin

Riboflavin {$C_{17}H_{20}N_4O_6$; 7, 8-dimethyl-10-[(2S,3S,4R)-2,3,4,5-tetrahydoxyp-entyl] benzo [g] pteridine-2,4-dione} is a signalling activator and systemic resistance elicitor in plants. It induces the expression of pathogenesis-related (PR) genes in plants and can trigger a signal transduction pathway of systemic resistance. It influences the outcome of plant and pathogen interactions. Emmanouil and Wood [52] reported that treatment of leaves of pepper, tomato, and eggplant with riboflavin before fungal inoculation caused a significant reduction in fungal load and overall disease symptoms of the plants. Riboflavin induced systemic resistance in chickpeas against charcoal rot and *Fusarium* wilt of chickpeas. Riboflavin (1.00 mM) treated chickpea inoculated with pathogens showed induction of resistance two days after treatment and reached the highest level between 5-7 days and then decreased. It induced a higher accumulation of phenols, POX and PAL activities after challenged inoculation with pathogens in riboflavin pre-treated plants. Dong and Beer [53] reported that riboflavin helps in the protection of various hosts from bacterial, viral, and fungal pathogens without phytotoxicity. Liu and coworkers [54] observed that the defence responses elicited by riboflavin in tobacco cells induce an oxidative burst, expression of defence-related genes, accumulation of phenolic compounds, and lignin. Results showed that riboflavin induced disease resistance based on MAPK-dependent priming for the expression of PR1gene.

Thiamine

Thiamine {$C_{12}H_{17}N_4OS$; 2-[3-[(4-amino-2-methylpyrimidin-5-yl) methyl]-4-methyl-1, 3-thiazol-3-ium-5-yl] ethanol} is an essential component of all living organisms' primary metabolism. It also plays a role in the processes of protection of plants against biotic and abiotic stresses. Ahn and coworkers [55] observed the effects of thiamine on disease resistance and defence-related gene expression. Peroxidase, polyphenol oxidase, and phenylalanine ammonia lyase activity were found to be higher in inoculated and non-inoculated plants treated with thiamine and riboflavin than in control plants. Abdel-Monaim [56] reported that treatment with riboflavin (2.5 mM) and thiamine (5 mM) showed maximum induction of systemic resistance against charcoal rot of soybean at six days after inoculation with *M. phaseolina* of soybean. It also increased the number and fresh and dry weight of nodules per plant compared to the untreated control. Muthukrishnan and coworkers [57] observed that the application of thiamine loaded chitosan nanoparticles increased seed germination and seedling vigour and decreased cell death in the roots of chickpeas.

Impact of Non-Conventional Chemicals on Phenol Content

The growth of pathogens is restricted due to the rapid accumulation of phenol content at the site of infection. The plant has the ability to synthesize phenols or their derivatives. Phenol content has antifungal, antiviral, and antibacterial properties. The accumulation of phenol content is higher in resistant than susceptible genotypes in different crop plants. It is involved in disease resistance in plants by hypersensitive responses or lignifications of the cell wall. The phenolic enzymes, *viz.*, peroxidase, polyphenol oxidase, and phenylalanine ammonia lyase showed increased activity in mustard leaves after spraying salicylic acid (SA) at 50 mg/l and 75 mg/l concentrations before the onset of disease [58]. Phenol also imparted resistance against soft rot pathogens in potato crops and showed significantly higher total phenols in resistant cultivars as compared with susceptible [59]. Geat and coworkers [60] evaluated salicylic acid, indole acetic acid, indole butyric acid, and carbendazim against *Colletotrichum capsici*, the causal agent of fruit rot of chili, in which salicylic acid treatment increased the level of total phenol in the resistant variety Sadabahar than in the susceptible Pusa Jwala.

Peroxidase Content

Peroxidase (POX) belongs to a versatile and ubiquitous group of enzymes that oxidise phenols to form more toxic quinols and play an integral role in disease resistance. Peroxidase activity was higher in the leaves of wilt-resistant chickpea cultivars than in susceptible ones after an attack of the pathogen [61]. Nagy and coworkers [62] found that peroxidase activity was higher in the root of Norway spruce when infected with *Rhizoctonia* species than in control. Ngadze and coworkers [59] studied the role of peroxidase in imparting resistance against the soft rot pathogen in potatoes and found significantly higher peroxidase activity in resistant varieties than susceptible ones. Ondobo and coworkers [63] estimated the peroxidase and polyphenol oxidase activities in *Theobroma cacao* after inoculation with a pathogen.

Polyphenol Oxidase Content

Polyphenol oxidase (PPO) causes browning, following cell damage by the attack of pests and pathogens, wounding, and senescence. Mandavia and coworkers [64] observed the activities of different enzymes in tolerant and susceptible cumin plants and found higher levels of polyphenol oxidase in tolerant varieties than in susceptible ones. Ashry and Mohammad [65] studied the changes in antioxidant enzymes in leaves and stems of eighteen flax lines resistant and susceptible to the pathogen, the role of polyphenol oxidase in imparting resistance to the pathogen, and significantly higher enzymatic activity of polyphenol oxidase in resistant varieties was observed.

Phenylalanine Ammonia Lyase Content

Phenylalanine ammonia lyase (PAL), an enzyme involved in the phenylpropanoid pathway of plants, catalyses the synthesis of secondary metabolites from L-phenylalanine such as lignin, phytoalexin and flavonoid pigments. Plants treated with abiotic elicitors and pathogens showed increased activity of PAL. It is more accumulated in infected areas undergoing lignification. A sudden increase in PAL activity following fungal infection indicates the synthesis of phenolic compounds for defence. Lignification of plant cell walls for resistance reactions is also confirmed in the absence of phytoalexin. Sharma and coworkers [66] found more induction of phenolics, peroxidase, polyphenol oxidase and phenylalanine ammonia lyase in resistant varieties of sesame than in susceptible ones after inoculation of *M. Phaseolina*.

Catalase Content

The activity of catalase was found to be higher in the leaves of treated plants with non-conventional chemicals (SA, IAA, and IBA) than in untreated controls [67]. EL-Fiki and coworkers [30] reported that catalase activity was higher when treating sesame seeds with SA, IBA, and IAA than untreated seeds. The activity of the defence enzyme catalase and defence inducing chemicals (total phenols) was found to be higher due to induced systemic resistance [68]. The changes in antioxidant enzymes in eighteen flax lines were observed. Infected plants significantly increased catalase activity in resistant flax lines than in susceptible ones.

Application of Phytoextracts and Fungicides

The use of phytoextracts and botanical pesticides for the management of the disease has gained momentum in view of environmental hazards caused by the excess use of pesticides. Savaliya and coworkers [69] evaluated phytoextracts of nine plant species under *in vitro* conditions by using a poisoned food technique against *M. phaseolina* of sesame. The extract of garlic cloves showed maximum inhibition of mycelial growth (77.65%) and scanty sclerotial formation, followed by onion bulb extract (63.98%) and the least growth inhibition (32.34%) by ginger rhizome extract. According to Khamari and Patra [70], clove oil inhibited 100% mycelial growth of pathogens at all concentrations, followed by neem oil (72.57%, 81.86%, and 80.66%) and mustard oil (71.82%, 77.76%, and 88.60%) at concentrations (2%, 3%, and 5%), respectively. Khaire and coworkers [71] reported that at 10% concentration, mycelial growth inhibition of *M. phaseolina* ranged from 1.26% (euphorbia) to 77.46% (garlic). However, significantly maximum inhibition of mycelial growth was observed with phytoextract of garlic (77.46%) followed by neem (62.36%) treatment. Lakhran and Ahir [72] evaluated six phytoextracts under *in vivo* conditions against dry root rot disease, in which neem (71.39%) was found most effective in controlling charcoal rot at 40 DAS of chickpea.

M. phaseolina can also survive on seed and plant debris. Seed treatment with fungicide is an effective and practical method of disease management. Gupta and Cheema [73] reported that sesame seed treatment with thiram (2.5 g/kg), captan (2 g/kg) and carbendazim (2 g/kg) resulted in increased seed germination, shoot length and a decreased 64 to 70 percent incidence of charcoal rot disease. Pineda and coworkers [74] found that treated seeds with carbendazim + thiram reduced the incidence of *Rhizoctonia bataticola* in sunflowers. Jaiman and Jain [75] used six fungicides (bavistin, raxil, topsin–M, captan, indofil M-45, and thiram) as seed dressers against dry rot of cluster beans and reported bavistin as the best suited fungicide for controlling disease under field conditions. Shumaila and Mujeebur [76] evaluated different fungicides and found that carbendazim was the most effective and mancozeb the least effective seed treating fungicide against charcoal rot of mungbean.

Use of Resistant Varieties

A resistant variety is the simplest, most effective, and most economical method of plant disease management. Evaluation and identification of resistant sources is a prerequisite for a successful breeding programme to develop promising varieties against the devastating disease. Avila [77] screened thirteen sesame varieties and cultivars, *viz.*, UCLA-1, EXP-1, and DV-9, and found them moderately resistant to charcoal rot disease. Thiyagu and coworkers [78] identified three genotypes of gingelly, namely ORM 7, ORM 14 and ORM 17, resistant to root rot disease. Shabana and coworkers [79] identified the sesame lines P5 (NM59), C6.3, C1.10, and C3.8 as resistant to charcoal rot disease. Farooq and coworkers [80] screened a total of nine lines of sesame and three lines, *viz.*, 87008, 87502 and 95002, were found moderately resistant to the disease.

CONCLUDING REMARKS

Soil microbes play a vital role in sustaining crop production in various ways. The soil microbes regulate the different soil reactions to maintain their ideal conditions in favour of plant growth and development, which ultimately reflects increased crop yield. There are certain soil-inhabitant beneficial fungi, better known as antagonists, which are efficient in controlling the soil-borne fungal phytopathogens that cause diseases such as seed rot, damping-off, root rot, and wilt. *Trichoderma* species are the promising representative candidates of antagonists. These are naturally available in different habitats variably and can be developed in the form of product formulation by various methods for use in agriculture for seed treatment and so many other purposes to take care of plant disease. Hun dreds of manufacturers are actively involved in manufacturing its various usable formulations in India. Such formulations have great scope in agriculture, particularly "organic agriculture", as far as current and future agricultural perspectives are concerned. Such products are much safer for the associated environment, being natural in origin and free from health hazards. In the current agricultural scenario, such bioagents should be popularized on a large scale to create a healthy environment. Product quality is an important issue that determines its performance under field conditions. Secondly, the dose of application for various purposes seems quite variable among the products manufactured by various manufacturers, creates confusion. Therefore, it should be unified like other agrochemicals.

ACKNOWLEDGEMENTS

Authors are grateful to CCS HAU, Hisar, for their support.

REFERENCES

[1] Tassi FI. Novae Micromycetum species. Series III Bull, del Labor, ed Orto Bot, di Siena, Anno 1901; 4: 7-12.

[2] Taubenhaus JJ. The black rots of the sweet potato. Phytopathology 1913, 3: 161-4.

[3] Petrak F. Mykologische Notizen VI. Annal Mycol 1923; 21: 314-5.

[4] Butler EJ. Identification of the Fungus. In: Briton-Jones HR, Ed. Mycological Work in Egypt During the Period 1920-1922. Tech Sci Serv Bull No 49. Botanical Section 1925; pp. 64-5.

[5] Reichert I, Hellinger E. On the occurrence, morphology and parasitism of *Sclerotium bataticola*. Pales J Bot 1947; 6: 107-47.

[6] Khare MN, Sharma HC, Kumar SM, *et al.* Current plant pathological problems of soybean and their control. Proceedings of the Fourth All Indian Workshop on Soybean, GBPUAT, Pantnagar 1973; pp. 55-67.

[7] Chaudhary MHZ, Sarwar N, Chaughati FA. Biochemical changes in chickpea plant after induction treatment with simple chemical for systemic resistance against *Ascochyta* blight in the field. J Chem Soc Pak 2001; 23: 182-6.

[8] Crous PW, Slippers B, Wingfield MJ, *et al.* Phylogenetic lineages in the Botryosphaeriaceae. Stud Mycol 2006; 55: 235-53.
[http://dx.doi.org/10.3114/sim.55.1.235] [PMID: 18490983]

[9] Akhtar KP, Sarwar G, Arshad HMI. Temperature response, pathogenicity, seed infection and mutant evaluation against *Macrophomina phaseolina* causing charcoal rot disease of sesame. Arch Phytopathol Pflanzenschutz 2011; 44: 320-30.
[http://dx.doi.org/10.1080/03235400903052945]

[10] Chowdhury S, Basu A, Kundu S. Biotrophy-necrotrophy switch in pathogen evoke differential response in resistant and susceptible sesame involving multiple signaling pathways at different phases. Sci Rep 2017; 7: 17251.
[http://dx.doi.org/10.1038/s41598-017-17248-7] [PMID: 29222513]

[11] Watanabe T, Smith RS, Snyder WC. Population of *Macrophomina phaseoli* in soil as affected by fumigation and cropping. Phytopathology 1970; 60: 1717-9.
[http://dx.doi.org/10.1094/Phyto-60-1717]

[12] Meyer WA, Sinclair JB, Khare MN. Factors affecting charcoal rot of soybean seedlings. Phytopathology 1974; 64: 845-9.
[http://dx.doi.org/10.1094/Phyto-64-845]

[13] Reuveni R, Nachmias A, Krikun J. The role of seedborne inoculum on the development of *Macrophomina phaseolina* on melon. Plant Dis 1983; 67: 280-1.
[http://dx.doi.org/10.1094/PD-67-280]

[14] Abawi GS, Pastor Corrales MA, Eds. Root rot of beans in Latin American and Africa: Diagnosis, Research, Methodologies, and Management Strategies. Centro Internacional de Agricultura Tropical. Cali, CO: Centro Internacional de Agricultura Tropical (CIAT) Cali, CO: CIAT 1990; p. 114.
[https://hdl.handle.net/10568/54258]

[15] Olaya G, Abawi GS. Effect of water potential on mycelial growth and on production and germination of sclerotia of *Macrophomina phaseolina*. Plant Dis 1996; 80: 1347-50.
[http://dx.doi.org/10.1094/PD-80-1347]

[16] Dubey RC, Upadhyaya RS. Survival and Control of *Macrophomina phaseolina* (Tassi) Goid. In: Maheshwari DK, Dubey RC, Eds. Innovative Approaches in Microbiology. Dehradun, India: Bishan Singh Mahendrapal Singh 2001; pp. 169-90.

[17] Baggio JS, Cordova LG, Peres NA. Sources of inoculum and survival of *Macrophomina phaseolina* in Florida strawberry fields. Plant Dis 2019; 103: 2417-24.
[http://dx.doi.org/10.1094/PDIS-03-19-0510-RE] [PMID: 31322978]

[18] Mehta PR. Observations on new and known diseases of crop plants of the Uttar Pradesh. Plant Protect Bull 1951; 3: 7-12.

[19] Jain AC, Kulkarni SN. Root and stem rot of sesame. Indian Oilseed J 1965; 9: 201-3.

[20] Tikhonov OI, Nedelko OK, Persestova TA. Methods for pathogenicity tests for seed borne *Macrophomina phaseolina* isolated from different hosts. Phytopathol Z 1976; 88: 234-7.
[http://dx.doi.org/10.1111/j.1439-0434.1977.tb03973.x]

[21] Murugesan M, Shanmugam N, Menon PPV, *et al.* Statistical arrangement of yield loss of sesame due to insect pests and diseases. Madras Agric J 1978; 65: 290-5.

[22] Yu SH, Park JS. *Macrophomina phaseolina* detected in seeds of *Sesamum indicum* and it's pathogenicity. Korean J Pl Prot 1980; 19: 135-40. [https://www.online-rpd.org/upload/pdf/JPP019--3-01.pdf]

[23] Vyas SC. Diseases of sesamum and niger in India and their control (sesame and *Guizotia abyssinica*). Pesticides 1981; 15: 10-5.

[24] Maiti S, Hegde MR, Chattopadhyay SB, Eds. Handbook of Annual Oilseed crops. New Delhi: Oxford and IBH Publication Company Private Limited 1988; p. 325.

[25] Dinakaran D, Mohammed SEN. Identification of resistant sources to root rot of sesame caused by *Macrophomina phaseolina* (Tassi.) Goid. Sesame Safflower News 2001; 16: 68-71.

[26] Chattopadhyay C, Sastry RK. Combining viable disease control tools for management of sesame stem root rot caused by *Macrophomina phaseolina* (Tassi) Goid. Indian J Plant Prot 2002; 30: 132-8.

[27] Deepthi P, Shukla CS, Verma KP, *et al.* Identification of charcoal rot resistant lines of *Sesamum indicum* and chemical management of *Macrophomina phaseolina*. Medicinal Plants - Int J Phytomed Related Indus 2014; 6: 36-42.
[http://dx.doi.org/10.5958/j.0975-6892.6.1.005]

[28] Min YY, Toyota K. Occurrence of different kinds of diseases in sesame cultivation in Myanmar and their impact to sesame yield. J Exper Agric Int 2019; 38: 1-9.
[http://dx.doi.org/10.9734/jeai/2019/v38i430309]

[29] Mayek-PÉrez N, GarcÍa-Espinosa R, LÓpez-CastaÑeda CÁ, *et al.* Water relations, histopathology and growth of common bean (*Phaseolus vulgaris* L.) during pathogenesis of *Macrophomina phaseolina* under drought stress. Physiol Mol Plant Pathol 2002; 60: 185-95.
[http://dx.doi.org/10.1006/pmpp.2001.0388]

[30] Mohamad FG, EL-Deeb AA, Khalifa MMA. Some applicable methods for controlling sesame charcoal rot disease (*Macrophomina phaseolina*) under greenhouse conditions. Egypt J Phytopathol 2004; 32: 87-101.

[31] Gupta GK, Chauhan GS. Symptoms, identification and management of soybean diseases. Indore: Technical Bulletin no. 10, NRC Soybean 2005; 10: 89-92.

[32] Moi S, Bhattacharyya P. Influence of biocontrol agents on sesame root rot. J Mycopathol Res 2008; 46: 97-100.

[33] Mahmoud A, Budak H. First report of charcoal rot caused by *Macrophomina phaseolina* in sunflower in Turkey. Plant Dis 2011; 95: 223.
[http://dx.doi.org/10.1094/PDIS-09-10-0631] [PMID: 30743418]

[34] Kumar S, Aeron A, Pandey P, *et al.* Ecofriendly Management of Charcoal Rot and Wilt Diseases in Sesame (*Sesamum indicum* L.). In: Maheshwari D, Ed. Bacteria in Agrobiology: Crop Ecosystems. Berlin, Heidelberg: Springer 2011; pp. 387-405.
[http://dx.doi.org/10.1007/978-3-642-18357-7_14]

[35] Khalili E, Javed MA, Huyop F, *et al.* Evaluation of *Trichoderma* isolates as potential biological control agent against soybean charcoal rot disease caused by *Macrophomina phaseolina*. Biotechnol Biotechnol Equip 2016; 30: 479-88.
[http://dx.doi.org/10.1080/13102818.2016.1147334]

[36] Gemawat PD, Verma OP. Root and stem rot of *Sesamum* in Rajasthan. Evaluation of varieties (*Macrophomina phaseolina*). Indian J Mycol Plant Pathol 1974; 4: 76-7.

[37] Patel KK, Patel AJ. Meteorological correlations of charcoal rot of sesame. Indian J Mycol Plant Pathol 1990; 20: 64-5.

[38] Ratnoo RS, Bhatnagar MK. Effect of straw and oilcakes on ashy gray stem blight *Macrophomina phaseolina* (Tassi) Goid of cowpea. Indian J Mycol Plant Phytopathol 1993; 23: 186-8.

[39] Singh R, Sindhan GS. Effect of fungicides on the incidence of dry root rot and biochemical status of chickpea plants. Plant Dis Res 1998; 13: 14-7.

[40] Parmar H, Kapadiya HJ, Bhaliya CM, *et al.* Effect of media and temperature on the growth and sclerotial formation of *Macrophomina phaseolina* (Tassi) Goid causing root rot of castor. Int J Curr Microbiol Appl Sci 2018; 7: 671-5.
[http://dx.doi.org/10.20546/ijcmas.2018.702.083]

[41] Satpathi AK, Gohel NM. Role of meteorological factors on development of stem and root rot of sesame. J Pharmacogn Phytochem 2018; 7: 1922-4. [https://www.phytojournal.com/archives/2018/vol7issue6/PartAH/7-6-303-719.pdf]

[42] Kuć J. Concepts and direction of induced systemic resistance in plants and its application. Eur J Plant Pathol 2001; 107: 7-12.
[http://dx.doi.org/10.1023/A:1008718824105]

[43] Pieterse CMJ, Leon-Reyes A, Van der Ent S, *et al.* Networking by small-molecule hormones in plant immunity. Nat Chem Biol 2009; 5: 308-16.
[http://dx.doi.org/10.1038/nchembio.164] [PMID: 19377457]

[44] Glazebrook J. Contrasting mechanisms of defense against biotrophic and necrotrophic pathogens. Annu Rev Phytopathol 2005; 43: 205-27.
[http://dx.doi.org/10.1146/annurev.phyto.43.040204.135923] [PMID: 16078883]

[45] Hayat S, Mori M, Fariduddin Q, *et al.* Physiological role of brassinosteroids: An update. Indian J Plant Physiol 2010; 15: 99-109.

[46] Ojaghian MR, Almoneafy AA, Cui Z, *et al.* Application of acetyl salicylic acid and chemically different chitosans against storage carrot rot. Postharvest Biol Technol 2013; 84: 51-60.
[http://dx.doi.org/10.1016/j.postharvbio.2013.04.006]

[47] Klessig DF, Tian M, Choi HW. Multiple targets of salicylic acid and its derivatives in plants and animals. Front Immunol 2016; 7: 206.
[http://dx.doi.org/10.3389/fimmu.2016.00206] [PMID: 27303403]

[48] El-Hai KMA, El-Metwall MA, Mohamed NT. Hydrogen peroxide and acetylsalicylic acid induce the defense of lupine against root rot disease. Plant Pathol J 2016; 15: 17-26.
[http://dx.doi.org/10.3923/ppj.2016.17.26]

[49] Hahn K, Strittmatter G. Pathogen-defence gene *prp1-1* from potato encodes an auxin-responsive glutathione S-transferase. Eur J Biochem 1994; 226: 619-26.
[http://dx.doi.org/10.1111/j.1432-1033.1994.tb20088.x] [PMID: 8001577]

[50] Martínez Noël GMA, Madrid EA, Bottini R, *et al.* Indole acetic acid attenuates disease severity in potato-*Phytophthora infestans* interaction and inhibits the pathogen growth *in vitro*. Plant Physiol Biochem 2001; 39: 815-23.
[http://dx.doi.org/10.1016/S0981-9428(01)01298-0]

[51] Awasthi DP, Mishra NK, Dasgupta B. Use of non-conventional chemicals against stem rot of groundnut (*Arachis hypogaea* L.) caused by *Sclerotium rolfsii* Sacc. Int J Curr Microbiol Appl Sci 2018; 7: 1288-92.
[http://dx.doi.org/10.20546/ijcmas.2018.702.157]

[52] Emmanouil V, Wood RKS. Induction of resistance to *Verticillium dahliae* and synthesis of antifungal compounds in tomato, pepper and eggplant by injecting leaves with various substances. J Phytopathol 1981; 100: 212-25.
[http://dx.doi.org/10.1111/j.1439-0434.1981.tb03294.x]

[53] Dong H, Beer SV. Riboflavin induces disease resistance in plants by activating a novel signal

transduction pathway. Phytopathology 2000; 90: 801-11.
[http://dx.doi.org/10.1094/PHYTO.2000.90.8.801] [PMID: 18944500]

[54] Liu F, Wei F, Wang L, *et al.* Riboflavin activates defense responses in tobacco and induces resistance against *Phytophthora parasitica* and *Ralstonia solanacearum*. Physiol Mol Plant Pathol 2010; 74: 330-6.
[http://dx.doi.org/10.1016/j.pmpp.2010.05.002]

[55] Ahn IP, Kim S, Lee YH. Vitamin B_1 functions as an activator of plant disease resistance. Plant Physiol 2005; 138: 1505-15.
[http://dx.doi.org/10.1104/pp.104.058693] [PMID: 15980201]

[56] Montaser FAM. Role of riboflavin and thiamine in induced resistance against charcoal rot disease of soybean. Afr J Biotechnol 2011; 10: 10842-55.
[http://dx.doi.org/10.5897/AJB11.253]

[57] Muthukrishnan S, Murugan I, Selvaraj M. Chitosan nanoparticles loaded with thiamine stimulate growth and enhances protection against wilt disease in chickpea. Carbohydr Polym 2019; 212: 169-77.
[http://dx.doi.org/10.1016/j.carbpol.2019.02.037] [PMID: 30832844]

[58] Sangha MK, Atwal AK, Sandhu PS, *et al.* Salicylic Acid Induces Resistance to *Alternaria* Blight in Crop *Brassica* Species. In: Shengyi L, Guoqing L, Junbin H, Eds. Proceedings of the 12th International Rapeseed Congress Volume IV - Plant Protection. USA: Science Press USA Inc 2007; 14: pp. 137-9. [https://www.gcirc.org/fileadmin/documents/Proceedings/IRCWuhan2007vol4/137-139.pdf]

[59] Ngadze E, Icishahayo D, Coutinho TA, *et al.* Role of polyphenol oxidase, peroxidase, phenylalanine ammonia lyase, chlorogenic acid, and total soluble phenols in resistance of potatoes to soft rot. Plant Dis 2012; 96: 186-92.
[http://dx.doi.org/10.1094/PDIS-02-11-0149] [PMID: 30731807]

[60] Geat N, Singh D, Khirbat SK. Effect of non-conventional chemicals and synthetic fungicide on biochemical characteristics of chilli against fruit rot pathogen *Colletotrichum capsici*. J Plant Pathol Microbiol 2016; 7: 328.
[http://dx.doi.org/10.4172/2157-7471.1000328]

[61] Singh R, Sindhu A, Singal HR, *et al.* Biochemical basis of resistance in chickpea (*Cicer arietinum* L.) against *Fusarium* wilt. Acta Phytopathol Entomol Hung 2003; 38: 13-9.
[http://dx.doi.org/10.1556/APhyt.38.2003.1-2.3]

[62] Nagy NE, Fossdal CG, Dalen LS, *et al.* Effects of *Rhizoctonia* infection and drought on peroxidase and chitinase activity in Norway spruce (*Picea abies*). Physiol Plant 2004; 120: 465-73.
[http://dx.doi.org/10.1111/j.0031-9317.2004.00265.x] [PMID: 15032844]

[63] Ondobo ML, Onomo PE, Djocgoue PF, *et al.* Influence of *Phytophthora megakarya* inoculation on necrosis length, phenolic content, peroxidase and polyphenol oxidase activity in cocoa (*Theobroma cacao* L) plants. Syllab Rev Science Series 2013; 4: 8-18.

[64] Mandavia MK, Gajera HP, Khan NA, *et al.* Inhibitory action of phenolic compounds on cell wall degrading enzymes: host pathogen interaction in *Fusarium* wilt of cumin. Indian J Agric Biochem 2003; 16: 39-42.

[65] Ashry NA, Mohamed HI. Impact of secondary metabolites and related enzymes in flax resistance and susceptibility to powdery mildew. Afr J Biotechnol 2012; 11: 1073-7.
[https://www.ajol.info/index.php/ajb/article/view/100252] [10.5897/AJB11.1023]

[66] Sharma A, Sharma S, Joshi N, *et al.* Alteration in biochemical response in *Sesamum indicum* upon different plant-pathogen interactions. J Agric Sci Technol 2011; 1: 68-75.

[67] Shalaby IMS, El-Ganainy RMA, Botros SA, *et al.* Efficacy of some natural and synthetic compounds against charcoal rot caused by *Macrophomina phaseolina* of sesame and sunflower plants. Assiut J Agric Sci 2001; 32: 47-56.

[68]	Anand T, Chandrasekaran A, Kuttalam S, *et al.* Association of some plant defense enzyme activities with systemic resistance to early leaf blight and leaf spot induced in tomato plants by azoxystrobin and *Pseudomonas fluorescens.* J Plant Interact 2007; 2: 233-44.
[http://dx.doi.org/10.1080/17429140701708985]

[69]	Savaliya VA, Bhaliya CM, Marviya PB, *et al.* Evaluation of phytoextracts against *Macrophomina phaseolina* (Tassi) Goid causing root rot of sesame. J Biopesticides 2015; 8: 116-9.
[http://www.jbiopest.com/users/lw8/efiles/vol_8_2_116-119.pdf]

[70]	Khamari B, Patra C. Evaluation of antifungal potency of natural products against stem and root rot of sesame. J Pharmacogn Phytochem 2018; 7: 156-8. [https://www.phytojournal.com/archives/2018/vol7issue6/PartC/7-5-544-821.pdf]

[71]	Khaire PB, Hingole DG, Padvi SA. Efficacy of different phytoextracts against *Macrophomina phaseolina.* J Pharmacogn Phytochem 2018; 7: 1124-6. [http://www.jbiopest.com/users/lw8/efiles/vol_8_2_116-119.pdf]

[72]	Lakhran L, Ahir RR. *In vivo* evaluation of different fungicides, plant extracts, bio- control agents and organic amendments for management of dry root rot of chickpea caused by *Macrophomina phaseolina.* Legume Res 2020; 43: 140-5. [https://arccjournals.com/journal/legume-research--n-international-journal/LR-3939] [DOI: 10.18805/LR-3939]

[73]	Gupta IJ, Cheema HS. Effect of microsclerotia of *Macrophomina phaseolina* and seed dressers on germination and vigour of sesame seed. Seed Res 1990; 18: 169-72.

[74]	Pineda JB, Colmenares O, Avila JM. Evaluation of sunflower (*Helianthus annuus* L.) hybrid seed production in relation to incidence of plant pathogens. Agro Tropic (Venezuela) 1991; 41: 215-24.

[75]	Jaiman RK, Jain SC. *Macrophomina phaseolina* in cluster bean (*Cyamopsis tetragonoloba*) seeds and its control. J Mycol Plant Pathol 2008; 38: 403-4.

[76]	Shahid S, Khan MR. Management of root rot of mungbean caused by *Macrophomina phaseolina* through seed treatment with fungicide. Indian Phytopathol 2016; 69: 128-36.

[77]	Melean JA. Resistance of white seeded sesame (*Sesamum indicum* L.) cultivars against charcoal rot (*Macrophomina phaseolina*) in Venezuela. Sesame Safflower New 2003; 18: 72-6.

[78]	Thiyagu K, Kandasamy G, Manivannan N, *et al.* Identification of resistant genotypes to root rot disease (*Macrophomina phaseolina*) of sesame (*Sesamum indicum* L.). Agric Sci Dig 2007; 27: 34-7.

[79]	Shabana R, Abd El-Mohsen AA, Khalifa MMA, *et al.* Quantification of resistance of F6 sesame elite lines against charcoal-rot and *Fusarium* wilt diseases. Adv Agric Biol 2014; 1: 144-50.
[http://dx.doi.org/10.15192/PSCP.AAB.2014.1.3.144150]

[80]	Farooq S, Mohyo-Ud-Din A, Naz S, *et al.* Screening of sesame (*Sesamum indicum* L.) germplasm for resistance against charcoal rot disease caused by *Macrophomina phaseolina* (Tassi) Goid. Int J Biol Biotechnol 2019; 16: 407-10. [https://www.ijbbku.com/assets/custom/journals/2019/2/Screening%20of%20sesame%20(Sesamum%20indicum%20L.)%20germplasms%20for%20resistance%20against%20charcoal%20rot%20disease%20caused%20by%20Macrophomina%20phaseolina%20(Tassi)%20Goid.pdf]

CHAPTER 13

Beauveria Bassiana: An Ecofriendly Entomopathogenic Fungi for Agriculture and Environmental Sustainability

Purnima Singh Sikarwar[1] and **Balaji Vikram**[1,*]

[1] *Department of Horticulture, AKS University, Satna (M.P.) India*

Abstract: In the present day perspective, with the increasing cost of chemical pesticides along with increasing incidences of pesticide toxicity, the application of microbial pesticides holds good promise for crop protection around the world. *Beauveria bassiana* is a common soil fungus, having a broad host range and therefore is used for biological control of soil-dwelling insect pests. As this fungus is epizootic, it is being used worldwide as a biopesticide to control several pests, such as termites, whiteflies, and malaria-transmitting mosquitoes. The use of this fungus in different crop protection systems significantly controls the Colorado potato beetle, codling moth, and several genera of termites and bollworms. As insecticides, the spores are sprayed on affected crops as an emulsified suspension or wettable powder. Generally, *B. bassiana* is considered a nonselective pesticide because it parasitizes a very high range of arthropod hosts. This entomopathogenic fungus is also applied against the European and Indian corn borer, pine caterpillar, and green leafhoppers. The ability of *B. bassiana* to antagonize, parasitize, and kill insects endorses it as an efficient biocontrol agent. Although *B. bassiana* has a good share in the total biopesticide market, there is still ample scope for further development of this superior strain through advertisement among the farming community.

Keywords: *Beauveria bassiana*, Biopesticide, Soil fungus, Soil inhabiting organisms.

INTRODUCTION

Beauveria bassiana is an important and effective entomopathogenic fungus that causes a disease called white muscardine that occurs in a range of insects, including whiteflies, aphids, thrips, grasshoppers, and some types of beetles. It does not need to be ingested by the host; it only requires the spores to come into contact with one host. After infection, the fungus rapidly grows inside the insect. *B. bassiana* feeds on the nutrients present in the insect's body and produces toxins

---

* **Corresponding author Balaji Vikram:** Department of Horticulture, AKS University, Satna (M.P.) India; E-mail: balaji.vikram55@gmail.com

Shampi Jain, Ashutosh Gupta and Neeraj Verma
All rights reserved-© 2023 Bentham Science Publishers

in the process. This causes the insect to die, and as the host dies, *B. bassiana* covers the carcass in a layer of white mould and produces more infective spores. *B. bassiana* is found naturally in the soil of many regions around the world and is primarily used to target foliar pests. Many soil-dwelling insects have developed a natural resistance to *B. bassiana* because it is native to many such areas. Therefore, the pest (s) should be closely examined for correct identification before being applied to the soil.

B. bassiana can be used for hard crops at the time of sowing and is not harmful to crop residues, although it is always good to wash the harvested material before consumption. All safety precautions set out on the label must be followed at all times when using *B. bassiana* products. *B. bassiana* is generally considered a low-risk pesticide, yet it is recommended that long-sleeved shirts and pants be worn with a mask and/or goggles when applying products containing the fungus so that the chances of getting into any kind of accident are very small [1, 2].

B. bassiana is a fungus that occurs naturally in soils around the world and grows in and acts as a parasite on various arthropod species, causing white muscardine disease. It is thus believed to be related to entomopathogenic fungi. It has been used for a long time as an organic pesticide to control many pests, such as termites, thrips, whiteflies, aphids, and various beetles. It is also used to control bedbugs and mosquitoes that spread malaria. *B. bassiana* was formerly also known as *Tritirachium shiotae*. The species is named after the Italian entomologist Agostino Bassi, who discovered it in 1815 as the cause of muscardine disease, also known as white muscardine disease. When microscopic spores of the fungus come into contact with the body of the insect host, they germinate on the insect itself, penetrate the cuticle, grow inside the insect, and take nutrients from the insect, weakening the insect and slowly dying. This spreads and colonizes the entire insect and thus deprives the insect of nutrients. The infected insects eventually die within a few days. Later, a white mould emerges from the carcass and produces new spores. A specific isolate of *B. bassiana* can also attack a wide range of insects. The fungus rarely infects humans or other animals, so it is generally considered safe as an insecticide.

Taxonomy

The systematic position of *B. bassiana* is given below [3]:

- Kingdom-Fungi
- Phylum-Ascomycota
- Class- Sordariomycetes
- Order-Hypocreales

- Family-Clavicipitaceae or Cordicipitaceae or ophiocordicipitaceae
- Genus-*Beauveria*
- Species- *bassiana*

MORPHOLOGY

B. bassiana (Bals.) Vuill. is saprophytic, ubiquitous and pathogenic for many insects belonging to different orders, such as Lepidoptera, Hemiptera, Coleoptera, Hymenoptera, Homoptera, Hemiptera, and Orthoptera. The hyphal diameter in these different types of insects varies from 2.5-25 µm. Conidiophores of long scaly transparent and septal filaments bear white to yellowish conidia (asexual spores).

Different types of conidia can be produced by different strains of *B. bassiana*, depending on environmental conditions. The fungus continues to produce spherical (1–4 µm in diameter) or oval (1–3 µm in diameter x 1–3 µm) conidia under aerobic conditions, but anaerobically it produces oval-shaped infective blastospores (2–2 in diameter x 1–3 µm) and conidia (3 µm), whose length can be up to 7 µm [4].

MODE OF ACTION AGAINST INSECTS

Conidia are wind-borne, although rain splashes or arthropod vectors can also help them establish infection in susceptible hosts. In the bodies of invertebrates, the *B. bassiana* infection cycle has been studied in detail by Douro and colleagues [5]. Host infection occurs mainly in four stages: (a) Adhesion, (b) Germination and differentiation, (c) Penetration, and (d) Proliferation.

Adhesion

This represents the first step in the recognition and compatibility mechanism of conidia to the host cuticle [6], which is attached to the insect's cuticle by electrostatic and chemical forces. The induced epicuticular modification led to the germination of conidia, possibly through the production of mucilage [3].

Germination and Differentiation

Environmental conditions and host physiology, such as the biochemical composition of the host cuticle, continue to influence the process of germination. Depending on these factors, the germination of spores can often be stimulated or inhibited. Under suitable conditions, conidia (blastospores) germinate to form a germ tube. This differentiation is characterized by the formation and establishment of appressorium (penetration peg), softening of the cuticle, and promoting penetration. Appressorium degrades host integument to allow hyphae development, on which it relies for survival [5]. However, the nutritional value of the host cuticle plays an important role in appressoria production that encourages mycelial growth rather than penetration [3].

Penetration

With the reaction of hydrolytic action of a variety of enzymes, *e.g.*, mechanical pressure from protease, chitinase, lipase (the most important proteases), and oppressor, as well as other factors (such as oxalate), the fungus penetrates all cuticle layers until it reaches the nutrient-rich environment, *i.e.*, in insect hemolymph [3].

Dissemination within and to Another Host

In hemolymph, fungi undergo a morphogenetic differentiation from filamentous growth into single-celled, yeast-like hyphal bodies or blastospores that strategically exploit nutrients as well as colonize internal tissues and the host immune system. The fungus may then secrete toxic metabolites that help overcome the insect's immune defense mechanisms for its successful colonization. Several experiments have also reported the secretion of non-enzymatic toxins (beauvericin, beaverolides, basianolides, and isrolides) by some strains to speed up the infection process [3]. These events eventually lead to the death of the host insect, which is then mummified. After the insect's death, the fungus produces an antibiotic called "osporin," which removes the competition from the bacteria produced in the insect's gut [6]. Then, the fungal hyphae cross the entire part of the insect and turn white in colour. After a few days, conidiophores begin to appear on the mummified carcasses and bear new infective conidia for dispersal (passive propagation) [3].

Arthropods in *B. bassiana* and some plant groups show multitrophic interactions between the soil as well as some other microbes [5]. The scenario depicts a conceptual summary of a community scenario. The fungus can also affect soil-dwelling insects and then transfer nitrogen from dead insects to plants through the establishment of root endophytic colonies in those insects [3].

The endophytism of *B. bassiana* is capable of forming an important life system within the plant. It can colonize the roots, aerial parts of plants, and seeds without any harm to the plant. Insects that eat these parts of the plant can become infected with fungal spores. Sometimes dead insects can sporulate, becoming a new contamination source through their carcasses, aerogenic spores, or endophytic colonization for other organisms, including predators and parasites. Such organisms may also come into contact with other fungi to transmit their spores (vector). Somewhere, the teleomorph trait of the fungus has been reported to be related to *Cordyceps* sp. It appears to be found only in Asia, where it is commonly used in considerable quantities in Chinese medicine [3].

APPLICATIONS OF *B. BASSIANA* IN INSECT PEST MANAGEMENT

The history of success of *B. bassiana* applications in the management of insect pests is due to the development of control programs that include *B. bassiana* applications as well as a commercial endophytic fungus. There are many reports about the insecticidal potential of *B. bassiana* [7]. All studies are found to involve the direct application of entomopathogens by inoculation of the insect host plant to target the insect or indirectly by inoculation of the insect host plant. In the endophytic colonization strategy, methods of inoculating plants include seed coating, plant injection, immersion of radicals, *etc* [3].

Efficacy of *B. bassiana* Against Thrips

At 1×10^4–10^7 conidia per ml, *B. bassiana* strain RSB is highly effective against western flower thrips (*Frankliniella occidentalis*), causing 69%–96% mortality of first instars 10 days after inoculation [3]. When this strain is used on broccoli leaves, it has shown a significant reduction in adult and larval numbers in laboratory and greenhouse tests. Second instar larvae and pupae of thrips that were attacked by predatory mites are more susceptible to *B. bassiana* infection [3]. Tofa-Mahinto and coworkers [8] evaluated 20 strains of *B. bassiana* to control onion thrips (*Thrips tabaci*) in the laboratory and in the greenhouse. Strain SZ-26 was found to be the most virulent, causing 83%-100% mortality in adults 4–7 days after 1×10^7 conidia ml^{-1} application. Further, under greenhouse experiments, strain SZ-26 significantly reduced the number of adults and larvae [3].

Furthermore, by using a combination of sub-lethal doses of neem tree extract with *B. bassiana*, one can improve the control of *T. tabaci*. Efficacy was also found to improve in controlling pests, while the amount of pesticide used in palm trees and stored grain was also reduced. *B. bassiana* has been tested with positive effects against palm weevil (*Rhynchophorus ferrugineus*) when it has been treated with three methods, *i.e.*, injection of the fungus into naturally infected palm trees, periodic dust application of fungal spores on plants and palm weevil with the fungus, and release of contaminated males of red palm weevil [3]. In naturally infected palm trees, *B. bassiana* was found to be highly effective, reducing 90% of the mite population. Under laboratory conditions, Dembilio and coworkers [9] observed that *B. bassiana* CECT-20752 is naturally infected with *R. ferrugineus* pupae, eggs, larvae, and adults with an average lethal concentration (LC50) of 6.3×10^7-3.0×10^9 conidia ml^{-1} that could be beneficial to crops.

Efficacy of *B. bassiana* Against Insect Pests of Stored Grains

Efficacy of fungal strains (IBCB 74, ICB 87, and ICB 146) with a mortality rate of 54% to 66% was observed when beetles (*C. sordidus*) were immersed in a fungal suspension with a concentration of 1.12×10^9 conidia ml^{-1}. The sporulation rates were found to be similar for these three strains. Olatinvo and coworkers [10] reported that *B. bassiana* application resulted in up to 92% mortality in weevil (*Ips typographus*).

For the management of *Polyphylla fullo* (L.), a combination of *B. bassiana* strain PPRI5339 and *Metarhizium anisopliae* has been suggested. The first and second instar larvae were more prone to infection than older (3rd instar) [3] and resulted in 79.8% and 71.6% mortality in young and old larvae, respectively.

In another experiment, it was observed that spraying of Botanigard 22WP-*B. bassiana* strain GHA in water at the rate of 3-57 gm per 1.5L at consecutive weekly intervals from two weeks after planting against two flea beetle susceptible varieties of okra (NH99/DA and LD88/1-8-5-2) not only effectively reduced the number of *Podagrica* spp. in six weeks but also improved the yield of okra. In addition, a synergistic interaction between spores and oil formation of *B. bassiana* was also observed to control pollen beetles (*Meligethes* spp.) [11]. *B. bassiana* was also found to be highly effective against *Ips avulsus*, causing about 84% adult mortality [3].

Efficacy of *B. bassiana* Against Crop Pests

During research, *B. bassiana* has been used against crop pests of Diptera, Lepidoptera, and Hemiptera and has been found to be effective against crop pests. It caused 37–80.1% mortality after 10 days of application against *Thaumastocoris peregrinus*, one of the pests of *Eucalyptus camaldulensis* [3], under laboratory conditions.

B. bassiana, *Metarhizium anisopliae*, and *M. flavoviride* were applied against citrus pests (*Ceratitis rosea*, *C. capitata*, and *Thaumatotibia leucotreta*) at a concentration of 1×10^7 conidia ml^{-1}, resulting in a significant decrease of the adult population. Their effect on the pupa under laboratory conditions was also observed. The LC50 values for the three fungal species ranged from 6.8×10^5-2.1×10^6 conidia ml^{-1} [12].

Three strains of *B. bassiana* viz., B-2, B-14, and B-9 have been used against *Dendrolimus punctatus* at LT50 of 7.63, 7.62, and 7.88 d, respectively. These strains were found to be more toxic with lethal concentrations (LC50) of 0.63×10^6, 0.96×10^6, and 0.78×10^6 conidia ml^{-1}, respectively [3]. Dried conidia of

B. Bassiana strs. SBT@11 and SBT@16 also affected chrysalids of *Spodoptera litura*, with 100% mortality under laboratory conditions. Conidial concentration turned out to be the major factor. SBT@11 was more virulent under laboratory conditions [3]. At 2 gl^{-1} water concentration, *B. bassiana* and *Verticillium lecanii* affected *S. exigua* larvae by up to 90% [13].

Being an entomopathogen, *B. bassiana* colonizes endophytically and plays an important role in protecting plants from herbivore attacks and diseases. Spraying *B. bassiana* either as a foliar spray or into the soil at a concentration of 10^8 conidia ml^{-1} (in water) reduces pest and disease attacks in *Phaseolus vulgaris* [3].

CONSTRAINTS TO USE *B. BASSIANA*

Environmental Conditions

Environmental conditions also play an important role in the influence of *B. bassiana* in the pest control program. The fungus can kill insects only 3 to 6 days after infection under laboratory conditions. Under sub-optimal field conditions, the lethal effect of fungi on insects can sometimes be much longer. This may be due to the delayed onset of the disease and the time taken for infection [3]. Abiotic factors such as humidity, temperature, precipitation, and ultraviolet (UV) radiation primarily affect the activity and efficacy of *B. bassiana*. Talking mainly about mycelium, the favourable temperature for its growth is between 13°C-36°C. When temperatures reach extremes (8°C and 40°C), mycelium stops growing [3].

Temperature

Several laboratory tests found the optimum temperature for spore germination, mycelial growth, and effective infection with *Dendrolimus punctatus* to be 24°C. For normal growth, the upper temperature was found to be between 34°C-36°C. High temperatures invariably reduce fungal spore production and its ability to infect [3]. Also, a temperature of 30°C was found to be optimal for the maximum production of spores. As well as for mycelial growth and spore germination, relative humidity of 100% was found to be most suitable. It was also found that spores of some strains of *B. bassiana* can germinate at 56.8% relative humidity. Whereas, in most experiments, it was discovered that sporulation may be favoured by a low percentage of relative humidity (25%–50%).

Humidity

B. bassiana has been found to tolerate relatively low humidity. It has been found to proliferate at suitable moisture content inside or outside the insect's body [15]. Excessive rainfall, on the other hand, has been shown in some studies to cause significant loss of *B. bassiana* conidia that adhere to the leaves of some monocot and dicot plants. This resulted in a decrease in the effectiveness and persistence of *B. bassiana* on treated plant leaves [3].

Other Constraints

Climatic conditions may not always affect only the physiology, infectivity, fungal progression in the insect's body, and cadaver sporulation, dispersal, and efficacy of conidia, but also host susceptibility or resistance [3]. To improve the effectiveness of a biological control agent depending on the situation, chemical pesticides can be used at low rates in combination with *B. bassiana*. In a synergistic interaction, such a combination can enhance the effectiveness of the biological pesticide and reduce the side effects of chemical pesticides to a great extent. It is often observed that synergistic interactions are not possible in all cases, and sometimes it is possible to get an antagonistic effect with *B. bassiana*. This has been observed with several insecticides that can affect different strains of *B. bassiana*. Lufenuron exhibits incompatibility with *B. bassiana* str. MTCC-984, even at very low doses. Chemical insecticides such as imidacloprid, flufenoxuron, teflubenzuron+ fuzalon, endosulfan, and amitraz have been studied for their effect on *B. bassiana* conidial germination, vegetative growth, and sporulation [3].

Timing and synchronization are other constraints often observed for the application of *B. bassiana* with other chemical insecticides. It has been observed that there is no effect on the efficacy of *B. bassiana* against adults of *Lygus lineolaris* if applied four days prior to the application of a chemical fungicide. But when fungicide is applied prior to the use of *B. bassiana*, it results in a decrease in the mortality rate of the adult population of *L. lineolaris* [3]. There are very few reports on the compatibility of herbicides and plant growth regulators with various entomopathogenic fungi. However, use of glufosinate ammonium inhibited the growth and sporulation of mycelia of *B. bassiana*, and it was observed to be inconsistent when applied to control potato beetles. However, diquat has been found to induce the insecticidal activity of *B. bassiana*, resulting in 50%–76.6% death of the adult Colorado potato beetle (*Leptinotarsa decemlineata*) [3].

Additionally, antagonism among microorganisms is one of the major limitations in the combination of biological control agents. It has been found that a biological pesticide may be ineffective for controlling a particular pest in the field in the presence of an antagonistic agent (another biological agent). For example, some pathogenic fungi like *Penicillium urticae* inhibit the germination of conidia and growth of mycelia of *B. bassiana* [3]. The application method of biological agents in fields or on plants may also affect their interactions with each other depending on their properties. It has been observed that adults of onion thrips, when treated with a mixture of *B. bassiana* and neem tree extracts (at a sub-lethal dose), are found to exhibit a higher mortality rate than when these insects are treated with fungicides alone [3]. However, when the neem tree extract was used across the field, an antagonistic effect was observed. But drenching of neem tree extract (at sublethal dosage) and topical application of *B. bassiana* were found to show a higher mortality rate in treated insects, showing a synergistic effect with each other [3].

In the same sequence, a combination of *Bacillus thuringiensis* and *B. bassiana* was found to increase the mortality rate of corn borer larvae (*Ostrinia nubilalis*) in maize fields. Inglis and coworkers [15] reported a greater control of larvae when *B. bassiana* was applied to the lower canopy in comparison to applications to the upper canopy. Applications against late-instar larvae failed to provide useful control of these larvae. However, almost all treatments in use against late instars (spray at weekly intervals and above the canopy) showed a significant reduction (53-84%) in the population of adult beetles [3].

ADVANTAGES OF USING *B. BASSIANA*

Based on many experiments, it can be said that it is one of the effective bioagents as widely described. A wide range of insects (707 species belonging to 15 orders) and mites (13 species) have been identified as being affected by different isolates of B. *bassiana* [16] on which it is completely dominant. It is a biocontrol agent that is safe for the environment but not for humans.

Under laboratory conditions, the cost of producing an effective chemical insecticide is much higher than that of *B. bassiana* strains, which can be grown easily and are cost-effective. Moreover, its spores in solution form can be easily used in the field. Under appropriate environmental conditions and insect physiology, mere contact of *B. bassiana* with susceptible insect bodies is sufficient to trigger infection, as compared to other biocontrol agents acting by ingestion. Its insecticidal activity with longevity exerts its effect much faster than other entomopathogenic agents [3]. Also, the fungus and its conidia have a very limited effect on beneficial insects and other non-target organisms. Theoretically, due to the multiple modes of action of this fungus, *B. bassiana*, on its own, can adapt to various host changes. Various experiments have shown that *B. bassiana* can be used as a biological agent as a potential alternative to chemical insecticides [3].

Improved Plant Nutrient Uptake and Plant Growth

B. bassiana and other hypocrealean fungi are soil-borne and have developed a mutualistic relationship with plants. At the same time, they are now known to colonize several species of plants endophytically (living inside the plant). When fungi are applied through soil digging, foliar application, or seed or transplant treatments, they have been found to colonize and benefit plants in one or more ways. Examples of plants colonized by *B. bassiana* and other EPFs include barley, cassava, cocoa, forbs, grasses, green beans, corn, soybeans, tobacco, wheat, strawberries, and many other crop plants. Several studies have shown that endophytic *B. bassiana* and other EPFs improve nutrient uptake and promote plant growth, leading to better plant growth. Endophytic *B. bassiana* has also been found to be useful in improving iron availability, chlorophyll content, root length, and fine root abundance in sorghum grown on calcareous substrate.

It has always been observed that soil application appears to be a better treatment method than seed treatment or foliar application. In California, cabbage was grown as a potted plant under simulated drought conditions. It was observed that EPF, specifically *B. bassiana*, improved plant growth, plant condition, shoot/root ratio, plant biomass, and nutrient absorption. Fungus-treated vines produced nine anti-insect compounds, compared to five anti-insect compounds in untreated vines. EPFs colonized with plant roots act as root extensions and possibly improve nutrient and water absorption and help plants cope with stress factors. A similar effect has also been observed in cotton plants inoculated with *B. bassiana* with increased water holding capacity and root length. In a study in Denmark, seed treatment of corn (monocot plants) with *B. bassiana* showed increased plant growth when nutrients were abundant but not under low-nutrient conditions. This supported its role in nutrient absorption. In a Brazilian study, *B. bassiana* treated seeds of green beans showed improvement in the fresh and dry weight of beans, the length of their roots and aerial parts.

In an Argentine study, the application of *B. bassiana* to soybeans showed significant improvements in plant height, number of branches, weight per branch and plant pods, number of pods per branch and plant, branch and seed count per plant, and also seed weight and yield. In a study in Jordan, it was observed that seed treatment of broad beans with *B. bassiana* and *Metarhizium brunum* increased seedling emergence, plant height, number of leaf pairs, and fresh seedling and root weight. In a three-month raised bed study in Southern California, transplant plug treatment with *B. bassiana* was found to improve plant health and growth, leading to better results than a commercial product with beneficial soil microbes. In cotton, seed treatment with *B. bassiana* affected the breeding of the cotton aphid (*Aphis gossypii*) in greenhouses, and their populations were found to be reduced under field conditions.

Using *B. Bassiana* For Insect Management

It is a well-known fact that *B. bassiana* causes white muscardine disease in insects. When the spores of this fungus come into contact with the cuticle (skin) of susceptible insects, they germinate and move very rapidly through the cuticle directly into the internal body of their host [6]. The fungus spreads throughout the insect's body, producing toxins and depleting the insect of nutrients, eventually killing it. Therefore, unlike bacterial and viral pathogens of insects, *Beauveria* and other fungal pathogens infect the insect by contact and do not need to be consumed by their host to cause infection. Once the fungus kills its host, it grows back through the softer parts of the cuticle, covering the insect with a layer of white mold (hence the name white muscarine disease). This downy mould produces millions of infectious spores that are released into the environment.

Beauveria is a naturally occurring soil-borne fungus found around the world that has been reported for the control of soil-borne insects. However, many soil insects may have a natural tolerance to this fungus, which is not present in many foliar insects. Therefore, commercial development of this fungus for biological control has been targeted primarily against foliar insects.

Environmental protection is also very important for biological control. These products are generally non-toxic to beneficial insects, yet applications should be avoided in areas where bees are present. *Beauveria* products should also not be applied to water, as they can also be potentially toxic to fish. For greenhouse management, it is recommended that treatment should begin as soon as pests are detected. In potatoes, treatment should be started when the limit of 1.5 larvae per plant is exceeded.

After mixing with water, *Beauveria* products should be sprayed as soon as possible as the spores of the fungus become ineffective very quickly, and the material loses its viability overnight. Foliar spray should be done until the plants are completely wet. Devices that move the material to the undersides of leaves will result in prolonged activity as the spores are inactivated by sunlight. Evening time applications are highly desirable.

The frequency of applications also depends on the pest and the crop. For greenhouse pest problems, it is recommended to apply every 5-7 days. Again, all applications should be based on monitoring insect populations. The rate at which *Beauveria* spores kill their hosts depends on the temperature. At a constant

22.2°C, the tiny potato beetle larvae die in 3-5 days. It may take 7 to 10 days to kill the larvae, depending on farm conditions.

Using *B. Bassiana* as an Insect Management of Field Crops

Soil application (for root grub): Mix bio-power @ 3 Ltr or 4 kg/ha in 500 L of water (*i.e.*, 6 ml or 8 g/l of water). Bio-power can be sprinkled around the root zone and incorporated into the soil either mechanically or through the watering of plant roots. At the recommended rate, Bio-Power can also be mixed with 1000 kg of organic fertilizer or field soil at the recommended rate and can be applied uniformly.

To Control Termites and Other Soil Insects

1. Take *B. bassiana* (biopesticide) 1.0% @ 2.5 kg /hectares.
2. Mix it thoroughly with 60-75 kg of digested dung manure.
3. Keep the mixture in the shade for about 7-8 days in the summer and 10-12 days in the winter season.
4. Cover the mixture with jute bags and sprinkle water to keep the mixture moist as the fungus *B. bassiana* grows in high humidity.
5. Spread the prepared mixture into the soil at the time of the last ploughing, before sowing the seed.

Drip System

Bio-Power is mixed at a rate of 6 ml or 8 g/litre of water. After filtering, it is incorporated into the soil through a drip irrigation system either before planting or during the post-planting phase. But the best practice is to start foliar applications early in the season, when you get the plants, or when cuttings are done before the pests first appear.

Preparation of Spray Solution

Suspend bio-power (3 L or 4 kg/ha) in a clean, dry container and add water (20 L). After that, stir the ingredients well and make up to the final volume (500 liters) with water or 1% neem solution while preparing the spray liquid. Apply it, preferably early in the morning or late in the evening. Use the spray solution as a direct spray targeting pests on the undersides of leaves. Relative humidity of more than 60% should be maintained during foliar applications.

The high temperature reduces the viability of the fungal spores. The optimum application temperature lies between 18°C-29°C. Spores develop slowly below 15°C and become ineffective above 33°C. *B. bassiana* spores are highly sensitive to hcat, low humidity, and poor storage conditions. It has played an important role in the management of many arthropods and agricultural, veterinary, and forestry pests. It is better to be used as a preventive bioagent rather than a cure. Due to the long incubation period, it takes several days to kill the pests.

DIFFICULTIES FACED IN FORMULATION OF BIOPESTICIDE BASED ON *B. BASSIANA*

Biopesticides tested under laboratory or controlled conditions should always be commercially prepared to be used as a powder or as a solution in water for spray. Stabilization of the bioagent, handling of the bioagent, protection from adverse environmental conditions, and longevity are some of the primary objectives for preparing biopesticide at a commercial level. These should also be safe to use and cost-effective as compared to synthetic insecticides. It should be kept in mind that a biopesticide should be eco-friendly.

Still, due to difficulties associated with biopesticide formulation based on *B. bassiana*, the search for good effective material to obtain viable and stable conidia is in progress. Progress in technology and development in the approach of pest management research and outcomes have also led to large-scale production. Recent developments in integrated pest management strategies, including the use of *B. bassiana*, may provide a better alternative to effectively and efficiently controlling harmful pests.

CONCLUDING REMARKS

From the research work done on *B. bassiana*, it is clear that our soil health and all-round life cycle can be maintained by using it in agriculture. *B. bassiana* can be used for the control of members of Lepidoptera, such as chick-beetle, hairy beetle, sucking moth, woolly aphid, whitefly, termite, and spider mite, infecting various crops. At high humidity and normal temperature, *B. bassiana* is more effective.

In the presence of suitable temperature and humidity, the spores of this fungus stick to the chitin-covered coating of insects and germinate, and the germination tubes enter various pores of insects, such as respiration and germination, and spread infection in them.

It can be used as a spray on crops because its action kills the sucking insects by sticking them to the leaves of the crop. *B. bassiana* can be used in termite and soil treatment as it penetrates the soil and controls the pests present in it. Despite having different types of soil composition, the use of *B. bassiana* has always proved beneficial. Its use is highly effective in the pest control at any stage of the plant.

REFERENCES

[1] De-Oliveira RC, Neves PMOJ. Compatibility of *Beauveria bassiana* with acaricides. Neotrop Entomol 2004; 33: 353-8.
[http://dx.doi.org/10.1590/S1519-566X2004000300013]

[2] Inglis GD, Goettel MS, Johnson DL. Persistence of the entomopathogenic fungus, *Beauveria bassiana*, on phylloplanes of crested wheatgrass and alfalfa. Biol Control 1993; 3: 258-70.
[http://dx.doi.org/10.1006/bcon.1993.1035]

[3] Dannon HF, Dannon AE, Douro-Kpindou OK, *et al.* Toward the efficient use of *Beauveria bassiana* in integrated cotton insect pest management. J Cotton Res 2020; 3: 24.
[http://dx.doi.org/10.1186/s42397-020-00061-5]

[4] Sabbahi R. Use of the entomopathogenic fungus *Beauveria bassiana* in a strategy of phytosanitary management of the main insect pests in strawberry plantations. Doctoral dissertation in Biology. University of Quebec, Quebec, Canada, 2008

[5] Mascarin GM, Jaronski ST. The production and uses of *Beauveria bassiana* as a microbial insecticide. World J Microbiol Biotechnol 2016; 32: 177.
[http://dx.doi.org/10.1007/s11274-016-2131-3] [PMID: 27628337]

[6] De Kouassi M. The possibilities of microbiological control. VertigO - Electron J Environ Sci 2001; 2.
[http://dx.doi.org/10.4000/vertigo.4091]

[7] Jaber LR, Ownley BH. Can we use entomopathogenic fungi as endophytes for dual biological control of insect pests and plant pathogens? Biol Control 2018; 116: 36-45.
[http://dx.doi.org/10.1016/j.biocontrol.2017.01.018]

[8] Wu S, Gao Y, Smagghe G, *et al.* Interactions between the entomopathogenic fungus *Beauveria bassiana* and the predatory mite *Neoseiulus barkeri* and biological control of their shared prey/host *Frankliniella occidentalis*. Biol Control 2016; 98: 43-51.
[http://dx.doi.org/10.1016/j.biocontrol.2016.04.001]

[9] Dembilio Ó, Quesada-Moraga E, Santiago-Álvarez C, *et al.* Potential of an indigenous strain of the entomopathogenic fungus *Beauveria bassiana* as a biological control agent against the Red Palm Weevil, *Rhynchophorus ferrugineus*. J Invertebr Pathol 2010; 104: 214-21.
[http://dx.doi.org/10.1016/j.jip.2010.04.006] [PMID: 20398670]

[10] Olatinwo R, Walters S, Strom B. Impact of *Beauveria bassiana* (Ascomycota: Hypocreales) on the small southern pine engraver (Coleoptera: Scolytidae) in a loblolly pine bolt assay. J Entomol Sci 2018; 53: 180-91.
[http://dx.doi.org/10.18474/JES17-58.1]

[11] Kaiser D, Bacher S, Mène-Saffrané L, *et al.* Efficiency of natural substances to protect *Beauveria bassiana* conidia from UV radiation. Pest Manag Sci 2019; 75: 556-63.
[http://dx.doi.org/10.1002/ps.5209] [PMID: 30221461]

[12] Goble TA, Dames JF, Hill MP, *et al.* Investigation of native isolates of entomopathogenic fungi for the biological control of three citrus pests. Biocontrol Sci Technol 2011; 21: 1193-211.
[http://dx.doi.org/10.1080/09583157.2011.608907]

[13] Hasyim A, Setiawati W, Jayanti H, *et al.* Identification and pathogenicity of entomopathogenic fungi for controlling the beet armyworm *Spodoptera exigua* (Lepidoptera: Noctuidae). AAB Bioflux 2017; 9: 34-46. [http://www.aab.bioflux.com.ro/docs/2017.34-46.pdf]

[14] Faria M, Wraight SP. Biological control of *Bemisia tabaci* with fungi. Crop Prot 2001; 20: 767-78.
[http://dx.doi.org/10.1016/S0261-2194(01)00110-7]

[15] Inglis GD, Ivie TJ, Duke GM, *et al.* Influence of rain and conidial formulation on persistence of *Beauveria bassiana* on potato leaves and Colorado potato beetle larvae. Biol Control 2000; 18: 55-64.
[http://dx.doi.org/10.1006/bcon.1999.0806]

[16] Lambert N. Biological control of pests: Applicability in Quebec. Degree of Master in Environment. University of Sherbrooke, Quebec, Canada 2010.

CHAPTER 14

Ergot, Ergotism and its Pharmaceutical Use

Doomar Singh[1,*]

[1] *Department of Plant Pathology, Faculty of Agriculture Science & Technology, AKS University, Satna 485001, India*

Abstract: Many fungi are directly or indirectly toxic to humans and animals. Ergot, a fruiting body of the *Claviceps purpurea* fungus, contaminates grain after harvest and is toxic to humans and animals who consume contaminated grains. The lysergic acid diethylamide (LSD) that was widely used as a hallucinogen is best known as the ergot alkaloids. The main symptoms of the disease caused by consuming ergot-contaminated grain flour in humans and animals are blistering and reddening of the skin with a burning sensation. Ergot alkaloids such as agroclavine, ergovaline, ergotamine, ergonovine, lysergic acid, dopamine, *etc*., are the natural alkaloids produced by *Claviceps* spp. in many cereal crops (mainly wheat, barley, rye, bajra, jowar, and dallisgrass), but rye (triticale) is the most common host of this fungus. Contaminated grain may cause very harmful diseases to internal organs, the circulatory and nervous systems of animals and humans, and even they may die. Ergot alkaloids are very important in the pharmaceutical industry. Therefore, this soil-borne fungus, which can be used in the manufacturing of different types of medicines for human and animal welfare, is very important.

Keywords: *Claviceps purpurea*, Ergot, Ergotism, Ergot alkaloids, Holy fire, LSD.

INTRODUCTION

The French word "argot," meaning "a cock's spiny," is the fruit-like structure developed by *Claviceps purpurea* fungus instead of the seed of the plant [1] and contaminates the grain after harvest. Ergot is also the name of the disease of cereals and grasses caused by this and related fungi. Ergot, a plant disease, can reduce grain yields significantly, as each ergot completely replaces the kernel that it infects. Most of the damage to the crop, however, is because it makes the rest of the crop unfit for human or animal consumption unless the ergots are removed. The fungus, *Claviceps* spp., produces a large number of ergot alkaloids and infects a wide variety of grass species during the growing season.

---

[*] **Corresponding author Doomar Singh:** Department of Plant Pathology, Faculty of Agriculture Science & Technology, AKS University, Satna-48500, MP, India; E-mail: doomarsingh@yahoo.in

Shampi Jain, Ashutosh Gupta and Neeraj Verma
All rights reserved-© 2023 Bentham Science Publishers

Structurally, ergot alkaloids are classified into three groups, namely, clavines, lysergic acid amides, and peptides (ergopeptines) [2]. The most commonly produced ergot alkaloids from *Claviceps* species are ergometrine, ergotamine, ergocristine, ergokryptine, ergosine, and ergocornine and their epimeric forms ergotaminine, ergometrinine, ergocristinine, ergokryptinine, ergocroninine, and ergosinine [3 - 6]. Ergotamine and ergovaline are two common alkaloids examined in ergot. These alkaloids have a therapeutic effect on some forms of migraine, post-partum hemorrhages, mastopathy, and a sedative effect on the central nervous system [7]. The most well-known of the ergot alkaloids, lysergic acid diethylamide (infamous LSD), was widely used as a hallucinogen by the hippie culture of the 1960s.

Several powerful alkaloids such as ergotamine, ergometrine, and ergonovine are used medicinally to induce labour and prevent post-partum haemorrhage during childbirth. Due to the importance of these drugs, rye and wheat fields are artificially inoculated in Europe and other countries to increase sclerotia production, which is a valuable source of a farmer's income. Pharmaceutical companies are much more interested in a method of growing the fungus in liquid culture in vats.

Depending on the climatic factors and plant species (rye, wheat, pearl millet, barley, sorghum, *etc.*) and *Claviceps* species, the quantity of ergot in the field and the severity of the symptoms of ergotism may vary. Rye is the most common host of the ergot, while wheat is the least common host of the ergot. These two grains were the source of feed for animals as well as food for poor people and food for rich men, respectively. The alkaloids produced by ergot can compress blood vessels and cause dead tissue by infection or lack of blood flow in humans and animals who consume contaminated food with ergot. Doctors and midwives used the ground fruiting body to stop severe bleeding during severe accidents and childbirth.

History

The disease (ergotism) seems to have existed since ancient times. A religious order was issued in 1093 against a severe outbreak of the disease in Southern France to help the people suffering from it. Because the order was issued by Saint Anthony, the disease was known as "St. Anthony's Fire." The disease varied in severity and occurrence from year to year and appeared to affect poor people more often [8]. During the Middle Ages, a Roman historian named Lucretius gave the name 'Ignissacer', meaning "Holy Fire," to ergotism in 98–55 BC. In China, it was described in 1100 B.C., in Assyria in 600 B.C., and was reported to be very severe for the troops of Julius Caesar in one of his campaigns in France. Several epidemics of "holy fire" appeared in France, including ones of 857, of 994, (which killed between 20,000 and 50,000 people) and of 1093. About 20,000 soldiers in Peter the Great of Russia's army died of consuming bread made with severely contaminated wheat flour in 1722. Outbreaks of the "holy fire" have also occurred during the 20th century, *e.g.*, in 1926-1927 in Russia, about 10,000 people were affected by the disease; in England in 1927, more than 200 cases were reported; and about 200 people were affected in 1951 in the village of Poni-St. Esprit, France, by eating bread made from ergot-contaminated wheat flour [8]. Because of improper methods of cleaning the grains before grinding them for flour, human beings suffered from this horrible disease.

Some recent outbreaks occurred in India in 1975, where the effects were more of the nervous type symptoms of giddiness, drowsiness, nausea, and vomiting, and in Ethiopia in 1978, where gangrene and loss of limbs occurred.

PRODUCTION OF ERGOT ALKALOIDS

Ergot is a disease of grasses (Poaceae family) caused by several *Claviceps* sp. It is an important disease of crops like rye (*Secale cereale*), wheat (*Triticum aestivum*) and barley (*Hordeum vulgare*) caused by *C. purpurea* (Fr.) TuL., bajra (*Pennisetum typhoides*) caused by *C. fusiformis*, jowar (*Sorghum bicolor*) caused by *C. africana* and dallisgrass (*Paspalum dilatatum*) caused by *C. paspali*. Therye (triticale) is the most common host of this fungus, while oats (*Avena sativa*) are rarely affected. The fungus (*Claviceps*) belongs to the class of filamentous Ascomycetes (Pyrenomycetes) of the phylum Ascomycota. In India, the ergot disease of bajra was first observed in the early 1940s in the southern region of India as of minor importance. The first outbreak of the disease-causing considerable yield loss was reported in the Satara district of Maharashtra in 1956 [9]. Since then, the disease continued to appear in sporadic form until 1968, when it appeared in an epiphytotic form in the northern part of India. In the pearl millet growing regions of India, *C. purpurea* is a common cause of this disease, which was first recognized as a fungus in 1711 and whose life cycle was described in 1853 by Tulasne. On the host, the pathogen produces three distinct stages in its life cycle, *i.e.*, the sclerotial (ergot) stage, the perithecial stage, and the honeydew or sphacelia (conidial) stage (Fig. **1**). The sclerotia serve as resting structures in the offseason. In the next season, during blossoming, sclerotia produce 1–60 flesh-coloured stalks. An orange/pink coloured spherical head (stroma proper) is produced on the tip of each stalk. Several perithecia are developed at the periphery of the head. Each perithecium contains many cylindrical or clavate asci. Each ascus contains a bundle of eight long, slender, filiform, hyaline, multicellular ascospores, which are the source of primary infection, carried by insects or wind to newly opened flowers, where they cause infection in the ovaries directly or through stigma. Within a week, the fungus forms sporodochia in the infected flower. The conidia are produced in each sporodochium, which exudates creamy droplets (honeydew). This honeydew also contains alkaloids and attracts insects for secondary infection of healthy flowers or the conidia can also spread through splashing rain. Instead of grain formation, the infected ovary is filled with the mass of fungal hyphae (mycelium) through germinating conidia, which converts into hard sclerotium (ergot). At the time of maturity of healthy seeds, these sclerotia are also mature at the same time and fall to the ground or are mixed with the healthy seeds during harvesting and threshing (Fig. **2**). These sclerotia are hard, brown to purple-black, horn-like (Figs. **3** and **4**) and considerably longer than healthy grains, which can easily be separated by sieving. The ergot grains can easily be produced through the artificial inoculation of susceptible crops for pharmaceutical purposes.

St. Anthony's fire is known today as ergotism and is the result of people and animals consuming grain obtained from cultivated cereals and wild grasses infected with one of several ergot-producing fungi. The ergot and nutrients for the biosynthesis of the ergot alkaloids are absorbed by the fungus from the concerned host. Ergot is not a storage problem, but it can be present in stored grains resulting from the harvesting of ergots along with the healthy grain.

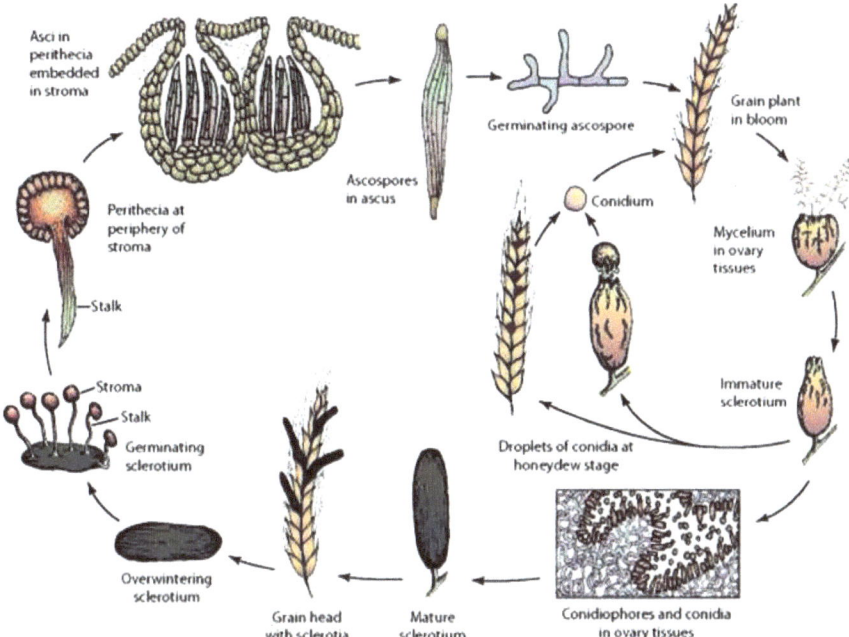

Fig. (1). Life cycle of *Claviceps purpurea* (Source: fungus https://www.apsnet.org/edcenter/disandpath/fungalasco/pdlessons/Pages/Ergot.aspx).

Fig. (2). Ergot sclerotia mixed with healthy seeds in wheat (Source: A Brief Summary, AUGUST 7, 2015 / BART MOSEMAN https://farmerscoop.wordpress.com/2015/08/07/ergot-in-wheat-a-brief-summary/).

Fig. (3). Ergots (the blackened claw-like objects) of *Claviceps purpurea* on wild oat (*Avena fatua*). (Source: http://www.davidmoore.org.uk/Sec04_03.htm).

Fig. (4). Ergot sclerotia (Source: https://www.shroomery.org/forums/showflat.php/Number/5399813).

ERGOT ALKALOIDS

The differences may be due to the different kinds of alkaloids present in the ergot, as variations in the amount and kinds of alkaloids can occur in the ergot sclerotia. Recent outbreaks have occurred in Ethiopia (1978), where gangrene and loss of limbs occurred, and in India (1975), where the effects were more of the nervous type symptoms of giddiness, drowsiness, nausea, and vomiting.

Ergometrine was discovered in Ethiopian sclerotia, while clavinet alkaloids such as agroclavine, elymoclavine, chanoclavine, penniclavine, and setoclavine were discovered in India. The ergot alkaloids found in these two outbreaks were produced by different *Claviceps* spp. An endophytic fungus called *Neotyphodium coenophialum* is found in most of the widely grown tall fescue grass in the United States. Ergovaline, an ergopeptine, produced by this endophytic fungus, causes ergot-like toxicosis in animals grazing on pastures containing fescue grass. Besides the two classical forms of ergotism—gangrenous and convulsive, another type of ergotism, manifested mainly as gastrointestinal disturbance, has been reported in India from areas of Maharashtra, Gujrat, and Rajasthan. Bajra (*P. typhoides*) ergot contains agroclavine, elymoclavine, chanoclavine, penniclavine, and setoclavine alkaloids, which are different from those found in rye and wheat ergot. A water-soluble alkaloid, ergometrine, is also present. The total amount of alkaloids in honeydew and sclerotia has been found to be up to 5 mg/100 g and 56 mg/100 g, respectively [10]. Cattle are frequently poisoned by grazing on grasses that contain the fungus's sclerotia or in fields where the sclerotia are present. Their legs, hoofs, and tails become gangrenous, and cows may abort their calves. This disease of animals is known as "ergotism." Ergot sclerotia contains up to 30%–40% fatty oils and 2% alkaloids [11]. Free amino acids, ergosterin, choline, acetylcholine, betaine, ergothioneine, uracil, guanidine, free aromatic, and heterocyclic amines (tyramine, histamine, agmatine) are also found in sclerotia (Fig. 5). Orange-red (endocrinin and clavorubin) and light yellow (ergochromes and ergochrysins) are acid pigments belonging to anthraquinolinic acid derivatives found in the outer shell of sclerotium. That is why the pigment of the sclerotium shell gives rise to the greyish-brown-violet colour. Albino ergot strains incapable of producing pigments also exist.

SYMPTOMS OF ERGOTISM IN HUMANS AND ANIMALS

Many grains, and sometimes other seeds and plant products such as bread, hay, purees, and rotting fruit, are often infected or contaminated with one or more fungi that produce toxic compounds known as mycotoxins. Animals or humans consuming such products may develop severe diseases of their internal organs, the nervous system, and the circulatory system and may eventually die. Many pasture grasses are infected with certain endophytic fungi that grow internally in the plant and, although they do not seem to seriously damage the grass plants, they produce toxic compounds that cause severe diseases in wild and domestic animals that eat those infected plants. Similarly, toxic and sometimes lethal to animals are some grasses whose seeds are infected with bacteria carried there by a nematode. These bacteria are often infected with a virus (bacteriophage) that induces the bacteria to produce compounds highly toxic to animals.

Fig. (5). Various alkaloids found in ergot grain.

Many grains, and sometimes other seeds and plant products such as bread, hay, purees, and rotting fruit, are often infected or contaminated with one or more fungi that produce toxic compounds known as mycotoxins. Animals or humans consuming such products may develop severe diseases of their internal organs, the nervous system, and the circulatory system and may eventually die. Many pasture grasses are infected with certain endophytic fungi that grow internally in the plant and, although they do not seem to seriously damage the grass plants, they produce toxic compounds that cause severe diseases in wild and domestic animals that eat those infected plants. Similarly, toxic and sometimes lethal to animals are some grasses whose seeds are infected with bacteria carried there by a nematode. These bacteria are often infected with a virus (bacteriophage) that induces the bacteria to produce compounds highly toxic to animals.

LSD is an illegal, semi-synthetic drug derived from ergot and a non-organic chemical called diethylamide. It stimulates serotonin production by activating serotonin receptors in the cortex and deep structures of the brain. Generally, the brain filters out irrelevant stimuli, but with LSD, this does not happen. The excessive stimulation of LSD causes changes in thought, attention, perceptions, and emotions in humans. These alterations appear as hallucinations. It takes about 60 minutes to onset hallucinations and can last for 6 to 12 hours. Physical stimulation caused by LSD includes pupil dilation, high blood pressure, increased heart rate, and temperature rise. Some of the short-term effects include dizziness and sleeplessness; reduced appetite; dry mouth; sweating; numbness; weakness; and tremors.

Individuals react differently to a drug, which is why the potency of any drug is unpredictable. The effects of a drug depend on the mindset of a user, their surroundings, stress level, expectations, thoughts, and mood at the time the drug is taken. At doses above 400 mcg, severe or life-threatening physical effects are seen, but the psychological effects can lead to unusual and risky behaviour or even death. Physical withdrawal symptoms do not occur after stopping the use of LSD, because it is not physically addictive, but psychological addiction can be common. In susceptible individuals, there is a danger of a long-term psychotic state or schizophrenia.

The reported symptoms of LSD are weakness, nausea, numbness, hypothermia or hyperthermia, goosebumps, elevated blood sugar, jaw clenching, heart rate increase, perspiration, saliva production, hyperreflexia, tremors, and mucus production. Visual hallucinations and illusions (colloquially known as "trips") are the most common immediate psychological effects of LSD, which can vary depending on how much is used and how the brain responds. Negative experiences, referred to as "bad trips," produce intense negative emotions, such as irrational fears and anxiety, panic attacks, paranoia, rapid mood swings, hopelessness, intrusive thoughts of harming others, and suicidal ideation. Good trips are stimulating and pleasurable and typically involve feeling as if one is floating, feeling disconnected from reality, feeling joy or euphoria (sometimes called a "rush"), decreasing inhibitions, and the belief that one has extreme mental clarity or superpowers.

LSD sensory effects include radiant colors, colour patterns behind closed eyelids, objects and surfaces appearing to ripple or breathe an altered sense of time, crawling geometric patterns overlaying walls and other objects, and morphing objects.

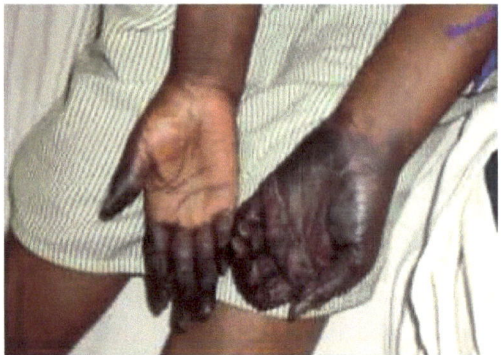

Fig. (6). Gangrenous fingers (Source: https://sites.google.com/site/135botany/home/lect13-ergot-of-ye-i?tmpl=%2Fsystem%2Fapp%2Ftemplates%2Fprint%2F&showPrintDialog=1).

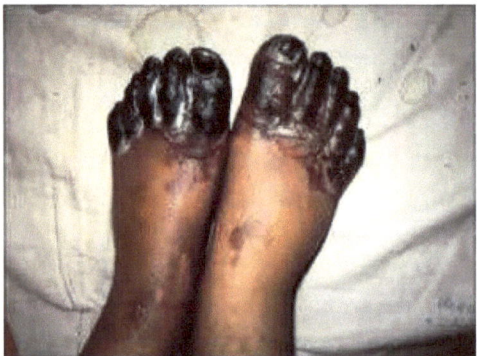

Fig. (7). Peripheral gangrene of toes (Source: https://www.idoj.in/viewimage.asp?img=IndianDermatol-OnlineJ_2013_4_3_228_115528_f1.jpg).

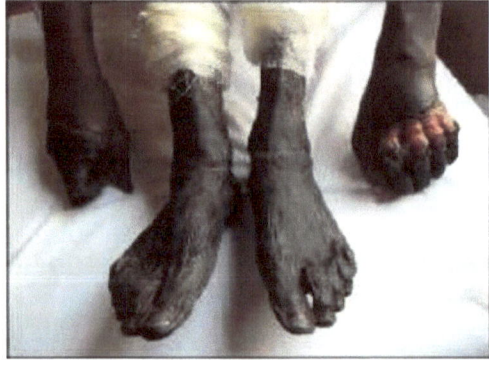

Fig. (8). Symmetrical peripheral gangrene of the hands and feet (Source: https://www.anmjournal.com/viewimage.asp?img=AnnNigerianMed_2014_8_2_98_153363_f2.jpg).

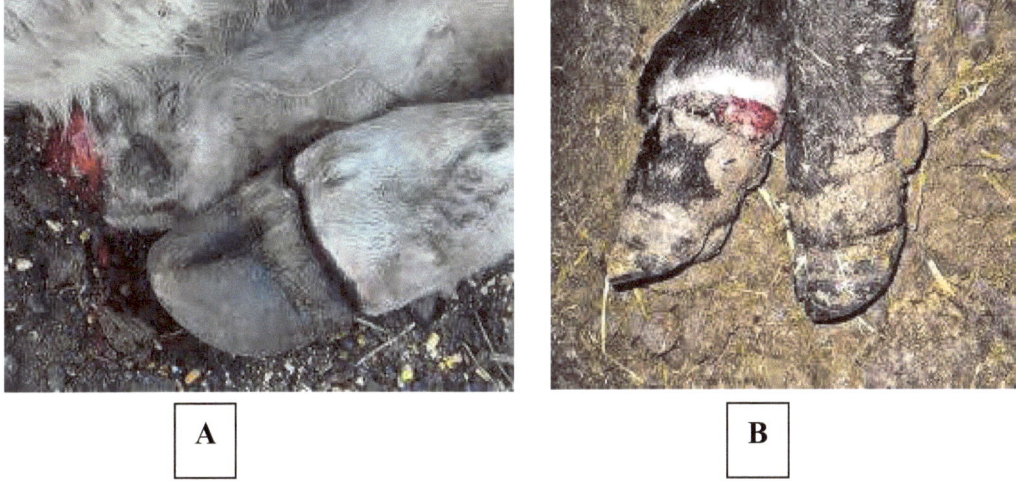

Fig. (9). (**A**) Symptoms of holy fire in animals hoof (Source: https://vetmed.iastate.edu/sites/default/files/vdpam/Extension/Ergot-Poisoning-in-Cattle.pdf) and (**B**) Cattle feet (Source: https://www.manito- bacooperator.ca/news-opinion/news/ergot-an-ongoing-problem-for-cattle-producers/).

PHARMACEUTICAL USE OF ERGOT ALKALOIDS

According to history, man has depended on nature for medicine. A huge source of pharmaceutically useful molecules is obtained from microorganisms such as fungi. The discovery of penicillin marked a new era. In 1995, six compounds out of the top 20 selling prescription medicines were derived from fungi [12]. The first, note that 'ergot can stimulate uterine contractions of labour by administering three sclerotia' was given by Adam Loncier in Germany in 1582 [13], and in 1822, Hosack in New York found that many stillbirths were due to uterine rupture resulting in maternal death. Rapid and sudden termination of labour, at delivery time lasting less than three hours, was treated with the most effective drug made by ergoted grain at that time.

A Swiss chemist, Albert Hofman, synthesized LSD for the first time in 1938 to treat respiratory depression. In 1943, he discovered the hallucinogenic properties of LSD when Hofman accidentally absorbed some LSD through his skin. The ergot alkaloids have high biological activity and a broad spectrum of pharmacological effects; hence, they are of considerable importance to medicine. They have adrenoblocking, anti serotonin, and dopaminomimetic properties. In 1967, LSD was banned from medical use and classified as a Schedule 1 drug. Its popularity has decreased since the 1970s. Ergot alkaloids are mainly used as medicines, and the same compounds have alleviated the suffering caused by treating symptoms of Parkinson's disease, migraine headaches, and stimulated uterine contractions and stemmed bleeding during childbirth. The toxicity and therapeutic potential of these compounds are due to their affinity for neurotransmitter receptors. The vasoconstriction leads to burning sensations associated with the stimulation of smooth muscle contraction. The effectiveness of ergot alkaloids in treating migraine and reducing bleeding lies in the same pharmacological mechanism.

Lysergic acid diethylamide (LSD) is an intermediate in the biosynthesis of ergot alkaloids. The artificial synthesis of most of the ergot alkaloids involves a common set of initial reactions, beginning with the addition of a pyrophosphate group to tryptophan, followed by a series of steps to produce a tetracyclic ergoline ring. Ergot alkaloids can further be divided into two structural groups, namely, amino acid alkaloids (*e.g.*, ergotamine) and amine alkaloids (*e.g.*, lysergic acid and ergonovine). Ergotamine and ergocristine (amino acid alkaloids) are physiologically the most active and of greatest importance. These alkaloids are potent vasoconstrictors and highly oxytocic (stimulate uterine smooth muscle) if given intravenously but not orally, and also inhibit nerves stimulated by sympathomimetic amines. Ergonovine (amine alkaloid) is the most potent and fastest acting, with a strong oxytocic effect and a weak vasoconstrictor. It is the most active and less toxic than ergotamine. These compounds have also been used for abortion (Fig. **10**).

Uterine motility is increased by ergot alkaloids, which have complex effects on cardiovascular function and suppress prolactin secretion. Nausea, vomiting, decreased circulation, rapid and weak pulse, and even coma conditions are the common effects of ergot alkaloids due to their toxicity. All natural ergot alkaloids significantly increase the motor activity of the uterus. Small doses increase contractions in force or frequency, but when the dose is increased, contractions become more forceful and prolonged. Although this characteristic precludes their use for induction or facilitation of labour, it is quite compatible with their use in postpartum or after abortion to control bleeding and maintain uterine contraction. The gravid uterus is very sensitive, and small doses of ergot alkaloids can be given immediately postpartum to obtain a marked uterine response, usually without significant side effects. Although all-natural ergot alkaloids have the same effect on the uterus. For these reasons, ergonovine and its semi-synthetic derivative (methylergonovine) have mostly been used as uterine-stimulating agents in obstetrics. The semi-synthetic derivative of ergonovine (methylergonovine) has fewer differences from ergonovine in its uterine actions.

Ergonovine and methylergonovine are rapidly and completely absorbed when they are administered orally and reach their peak concentrations in plasma within 60 to 90 minutes. They are ten times more active than an equivalent dose of ergotamine. A uterotonic effect can be observed within 10 minutes after oral administration of 0.2 mg of ergonovine to women postpartum. In terms of the relative duration of action, ergonovine is metabolized and/or eliminated more rapidly than ergotamine. In a study, the half-life of methylergonovine in plasma ranges between 0.5 and 2 hours, as was found by Hardman and Limbird [14]. While ergotamine, the natural amino acid alkaloid, constricts both arteries and veins, dihydroergotamine retains appreciable vasoconstrictor activity. Therefore, it is more effective in incapacitance than in resistance vessels. Investigation of its usefulness in the treatment of postural hypotension depends on this property. Ergot derivatives were first found as an effective anti-migraine agent in the 1920s, and since then they have continued as a major class of therapeutic agents for the fast relief of moderate or severe migraine.

However, ergot alkaloids are non-selective pharmacological agents because they interact with many neurotransmitter receptors. Ergot alkaloids can be extracted from many sources by various simple approaches. They can be separated without extensive purification and can be quantified by HPLC.

Fig. (10). Ergot Drugs (Source: https://pbs.twimg.com/media/CzttKOAXcAAGv4N.jpg; https://i.pinimg.com/originals/9b/1e/f8/9b1ef83f1a94b783768de7dbaf820c0d.jpg).

CONCLUDING REMARKS

Ergot is the fruiting body of *Claviceps* spp., produced in the ovary of a susceptible host. It is not only a disease that contaminates healthy grains after harvest, but it also causes many disabilities, disorders, and diseases in humans and animals for a long time after being taken up in the food. In ancient times, when humans were unaware of the toxic effects of ergot grains, they used these grains in their food unknowingly. In some cases, the consumption of ergot grains in small amounts was found to be beneficial, while in other cases, it was used to cure many diseases and disorders in humans and animals. Both the toxic and beneficial effects of ergot grains increased their importance in human society.

Ancient people started to use and sell drugs made from ergot grains. Most animal and human drugs and medicines are manufactured by plant metabolites, which are grown everywhere. By using these plant metabolites, any country can become self-sufficient in the field of medicine. Cereal crops are grown all over the world, which can be used for natural ergot grain production. These ergot grains can be used as raw materials for many types of ayurvedic medicines. It is also concluded that there is a further need for more detailed research about ergot alkaloids and their uses. The faith in ayurvedic and herbal medicines has increased during the COVID-19 pandemic. Therefore, many educated youths who have lost their jobs during this pandemic can start their own businesses in the field of medicine by isolating the required, specific alkaloids from ergot grains with minimum effort and technology. India is a hub of medicine, and therefore, Indian youth can prey in their future for a unique piece of medicine required by most of the world's population.

ACKNOWLEDGMENTS

The author will like to acknowledge AKS University, Satna, M.P., India for providing different facilities to prepare this manuscript.

REFERENCES

[1] Krska R, Crews C. Significance, chemistry and determination of ergot alkaloids: A review. Food Addit Contam Part A Chem Anal Control Expo Risk Assess 2008; 25: 722-31.
[http://dx.doi.org/10.1080/02652030701765756] [PMID: 18484300]

[2] Shahid MG, Nadeem M, Gulzar A, *et al.* Novel ergot alkaloids production from *Penicillium citrinum* employing response surface methodology technique. Toxins (Basel) 2020; 12: 427.
[http://dx.doi.org/10.3390/toxins12070427] [PMID: 32610508]

[3] Lee HJ, Ryu D. Worldwide occurrence of mycotoxins in cereals and cereal-derived food products: Public health perspectives of their co-occurrence. J Agric Food Chem 2017; 65: 7034-51.
[http://dx.doi.org/10.1021/acs.jafc.6b04847] [PMID: 27976878]

[4] Holderied I, Rychlik M, Elsinghorst P. Optimized analysis of ergot alkaloids in rye products by liquid chromatography fluorescence detection applying lysergic acid diethylamide as an internal standard. Toxins (Basel) 2019; 11: 184.
[http://dx.doi.org/10.3390/toxins11040184] [PMID: 30925708]

[5] Khaneghah AM, Farhadi A, Nematollahi A, *et al.* A systematic review and meta-analysis to investigate the concentration and prevalence of trichothecenes in the cereal-based food. Trends Food Sci Technol 2020; 102: 193-202.
[http://dx.doi.org/10.1016/j.tifs.2020.05.026]

[6] Kodisch A, Oberforster M, Raditschnig A, *et al.* Covariation of ergot severity and alkaloid content measured by HPLC and one ELISA method in inoculated winter rye across three isolates and three

European countries. Toxins (Basel) 2020; 12: 676.
[http://dx.doi.org/10.3390/toxins12110676] [PMID: 33114663]

[7] Boichenko LV, Boichenko DM, Vinokurova NG, *et al.* Screening for ergot alkaloid producers among microscopic fungi by means of the polymerase chain reaction. Microbiology 2001; 70: 306-10.
[http://dx.doi.org/10.1023/A:1010403427611]

[8] Agrios GN, Ed. Plant Pathology. 5th ed. Burlington, USA: Elsevier Academic Press 2005; pp. 37-9.

[9] Thakur RP, King SB. Ergot disease of pearl millet. Information Bulletin no.24. Patancheru, A.P., India: ICRISAT 1988; p. 24.

[10] Sharma PD, Ed. Plant Pathology. 2nd ed. Meerut, India: Rastogi Publications 2018; p. 636.

[11] Komarova EL, Tolkachev ON. The chemistry of peptide ergot alkaloids: Part 1. Classification and chemistry of ergot peptides. Pharm Chem J 2001; 35: 504-13.
[http://dx.doi.org/10.1023/A:1014050926916]

[12] Langley D. Exploiting the fungi — Novel leads to new medicines. Mycologist 1997; 11: 165-7.
[http://dx.doi.org/10.1016/S0269-915X(97)80094-8]

[13] van Dongen PWJ, de Groot ANJA. History of ergot alkaloids from ergotism to ergometrine. Eur J Obstet Gynecol Reprod Biol 1995; 60: 109-16.
[http://dx.doi.org/10.1016/0028-2243(95)02104-Z] [PMID: 7641960]

[14] Hardman JG, Limbird LE, Molinoff PB, *et al.*, Eds. Goodman & Gilman's The Pharmacological Basis of Therapeutics. 9th ed. Tx: McGraw-Hill 1996; p. 1905.

CHAPTER 15

Identification of Fungi on Rhizoplane and in Rhizosphere of Leguminous Crop by Adopting Different Techniques

Deepali Chaturvedi[1,*]

[1] *Department of Botany, Lucknow University, U.P., India*

Abstract: Microbial inhabitants of soil are in an active state and change in response to modifications in environmental conditions, amendments, management, *etc*. Although various methods have been developed to study the biological properties of a disturbed soil sample or with their precincts, adoption of a method should be done keeping in mind the objective, limitations and assumptions of the method. Reproducibility of results is important to have broad applicability and comparisons, which in turn will depend upon effectiveness. Soil is a diverse medium having varying physicochemical properties, and hence, there should be modifications while standardizing the method. It will help in getting reproductive results. Sampling, processing, and storage of samples are equally important to have a true picture of the soil and need appropriate care. The numbers and kinds of fungi on the root surface, *i.e.*, rhizoplane of the leguminous crop plant and in the rhizosphere (near the roots), have been compared with the number and kinds in root-free soil. The crops showed a typical rhizosphere effect, and there were more microorganisms in the rhizosphere than in root-free soil. A total of 34 different species of fungi were identified. The majority belong to the *Aspergillus* genus. Roots and the rhizosphere of moth bean (*Vigna aconitifolia*) yielded a higher proportion of fungi than did root-free soil.

Keywords: Hyphae, Inhibition, Rhizosphere, Soil Factor, Soil Dilution.

INTRODUCTION

The rhizosphere is the narrow region of soil that is directly influenced by root exudates and associated soil microorganisms known as the root microbiome, and the rhizoplane is the region where the root surface is in contact with soil and corresponds to the inner limit of the rhizosphere. The microflora on the root surface may be more characteristic of a plant species than that of the rhizosphere. In the soil, a specific root effect may be diluted out or possibly obscured by the

---

[*] **Corresponding author Deepali Chaturvedi:** Department of Botany, Lucknow University, U.P., India; E-mail: Dr.DeepaliChaturvedi1@yahoo.com

soil microflora. Organisms intimately associated with roots may also reflect a plant's growth and metabolism much more precisely than organisms in the rhizosphere soil [1 - 4].

Soil factors are known to affect both root disease development and rhizosphere microflora. These factors no doubt change with the growth of the plant. There is always more microbial activity in the soil in the vicinity of roots than in the soil away from them. Between these two zones is an area of transition (Fig. **1**) in which the root influence diminishes gradually with an increase in distance.

To study soil microorganisms, knowledge of suitable techniques and their methods is of utmost importance. There are some techniques for trapping specific microorganisms, while some are relatively common for others as well. Thus, a large number of methods have been devised to study microorganisms qualitatively and quantitatively as well as to isolate pure cultures of microorganisms from the soil.

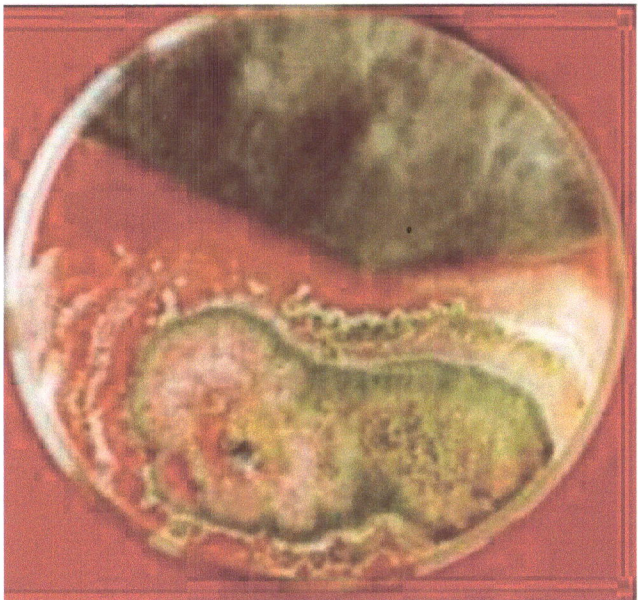

Fig. (1). A photomicrograph showing transition zone antagonistic reaction between microorganisms isolated from rhizosphere soil.

SOIL SAMPLING

The soil factors vary from place to place, so methods for collecting soil samples are not standardized. The microflora is rarely free in soil but occurs largely as colonies attached to clay, humus, and organic matter particles (*e.g.*, small rootlets). The microbial population of soil should be considered as a composite of these microorganisms' environments. It is generally presumed that similar types of soil samples from a locality at the same depth may contain a nearly equal number of microorganisms. It is generally suggested that 3 to 5 composite samples should be taken from each soil for each determination, after removing about 1 cm of the soil from the surface.

Collected samples are taken in clean and sterilized plastic bags or other suitable containers. It is always preferred to shorten the gap between sampling and the results of their studies. Although some workers believe that the soil samples can be stored at a low temperature, the multiplication of microbial cells is slowed down. Furthermore, it is recommended that prior to the use of samples, they should be partially dried (preferably air dried). Sieving and mixing of soil are also necessary. However, it has been reported by some workers that there is no necessity for taking unusual precautions, including aseptic conditions during sampling [5].

ISOLATION OF FUNGAL SPECIES FROM RHIZOSPHERE AND RHIZOPLANE

Methods

Various methods were applied to study the microflora of the rhizosphere and rhizoplane of moth bean (*Vigna aconitifolia*). The details of them are given below:

Direct Inoculation

It is for studying fungi that produce viable mycelia in soil. The period for incubation is kept short so that the fungal spores do not germinate and produce a mass of mycelium about 1 cm in diameter. Soil is placed in the centre of a petri dish containing solidified agar medium and incubated for 24 hours at 20–22°C. The mycelium, which radiates out from the soil piece, is cut and transferred to agar slants [6].

The Soil Plate Method

The method was designed to study the distribution of various species of fungi. The spore masses remain more intact in the soil plate than in the dilution plate. A small amount of soil (0.05 to 0.015g) will be examined in a petri dish with the help of a micro spatula. It is preferable to mix the soil particles with a drop of water before the melted and cooled agar medium is added and soil particles are dispersed throughout the agar [7].

Soil Dilution and Plate Count

This method is used for enumerating the pure cultures of a population of fungi, bacteria, actinomycetes, *etc.*, in soil, considering viable cells and mycelia fragments. Soil samples are sieved and the moisture content is estimated by drying a portion of soil in an oven (105–110°C). To a 25g air-dried soil sample, water is added to make the volume 250 ml, and the suspension is stirred and poured into a 1 litre Erlenmeyer flask. From this suspension, 10 ml is drawn with the help of a sterile pipette and transferred into 90 ml of sterile water (blank). Repeat this procedure until the desired final dilution is reached. The sample should not be allowed to remain in any dilution for more than 10 minutes. With this method, soil dilutions are prepared to yield dilutions of soil in water of 1 in 10, 1 in 100, 1 in 1000, 1 in 10,000, 1 in 1,00,000, 1 in 10,00,000 and so on. From each, 1 ml of the desired solution is poured into each petri dish, and an appropriate agar medium just above the solidifying temperature is added. The resulting colonies are counted.

Immersion Techniques

Several procedures have been developed, which consist essentially of placing agar media or other substances in the soil and permitting growth on the medium of fungi from specific loci in soil. The apparatus holding the medium is removed, and fungi colonizing the medium are isolated in the laboratory [8]. This method was further modified as:

Profile Plate Technique

It is utilized to study the distribution and association of actively growing fungi over a larger area. The plates used are made of polypropylene plastic and contain holes. These holes are filled with agar medium, which is brought into contact with soil to study the fungi isolated [9].

Immersion Tube

Immersion tubes are prepared from plastic centrifuge tubes having holes on their walls, which are filled with nutrient agar. These tubes are embedded in soil for 4-6 days, and the agar and fungal invaders are transferred to suitable media like Rose Bengal Agar or Sabouraud Agar for study [10].

ISOLATION OF FUNGAL HYPHAE FROM SOIL

Physical Dissection

This method permits the isolation of fungi that rarely appear on dilution plates as many of the fungal hyphae remain with the heavier soil particles of the residue soil. A crumb of 1.0-1.5 g is placed in water in a beaker and left to become saturated. More water may be added and the heavier particles are allowed to sediment while the supernatant fluid is removed. This process is continued until the liquid remains clear after standing for 1.0 to 1.5 minutes. The soil particles of the residue are distributed in small quantities among 3 sterile petri plates. The material is examined for the presence of fungal hyphae, which may be attached to mineral grains or humus particles and are removed by fine forceps and placed in clean, sterile water.

Hyphal masses need to be teased out since this usually consists of more than one species of fungi, and an agar medium is added. The hyphae are located and numbered on the bottom of the dish on examination under a microscope (120x). The tips of growing hyphae are cut out in agar blocks and transferred to fresh media. Careful examination is necessary to determine whether the colonies grow from the hyphal strands or humus particles. Since spores are likely to be attached to humus particles, the colonies suspected to be developing from spores should be cut out to prevent them from overgrowing unterminated or slowly growing hyphae by examining them.

RESULTS

In the plate counts, it may be noted that the numbers of organisms in the rhizosphere and control soils are calculated on an oven-dry basis and can be compared directly. For the dilution plate and soil plate methods, it was found that the soil plate method was superior for isolating more species. During the pipetting and soil water dilution series, larger soil particles were generally left in water blanks on which the fungal hyphae are normally attached.

The soil plate method is easier and less time-consuming as compared to the dilution plate method. In methods like immersion tubes *etc.*, the fungi trapped generally belong to selected fungi capable of growing out of the soil through an air space and on an agar surface. Thus, accurate or complete information about the distribution of fungi in soil cannot be obtained if other, unlike methods, are employed simultaneously for critical soil studies.

All colonies from the plates used for the fungal counts, or from suitable sectors of such plates, were picked to provide isolates for further studies (Fig. **2**). The present studies were undertaken to study the flora of the rhizosphere and rhizoplane (root surface) and also various aspects related to the rhizosphere zone, *viz.*, root exudates, secretions, soil factors, and rhizosphere effects, and were compared [11]. These studies proved quite effective in checking the population of the pathogen in the soil, leading to a reduction in disease and an increase in productivity. Katznelson and coworkers [12] have also shown that microorganisms predominate in the vicinity of the plant root. A total of 34 different species of fungi were isolated and identified from the rhizosphere and 12 from the rhizoplane of moth bean (*Vigna aconitifolia*) from six districts of Rajasthan, India, *viz.*, Newai, Ganganagar, Tonk, Sikar, Nagaur, and Pali (Table **1**).

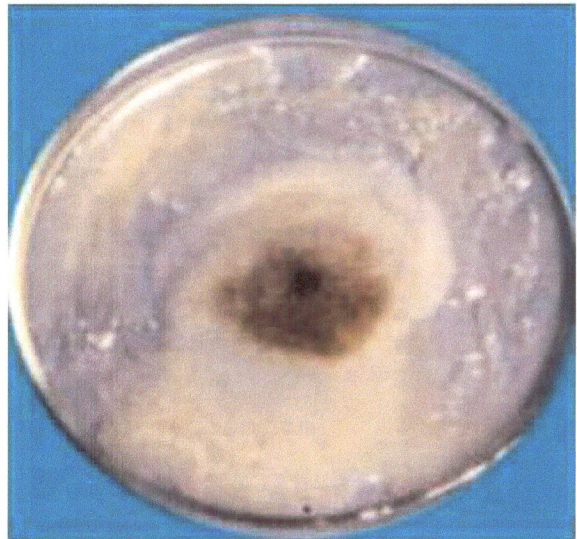

Fig. (2). A photomicrograph showing microorganisms isolated from rhizosphere soil through soil plate method.

Table 1. List of fungi identified in rhizosphere and rhizoplane of moth bean (*Vigna aconitifolia*).

S. No.	Fungi Identified in Rhizosphere		Fungi Identified in Rhizoplane	
	Genus	**Species**	**Genus**	**Species**
1	*Aspergillus*	*tamarii, terreus, variecolor, japonicus, flavus, aculeatus, niveus, nidulans,* and *quadrilineatus*	*Aspergillus*	*niger, terrus,* and *sydowi*
2	*Fusarium*	*javanicum, oxysporum, moniliforme, oxysporum, udum,* and *acuminatum*	*Fusarium*	*Solani* and *semitectum*
3	*Penicillium*	*oxalicum, rubrum, humicola,* and *notatum*	*Penicillium*	*notatum*

S. No.	Fungi Identified in Rhizosphere		Fungi Identified in Rhizoplane	
	Genus	Species	Genus	Species
4	*Trichoderma*	*harzianum, longibrachiatum, viride,* and *lignorum*	*Trichoderma*	*viride*
5	*Mucor*	*luteus, varians,* and *fragilis*	*Mucor*	*hiemalis*
6	*Curvularia*	*falcata* and *lunata*	-	-
7	*Sclerotium*	*oryzae*	-	-
8	*Rhizoctonia*	*solani*	-	-
9	*Helminthosporium*	*oryzae*	*Helminthosporium*	*oryzae*
10	*Rhizopus*	*nigricans*	*Rhizopus*	*nigricans*
11	*Cylindrospermum*	-	-	-
12	*Alternaria*	*alternata*	-	-
13	-	-	*Macrophomina*	*phaseolina*

DISCUSSION

An examination of the qualitative data showed that a determination of the rhizosphere and rhizoplane microflora may be a useful additional parameter for studying the influence of plant roots on soil microorganisms. The abundance of microorganisms in the rhizosphere appears to be under the influence of certain specific conditions prevailing in the rhizosphere. *Aspergillus* spp. were the most abundant in the rhizosphere and rhizoplane of the moth bean, followed by *Fusarium*, *Penicillium*, *Trichoderma*, and *Mucor* spp. In the rhizoplane, *Curvularia* sp., *Sclerotium* sp., *Rhizoctonia* sp., *Cylindrospermum* sp., and *Alternaria* sp. were not found. *Macrophomina phaseolina* was the only fungus that was found in the rhizoplane but not in the rhizosphere.

Secretions from the roots are known to be wholly or partly responsible for the increase in the number of microorganisms. These secretions consist of certain amino acids which influence the growth and spore germination of the microorganisms. Soil factors are known to affect both root disease development [13] and rhizosphere microflora. These factors no doubt change with the growth of the plant and, evidently, in turn are affected by the host.

CONCLUDING REMARKS

In the present study, methods to isolate microflora from soil have been studied concerning the rhizosphere and rhizoplane. It was found that the soil plate method was easier and less time-consuming. Each of the methods has its limitations, and during the study, 13 various species of fungi have been isolated from the rhizosphere and rhizoplane of moth bean (*Vigna aconitifolia*), with a majority of *Aspergillus* species being nine in number. It seems that the rhizosphere is the microorganism-rich area of moth beans, while the rhizoplane area is not suitable for the microorganisms. It may also be due to some secretions that may restrict the growth of the fungi in rhizoplane.

REFERENCES

[1] Graf G. Über den Einfluss des Pflanzenwachstums auf die Bakterien im Wurzelbereich Centbl Bakt II 1930; 82: 44-69.

[2] Clark FE. Effects of soil amendments upon the bacterial populations associated with roots of wheat. Trans Kans Acad Sci 1939; 42: 91-6.

[3] Sperber J, Rovira AD. A study of the bacteria associated with the roots of subterranean clover and wimmera rye-grass. J Appl Bacteriol 1959; 22: 85-95.
[http://dx.doi.org/10.1111/j.1365-2672.1959.tb04613.x]

[4] Kumar CPC, Balasubramanian KA. Phyllosphere and rhizosphere microflora of pearl millet with reference to downy mildew incited by *Sclerospora graminicola* (Sacc.) Schroet. Plant Soil 1981; 62: 65-80.
[http://dx.doi.org/10.1007/BF02205026]

[5] Johnson LF, Manka K. A modification of warcups soil- plate method for isolating soil fungi. Soil Sci 1961; 92: 79-84.
[http://dx.doi.org/10.1097/00010694-196108000-00001]

[6] Waksman SA. Soil fungi and their activities. Soil Sci 1916; 2: 103-56.
[http://dx.doi.org/10.1097/00010694-191608000-00001]

[7] Warcup JH. The soil-plate method for isolation of fungi from soil. Nature 1950; 166: 117-8.
[http://dx.doi.org/10.1038/166117b0] [PMID: 15439172]

[8] Chesters CGC, Thornton RH. A comparison of techniques for isolating soil fungi. Trans Br Mycol Soc 1956; 39: 301-13.
[http://dx.doi.org/10.1016/S0007-1536(56)80015-6]

[9] Anderson AL, Huber DM. The plate profile technique for isolating soil fungi and studying their activity in the vicinity of roots. Phytopathology 1965; 55: 592-4.

[10] Mueller KE, Durrell LW. Sampling tubes for soil fungi. Phytopathology 1957; 47: 243.

[11] McNear JrDH. The rhizosphere - roots, soil and everything in between. Nature Edu Knowl 2013; 4: 1.
[https://www.nature.com/scitable/knowledge/library/the-rhizosphere-roots-soil-and-67500617/]

[12] Katznelson H, Lochhead AG, Timonin MI. Soil microorganisms and the rhizosphere. Bot Rev 1948; 14: 543-86.
[http://dx.doi.org/10.1007/BF02861843]

[13] Bharti Mittu AC, Mittu B. Soil health - an issue of concern for environment and agriculture. J Bioremediat Biodegrad 2015; 6: 286.
[http://dx.doi.org/10.4172/2155-6199.1000286]

CHAPTER 16

Trichoderma: A Potential Arsenal for Industries

P.B. Khaire[1,*], **S.S. Mane**[1] **and S.V. Pawar**[2]

[1] *Department of Plant Pathology and Agril. Microbiology, PGI, MPKV, Rahuri, India*
[2] *Department of Plant Pathology, Sumitomo Chemical India Limited, Mumbai, India*

Abstract: The genus *Trichoderma* (fungi) is a very large group of microorganisms that play a significant role in the environment. This is omnipresent in the climate, particularly in soils. *Trichoderma* species could be easily isolated from the soil by all traditional methods available because of their rapid growth and abundant conidiation. These are used both as biofungicides for biological plant protection as well as for bioremediation. In addition, the genus *Trichoderma* includes edible and medicinal mushrooms but also human pathogens. Members of the *Trichoderma* genus are often used primarily in the processing of enzymes, antibiotics, and other metabolites, but also for biofuel in various branches of industry. Several researchers have confirmed, based on phylogenetic analysis, that *Trichoderma* and *Hypocrea* form a single holomorphic genus. In which two can be differentiated by large clades. Several *Trichoderma* spp. positively affect plants by stimulating plant growth and protecting plants from fungal and bacterial pathogens. *Trichoderma* has entered the genomic period at present, and sections of the genome sequences are open to the public. For this purpose, *Trichoderma* can be used to an even greater degree than before for human needs. *Trichoderma* species possess diverse biotechnological applications, such as acting as biofungicides to control various plant diseases and as biofertilizers to promote plant production. *Trichoderma* secretes various volatile compounds, including alcohols, aldehydes, ketones, ethylene, hydrogen cyanide, and monoterpenes, as well as non-volatile compounds known to exhibit antibiotic activity, including peptaibols, and diketopiperazine-like gliotoxins and gliovirins. Nonetheless, further studies are required to make the application of these fungi more effective and safe.

Keywords: *Trichoderma*, Bioactive Compounds, Antibiotics, Hydrolytic Enzymes, Brewing Industry, Bioremediation.

INTRODUCTION

Trichoderma-type fungi are commonly found in all climatic zones. The most common habitats include soil and rotting wood. They can use various substrates and are considered immune to specific toxic chemicals. These fungi can be found

---

[*] **Corresponding author P.B. Khaire:** Department of Plant Pathology and Agril. Microbiology, PGI, MPKV, Rahuri, India; E-mail: pravinkhaire26893@gmail.com

Shampi Jain, Ashutosh Gupta and Neeraj Verma
All rights reserved-© 2023 Bentham Science Publishers

in the soil environment on sclerotia and other propagating types of fungi. They are considered to dominate in all climatic areas, including agricultural, woodland, salt marsh, and desert soils. These were also separated from such rare sources as shellfish, sea bivalves, and termites. In addition, *Trichoderma* species are known to occur in rotting wood and soil due to their productive heterotrophic interactions, such as parasitism, decomposition, and even opportunistic endophytes. Such species produce different pigments, ranging from a greenish-yellow one up to a reddish tinge, but there are still some colourless specimens. The conidia can also have a variety of colours, from colourless to different shades of green, or even grey or brown tinges. This genus currently consists of more than 260 species, and approximately 35 known species are economically important either because of their ability to produce enzymes and antibiotics or because they are used as biocontrol agents. On the other hand, some members of the genus have been identified as emerging opportunistic human pathogens and as causative agents of green mould, a disease that causes significant losses in the production of agricultural edible mushrooms. This chapter covers the wild range of the biodiversity of a ubiquitous *Trichoderma* from different sources and its applications in industry as a producer of bioactive compounds and extracellular hydrolytic enzymes, and in agriculture as a prompter for plant growth and biocontrol.

BIOTECHNOLOGICAL APPLICATIONS

Production of Secondary Metabolites

Secondary metabolites are small organic compounds that do not participate directly in normal growth and reproduction but play an important role in the formation, signaling, and interaction with other species. The lack of secondary metabolites does not lead to the immediate death of the individual but results in a long-term impairment of the life, fecundity, or aesthetics of the organism or even no noticeable improvement at all. Often, there are some environmental conditions where secondary metabolites are essential for survival, such as siderophores that are necessary for growth where iron concentrations are poor. For plants, secondary metabolites are a significant defense against herbicides and other interspecies protection, whereas secondary metabolites are used by humans as medications, flavouring products, and therapeutic medicinal products [1]. Demain and Fang [2] reported that secondary metabolites are used as a strategic tool against other bacteria, fungi, amoebas, plants, insects, and animals for the transport of metals as well as agents for plant-to-organism symbiosis. Essential secondary metabolites are alkaloids, terpenoids, and phenolics.

Trichoderma spp. are economically important as they serve as biocontrol or biopesticide agents that inhibit the growth of phytopathogenic fungi. *Trichoderma* functions as a biocontrol agent due to the existence of a variety of extracellular lytic enzymes and secondary metabolites [3]. It produces numerous secondary metabolites as well as extracellular enzymes, including β-glucanase, chitinase, and proteinase [1]. They reveal a range of mechanisms of action for functioning as biocontrol agents in their antagonistic encounters with fungal pathogens, such as antibiotic activity, mycoparasitism, nutrient rivalry, cell wall lytic enzyme activity, and the development of systemic tolerance to plant pathogens [4, 5]. Antibiotic development by *Trichoderma* sp. is considered to play an important role in the case of biocontrol. A variety of antibiotics as well as antifungal toxins, such as trichodermin, gliovirin, and harzianic acid, have been known to be developed by *Trichoderma* species that have a direct effect on other organisms [6].

Table 1. Secondary metabolites /strains, bioactive compounds and their effects formed by *Trichoderma*.

Species	Bioactive Compounds	Effects	References
T. koningii and *T. viride*	Dermadin (U-21, 963)	Antimicrobial activity *S. aureus* and *Escherichia coli*	[7]
T. reesei	Cellulases	Degrade cellulase during root colonization to penetrate the plant tissue	[8]
T. longibrachiatum and *T. pseudokoningii*	Compactin	Act as a cholesterol-lowering agent	[9]
T. koningii	Koninginin A	Act as a regulator of plant growth	[10]
T. longibrachiatum	5-Hydroxyvertinolide	Antagonistic against the fungus *Mycena citricolor*	[11]
T. virens	Gliovirin	Antimicrobial against oomycetes and *Staphylococcus aureus*	[12]
T. longibrachiatum	Bisvertinolone	Antifungal properties	[13]
T. harzianum	Fleephilone	Inhibitory activity against the binding of REV-proteins to REV responsive element RNA	[14]
T. harzianum	Harziphilone	Cytotoxicity against the murine tumor cell line M-109	[14]
T. longibrachiatum	Trichodimerol	Inhibit tumor necrosis factor in human monocytes	[15]
T. harzianum	6-(1-pentenyl)-2H-pyran-2-one	Antifungal activity	[16]
T. virens	Trichocaranes A, B, D	Inhibit the growth of etiolated wheat coleoptiles	[17]

Species	Bioactive Compounds	Effects	References
T. viride	Viridepyronone	Antagonistic against *Sclerotium rolfsii*	[18]
T. harzianum	Harzianopyridone	Antifungal against *Botrytis cinerea, R. solani* and inhibitor of the protein phosphate type 2A	[19]
T. virens	Trichodermamide A	It has a weak cytotoxic effect on three cell lines P388, A-549, and HL-60	[20]
T. viride	Emodin	Antimicrobial and antineoplastic agent	[21]
T. harzianum	Trochosetin	Inhibited the growth of rice, tomato, and medicago	[22]
T. longibrachiatum, T. koningii, and T. viride	Ergokonin A	Antifungal activity against *Candida* sp.	[22]
T. virens	Trichodermamide B	Displays cytotoxicity against HCT-116 human colon carcinoma	[22]
T. virens	Wortmannolone	Inhibitor of the phosphatidylinositol 3-kinase with potential to attack human neoplasm in humans	[22]
T. virens	Virone	Inhibitor of the phosphatidylinositol 3-kinase	[22]
T. virens and T. viride	Heptelidic acid	Activity against *Plasmodium falciparum*	[22]
T. atroviride and T. virens	Indole-3-acetic acid	Growth and development regulator	[23]
T. atroviride and T. virens	Indole-3-acetaldehyde	Control root growth in *Arabidopsis thaliana*	[23]

Production of Antibiotics and Bioactive Compounds

Several fungi have the ability to produce toxins (antibiotics) that can destroy other microbes at low concentrations. The diversity of these antibiotics has shown a variety of actions against both prokaryotes and eukaryotes [24 - 26]. *Trichoderma* spp. produce antibiotics and the use of antibiotics was first described by Weindling [27], but Dennis and Webster [28] documented the role of antibiotics in the control of plant pathogens. Paracelsin was the first secondary antibiotic metabolite isolated from *Trichoderma* spp [29]. *Trichoderma* spp. contain a significant variety of antibiotic-like compounds such as alcohols, aldehydes, ethylene, hydrogen cyanide, monoterpenes, ketones, peptaibols, and diketopiperazine-like gliovirin and gliotoxin. The ability of *Trichoderma* spp. to produce antibiotics depends on certain conditions such as the quantity of microorganisms, pH, temperature, and the type of substrate. A single *Tricho-*

derma sp. can produce multiple antibiotic compounds and, similarly, another *Trichoderma* sp. can produce a defined antibiotic [30].

Multiple strains of *Trichoderma harzianum* contribute to the development of antibiotics, and therefore, these strains have the potential to limit the consumption of wheat [31]. Howell and Stipanovic [12] have demonstrated that antibiotics play a significant role in the antifungal function of *Trichoderma virens*. In addition, *Trichoderma* spp. are also known to develop several bioactive metabolites [32], including antimycobacterials such as aminolipopeptide trichoderine, antifungals including trichodermaketone A, and cytotoxics such as dipeptide, trichodermamid B, and antibacterial.

Production of Hydrolytic Enzymes

Both economically important crops are affected by pathogens, including fungi as the most destructive soil pathogen, and have also been thoroughly studied for crop damage [26, 33, 34]. *Botrytis*, *Fusarium*, *Pythium*, and *Rhizoctonia* are some of the major pathogenic fungal genera [35]. Pesticides have been extensively used to combat these diseases [36], but pesticide use has contributed to environmental and public health issues around the world [37, 38]. Therefore, any environmentally friendly alternatives are relevant.

Trichoderma spp. cell walls are known to contain hydrolytic enzymes, such as cellulase, chitinase, *etc.*, which play an important role in the degradation of biomass [39]. Chet [40] studied that the cell wall of phytopathogenic fungi is primarily composed of β-1, 3-glucans and chitin-like cellulose in some oomycetes, such as *Pythium* spp. But due to the involvement of hydrolytic enzymes, *Trichoderma* spp. interferes with pathogen development. Hydrolytic enzymes secreted by *Trichoderma harzianum*, which inhibit the growth of pathogens such as *Crinipellis perniciosa*, believed to be the cause of the cocoa disease (*Theobroma cacao*), have been shown to produce hydrolytic enzymes [41]. *Trichoderma* isolates produce hydrolytic enzymes such as chitinase, β-1, 3- and β-1, 6-glucanase, and proteases (β-1, 3-glucan), chitin or fungal cell walls as a carbon source [42, 43].

Trichoderma sp. is capable of using a wide range of chemicals, such as carbon and nitrogen, to concurrently secrete a number of enzymes to dissolve plant polymers into simple sugars for energy and growth. Due to the high cost of chemical inductors for these enzymes, it is important to find some inexpensive organic inductors from agricultural waste to maximize the mass production of *Trichoderma*. Studies on enzymes produced by *Trichoderma* are important to find more effective and low-cost enzymes that will be useful in various phases of the hydrolytic biomass degradation process.

ROLE OF *TRICHODERMA* IN WINEMAKING AND BREWERY INDUSTRY

The main advantage of biocatalytic processes is the prospect of developing less adulterating, biodegradable industrial products. The stimulation of *Trichoderma* hydrolytic enzymes with pectinase, cellulase, chitinase, and/or glucanase activity, as well as supra-extraction of Palomino fino grapes, was studied in the fermentation process, in the clarification of juices, and in wine characteristics. The greatest activity was found with the use of enzymes in contaminated concentrate. In addition, when comparing wines from juices subjected to various conditions (healthy and tainted, frozen or fresh), the highest difference in the wine characteristics was found. Supra-extracted juices produced wines with better acidity and higher alcohol content, such as methanol, propanol, and isobutanol [44].

The enzymes from *Trichoderma viride* WEBL0703 showed remarkable behaviour against a large variety of phytopathogens produced by solid-state fermentation using grape marc and wine lees. The yield of certain essential enzymes that play an important role in protecting plants from various diseases, such as chitinase, β-glucanase and pectinase, was 47.8 U/g IDS, 8.32 U/g IDS and 9.83 U/g IDS, respectively. The study proposed that it is possible to transform waste from the winery into a value-added and eco-friendly biocontrol agent [45].

ROLE OF *TRICHODERMA* IN BIOREMEDIATION

Trichoderma species [an anamorphic Hypocreaceae (class Ascomycetes)] have been used for the biological regulation of plant diseases, for the development of cellulolytic and hemicellulolytic enzymes, for the biodegradation of chlorophenolic compounds and soil bioremediation [46, 47]. *Trichoderma reesei* is assumed to be one of the most effective hypercellulase manufacturers used in the industry. The concentration of biomass at a point of time was constant with relatively fast, early growth on easily metabolized growth medium components (yeast extract), followed by a second slower growth period due to cellulose hydrolysis that follows a rise in the concentration of cellulose [48]. Bioremediation and phytoremediation in tandem with microbes are groundbreaking methods that can alleviate multiple environmental problems. The *Trichoderma* genus is genetically very diverse and has a wide variety of capabilities amongst various strains of agricultural and industrial value. It is also tolerant to a variety of recalcitrant toxins, such as heavy metals, contaminants, and polyaromatic hydrocarbons. *Trichoderma* has wide potential possibilities for the biological or phytobial remediation of environmental pollutants [49].

A newly isolated fungus, *Trichoderma lixii* F21, was investigated for the bioremediation of polar (Alizarin Red S) [50] and non-polar [Quinizarine Green SS (QGSS)] anthraquinone dyes. Results have shown that *T. lixii* F21 could be a strong candidate for the bioremediation of industrial effluents adulterated with anthraquinone dyes [51]. *Trichoderma* species may be a model fungus that preserves crop productivity as well as being commonly used as inoculants for biocontrol, biofertilization, and phytostimulation. *Trichoderma* spp. are reported to increase photosynthetic ability, increase nutrient uptake, and increase the effectiveness of the use of nitrogen in agriculture. On the other hand, they can be used to produce bioenergy, promote adaptation and reduce the detrimental effects of climate change [52]. *Trichoderma* species have broad biotechnological uses as biofungicides to combat plant disease and as biofertilizers to stimulate plant growth, resulting in high yields and productivity, maintaining food protection and environmental sustainability by minimizing the use of harmful agrochemicals, providing industrially important chemicals, and having potential for bioremediation. Bioremediation using effective *Trichoderma* can help to remove heavy metals that could pollute the climate. *Trichoderma* was isolated by serial dilution and spread technique on potato dextrose agar (PDA) with individual heavy metals, i.e. chromium (Cr), copper (Cu), lead (Pb), zinc (Zn) and nickel (Ni). The 29 fungal isolates of 4 species were chosen and the growth test showed that all *Trichoderma* isolates can tolerate high levels of Cr and Pb, but tolerance to Cu, Zn and Ni was unique to the species. The findings showed the ability of *Trichoderma* isolates for biological wastewater treatment in the mining industry [53].

ROLE OF *TRICHODERMA* IN AGRICULTURE

Trichoderma is one of the most studied and usable fungal biocontrol agents. Biological activity is linked to a variety of metabolites they produce that have been shown to inhibit specific pathogen development, and enhance disease tolerance and plant growth. *Trichoderma* sp. is excellent for the synthesis of enzymes with high xylanolytic activity. Diverse xylanases have been well characterized, identified and purified for their physicochemical, hydrolytic, and molecular properties. Cellulase-free xylanase formulations have also been successfully studied in modified or alternative industrial applications [54].

Economically important cultivated cotton plants were selected to observe the growth progression of *Trichoderma viride* and *Pseudomonas fluorescens* with and without pathogens, *Rhizoctonia solani*, and *Macrophomina phaseolina*. Of these, *Trichoderma viride* was found to be more effective in shoot and root length elongation than *P. fluorescens*, *T. viride*, and *P. fluorescens* greatly improved the percentage of seed germination, root longitude, shooting length, fresh weight, dry weight, and vigour index [55].

Trichoderma has shown promise for its diverse role in agriculture, and several strains of *Trichoderma* have been successfully evaluated for their constructive effects on soil fertility and plant health aspects, but we need a pollution-free ecosystem that focuses on multiple roles of *Trichoderma* to tackle different biotic and abiotic stresses and hazards. Strains of *Trichoderma atroviride* and *Trichoderma harzianum*, overgrown with biological material, have been used as soil bio-preparations in open-field growing lettuce, and their population levels have been tracked over time. This *Trichoderma* sp. was recognized in field soil; however, its occurrence was predicted to be comparatively low at 103 CFU compared to 105 CFU g^{-1} of dry soil, subsequently bio-prepared, and *Trichoderma* continued to do so after 2 years [56]. Another study used MiSeq sequencing to test the reactions of local microbial populations to three separate fertilization methods, involving the use of heavy chemical fertilizers (CF) and reduced chemical fertilizer formulations, complemented by organic fertilizer (OF) or *Trichoderma*-enriched organic fertilizers (BF) for tomatoes over a five-year span. The result showed stronger plant growth and soil fertility in BF treatment followed by OF and CF therapy. It was concluded that reduced chemical fertilizer plus *Trichoderma*-enriched organic fertilizer (BF) may be the most effective regime for regulating soil microbiome degeneration and maintaining the growth and health of tomato plants [57].

MASS PRODUCTION, FORMULATION AND ITS SCOPE FOR THE MANAGEMENT OF PLANT DISEASES

Primary biocontrol research is based on the use of *Trichoderma* spores specifically in plants. When scientific results are moved from the laboratory to the field, technologies become feasible. While *Trichoderma* has a very strong potential in disease control, it could not be used under faithful conditions as a spore suspension. Thus, in certain carriers, the *Trichoderma* culture should be immobilized and prepared as formulations for easy application, storage, commercialization, and field use.

Characteristics of *Trichoderma* for Formulation Development

To develop a successful *Trichoderma* formulation, it should possess the following qualities:

  i. High rhizosphere competence.
  ii. High competitive saprophytic ability.

iii. Able to enhance plant growth.
  iv. Ease for mass multiplication.
   v. Broad spectrum of action.
  vi. Excellent and reliable control.
 vii. Safe to the environment.
viii. Compatible with other bioagents.
  ix. Should tolerate desiccation, heat, oxidizing agents, and UV radiations [58].

Characteristics of an Ideal Formulation

An ideal formulation should possess the following properties:

   i. Should have increased shelf life.
  ii. Should not be phytotoxic to crop plants.
 iii. Should tolerate adverse environmental conditions.
  iv. Should be cost-effective and should give reliable control of plant diseases.
   v. Should dissolve well in water.
  vi. Carriers must be cheap and readily available for formulation development.
 vii. Should be compatible with other agrochemicals.

Model Methods for the Mass Multiplication of *Trichoderma*

The commercial application of *Trichoderma*, either for crop health or for the management of plant diseases, depends on the production of commercial formulations with effective carriers that, for a significant period, enable the survival of *Trichoderma*.

Formulation Based on Talc

T. viride formulations based on talc are available in India at the Tamil Nadu Agricultural University, Coimbatore, which was established for seed treatment of pulse crops and rice [59] and the Talc-based formulation of different biofertilizers at Mahatma Phule Krishi Vidyapeeth, Rahuri (M.H.) under the Plant Pathology section. *Trichoderma* is produced in a fluid solution, then mixed into talc powder in a 1:2 ratio and dried to 8% moisture level in the shade. *Trichoderma* powder formulations have a life span of 90-120 days. It has become quite common in India for a variety of reasons. Seed treatment at 4 to 5 g/kg is effective to control a variety of soil-borne pathogens in different crops. In India, many corporate factories are manufacturing talc products for agriculture. *Trichoderma* is expected to be needed at 5,000 tonnes per year to occupy 50% of India's agricultural land [60].

Formulation Based on Vermiculite and Wheat Bran

For ten days, *Trichoderma* is compounded in molasses yeast media. Over 3 days, 100 g of vermiculite and 33 g of wheat bran are sterilized in a 70°C hot air oven. After that, 20-gram fermentor biomass, 0.05 N medium, and diluted or whole biomass containing HCl are placed, gently mixed, and dried in the shade [61].

Formulation Based on Pesta Granules

Reactor biomass (52 L) is poured into 100 g of wheat flour with gloved hands to create a continuous mixture. So many times, the mixture is kneaded, rolled, and assembled by hand. Then one-millimeter thick sheets/pesta are ready and air-dried until crispy. The dough layer, after drying, was ground and then crystals were gathered after passing through an 18-layer [62].

Formulation Reliant on Alginate Prills

In one part, sodium alginate is diluted with distilled water (25 g/750 ml), while the food basis is dissolved in the other (50 g/250 ml). Such materials are sterilized by autoclaving and then merged with biomass once they have cooled [63]. Drop by drop, the aggregate is applied to the $CaCl_2$ solution, forming spherical beads that are air dried and maintained at 5°C.

Formulation Based on Press Mud

As a by-product of the sugar industry, press mud is abundant and can be used as a medium for mass multiplying of *Trichoderma*. The process consisted of equally integrating a 9-day-old *T. viride* culture formed in PD (potato dextrose) broth within 120 kg of press mud. Water is sprayed constantly to keep it wet. To allow airflow and hold moisture underneath, gunny bags are used to shield this. A nucleus culture with further multiplying is available in 4 weeks. The same has been thoroughly mixed into eight tonnes of press mud and reared for eight days in the shade before being used in the agriculture field. One can be able to introduce eight thousand times more culture media into the land than the prescribed levels of biocontrol agents due to a rapid and noticeable impact. According to Sabalpara [64], some compounds can also be used to multiply efficiently with various bioagents at the bulk stage.

Formulation Based on Coffee Husks

Sawant and Sawant [65] developed a *Trichoderma* formulation based on using the coffee husk, a waste product from the coffee processing manufacturing sector in Karnataka. The above material was found to be incredibly useful in treating black pepper *Phytophthora* foot rot disease and is commonly used in the Karnataka and Kerala states of India.

Formulations Based on Oil

They are prepared by combining solid-state or liquid-state ferment conidia with a robust emulsion formulation of plant oils. With the help of a ground active agent, microbial agents are dissolved in a water immiscible solvent such as a petroleum component or vegetable oil. That makes a solid emulsion and spreads this in water. EC needs a high concentration of an oil-soluble emulsifier to produce a homogeneous emulsion on mixing with water. Species, crops, people, and creatures must not be harmed by the oils utilized. *Trichoderma* products are being included as a foliar application. Oil-based products should be ideal for foliar spray in dry conditions and have a long lifespan. Since the spores are protected by oil, which prevents them from drying at 5°C, they can live for a long time on the surface of the plant, including dry conditions. Batta [66] formulated an emulsified composition of *T. harzianum* to combat *Botrytis cinerea*, a causal organism of apple post-harvest decay. Utilizing native components, an invert-emulsion mixture of *T. harzianum* with an eight-month prolonged storage period has been produced. This formulation was tested and reported to be successful against soil-borne peanut diseases by the former Project Directorate of Biological Control (PDBC) in India.

Formulations Focused on Banana Waste

Balasubramanian and coworkers [67] developed a mass multiplication strategy for *Trichoderma* sp. using banana waste. Urea, rock phosphate, *Bacillus polymyxa*, *Pleurotus sajor-caju*, and *Trichoderma viride* cultures were used in similar waste biomass. A pit of various banana waste, such as girdled pseudostems and bases, is diced into 5 to 8 cm lengths. A chamber is built, and various ingredients are layered in five layers. So each level contains 1 tonne of banana waste biomass, 5 kilogrammes of urea, 125 kilogrammes of rock phosphate, and 1 litre of *B. polymyxa*, *Pleurotus sajor-caju*, and *T. viride* broth culture is divided into five layers, each of which is made in the same way and thoroughly mixed. The banana waste decomposes in about 45 days, and the enriched media mass is ready for farm use.

Shelf Life of *Trichoderma* Formulations

For effective marketing, the shelf life of the manufactured product of a biocontrol agent plays a major part. The antagonists multiplied in an organic food base usually have a longer shelf life than the bases of inert or inorganic food. *Trichoderma*-based formulations of talc, peat, lignite, and kaolin have a shelf life of 3 to 4 months. The viable propagules of *Trichoderma* in talc formulation decreased by 50% after 120 days of storage [68]. Studies on the handling of *T. viride* formulations in polypropylene bags in different colours showed that the highest population of *T. viride* was contained in 100 micron thick milky white bags. At PDBC, work was carried out in Bangalore on increasing the shelf life of *Trichoderma* talc formulations using different ingredients (chitin and glycerol) in the processing medium, and heat shock was applied at the end of the log fermentation process, which can prolong the shelf life of talc.

TRICHODERMA DELIVERY FOR DISEASE MANAGEMENT

For successful disease control, delivery and establishment of *Trichoderma* to the site of action is very important. The most common methods of application of *Trichoderma* are as follows:

Treatment of *Trichoderma* to seeds

It has been proposed to treat seed with formulations based on talc and wheat bran at a rate of 4 g/kg of seed [69]. *Trichoderma* propagules germinate on the treated seed surface. When they are planted in the soil, the germinating propagules colonize the roots and rhizosphere of the seedlings [70]. Crops covered with *Trichoderma* is a simple and effective method of delivering the antagonist *Trichoderma* for the treatment of seed/soil-borne plant diseases. Just before planting seeds, the seeds are dusted with dry *Trichoderma* talc. The dry powder of antagonists is being used at 3 to 10 g per kg of seed for commercial gain, depending on agronomic traits [71]. Seed protectants against *Pythium* spp. and *R. solani* were found to be *T. harzianum*, *T. virens*, and *T. viride* by Mukherjee and Mukhopadhyay [72]. Rice seed treated with two antagonistic fungi, *T. viride* and *T. harzianum*, was found to be successful in controlling sheath blight disease and increasing crop productivity [73]. In some other research, *T. viride* was discovered to be an effective inhibitor of *R. solani* toxin production in the same illness [74]. Seed therapy with bio-controls such as *T. viride*, *T. harzianum*, and *Gliocladium virens* was proved to be efficient in battling loose smut of wheat by Singh and Maheshwari [75].

However, *Trichoderma* sp., like a growth promoter, aids in crop productivity, as indicated by using *T. harzianum* (Th3) in the watered and dried regions of Rajasthan's Kota and Jaipur regions of India, which is environmentally friendly [76]. Chavan and coworkers [77] found that seed application with *T. harzianum*, *Allium sativum*, and *Azadirachta indica*, in combination with an aerial/foliar application as spraying of mancozeb, was effective in preventing *Alternaria* blight disease of mustard caused by *A. brassicae* and *A. brassicicola*, as well as an increase in production. Oil-seed-disease causing fungi such as *Aspergillus flavus*,

Alternaria alternata, *Curvularia lunata*, *Fusarium moniliforme*, *Fusarium oxysporum*, *Rhizopus nigricans*, *Penicillium notatum*, and *Penicillium chrysogenum*, which damage oilseed plants such as soybean, sesame, and sunflower, were prevented by seed application of *Trichoderma* sp [78].

Treatment of *Trichoderma* as biopriming seeds

Bioprimming is the process of therapy of seeds by using bio-control agents and then incubating them in a humid and wet environment before the radical emerges. This technology offers the ability to outperform easy seed covering in terms of young plants' development speed and uniformity. *Trichoderma* conidia grow on top of the layer of bioprimed seeds and build a coating around them. Such seeds are more tolerant of a variety of soil conditions. Bio-priming can also cut down on the quantity of biological control agents used on the crop. In the Tarai area of Uttaranchal state, bio-priming of seeds has been found effective in tomato, brinjal, soybean, and chickpea crops [79]. In bio-primed seeds of chickpea and rajma in pots and farms, 3 rhizosphere capable microbial strains, *Pseudomonas fluorescens* OKC, *T. asperellum* T42, and *Rhizobium* species RH4, independently and in mixture, displayed greater plant growth percentages and improved growth parameters for both plants in comparison to non-bioprimed untreated control. It was also discovered that combining the microorganisms improved seed growth rate and plant growth rather than applying them separately. All the formulations that included *Trichoderma* performed better than the others, and the tri-microbial mixture performed better in terms of seed germination and plant development in each of these chickpeas and rajma [80].

Treatment of *Trichoderma* to the soil

All kinds of helpful and microbial bacteria can be found in soil. *Trichoderma* spp. delivered to the soil will improve the population biology of enhanced fungal enemies, preventing pathogenic microorganisms from establishing themselves on the contamination site. A few studies [81 - 83] show that bio-control compounds can be applied to the soil before or after cultivating the plants to manage a huge variety of soil-borne plant diseases. Application of *T. viride* to the soil, whether individually or in conjunction with other therapies, greatly decreased the incidence of *Colletotrichum falcatum*, causing the red rot of sugarcane [84]. According to Srivastava and coworkers [85], soil treatment with *T. viride* was reported to be the most effective in reducing jute seedling blight, colour rot, stem rot, and root rot fungus. Mustafa and coworkers [86] found that applying an organic *Trichoderma* formulation to the soil was successful in controlling seed-borne pathogenic fungi such as *Fusarium oxysporum*, *F. moniliforme*, *F. solani*, *Alternaria alternata*, and *R. solani*, as well as seed germination of *Dalbergia sissoo* Roxb. Since *Trichoderma* sp. can colonize farmyard manure (FYM), it is much more suitable and advantageous to apply colonized FYM to the soil. This is perhaps the most efficient way to use *Trichoderma*, especially for the treatment of soil emerging pathogens.

Treatment of *Trichoderma* to the Root

Seedling roots may be handled with opponent spore or cell solution while soil application of *Trichoderma* in raised nursery-beds or soaking roots in *Trichoderma* fluid prior to transplanting. The roots of rice, onion, brinjal, chili, and capsicum, dipped in antagonist solution, decrease disease severity and improve seedling development [87]. There have also been records of root dips of rice seedlings prior to transplantation reducing sheath blight disease [88].

Spraying the Leaves/Dressing for Wounds

Variations in microclimate have a major impact on the effectiveness of biological control agents for foliar and leaf diseases. Heat, relative humidity, dew, rainfall, air and radiation are all prone to diurnal and nocturnal, periodic and non-periodic variations in the phyllosphere. As a result, the water capacity of phylloplane microorganisms would be continuously changing. It can also differ on foliage or the canopy's perimeter, as well as on shielded leaves. In the shaded, compact area of the vine, relative humidity was stronger than in the periphery foliage. The deposition of dew is more prominent in the middle and outskirts. The amount of moisture exuded by stomata, lenticels, hydathodes, and injuries differs widely. In phylloplane, it influences the effectiveness and persistence of antagonists [89]. For such biological control of insects, a fluid solution of *Trichoderma* has been applied successfully to aerial plant components to manage the *Alternaria* leaf spot of *Vicia faba* [90]. *T. harzianum* and *T. virens* are effective in foliar application and powder formulations for rice sheath blight [91, 92].

Sharma and coworkers [76] conducted research experiments in Rajasthan on groundnut root rot affected by a multipathogen cluster, primarily *A. niger*, *A. flavus*, *Thielaviopsis basicola*, *R. rolfsii*, *T. solani*, and *Pythium aphanidermatum*. Treatment with *T. Harzianum* was tested in the fields as both dust and a fluid bio-formulation, and it was proven to be effective in controlling the disease. Citrus scab caused by *Elsinoe fawcettii* was managed by Singh and coworkers [93]. They discovered *T. harzianum* and *Epicoccum purpurascens* decreased the prevalence of the disease in the field by 17.8% and 10%, respectively, after spraying. Although foliar spray of *Trichoderma* reduces the incidence of diseases in the region, it is not theoretically feasible due to higher concentration and crop economy. As a result, the dose and treatment rate must be uniform depending on crop value, which may be a safe and realistic solution.

MASS PRODUCTION AND FORMULATION DEVELOPMENT

In commercial production processes, the first major problem is the achievement of sufficient growth of the biocontrol agent. In certain cases, due to the particular requirements of nutritional and environmental conditions for the growth of the organism, the biomass development of the antagonist is not straightforward. Through solid and liquid fermentation processes, mass processing is accomplished. The commercial success of biocontrol agents requires the following:

  i. Reliable and broad-spectrum action.
 ii. Stability and Safety.
iii. Increased shelf life.
 iv. Low capital requirements.
  v. Easy access to career materials.
 vi. Market demand that is both economical and viable.

Fermentation

Liquid and solid fermentation methods are used for the mass production of *Trichoderma*.

Liquid Fermentation

For processes where soluble materials in water are used for microbial formation, the term liquid state fermentation (LSF) is applied. The quantity of water should be greater for fermentation processes of this kind. In LSF, water is present in thick layers and is periodically absorbed within the substrates, both for microbial formation. This fermentation scheme has been introduced for the mass production of fungal biocontrol agents. The preferred medium should be inexpensive for mass production and readily available with acceptable nutrient quality. In general, potato dextrose broth, V-8 juice, molasses-yeast medium, and wheat bran are used by liquid fermentation technology for the mass production of *Trichoderma* spp [94].

Solid Fermentation

The term "solid state fermentation" (SSF) is applied to processes in which insoluble materials in water are used for microbial formation. The quantity of water in fermentative processes of this sort does not exceed the saturation potential of the solid bed in which the micro-organisms (fungi) emerge. For microbial growth, water is important in SSF; it is present in thin layers and sometimes absorbed within the substrates. Solid-state fermentation (SSF) is an important method of mass processing of fungal biopesticides since it provides a higher content of conidia to micropesticides. Various cheap cereal grains are used as substrates, such as sorghum, millets, and ragi [60, 61]. The grains are moistened, sterilized, inoculated with *Trichoderma*, and incubated by liquid *Trichoderma*, which produce a dark green spore coating on the grains for 10 to 15 days for the processing of *Trichoderma*. These grains can be thinly powdered and used for the treatment of seeds, or the grains can be used for the enrichment of FYM for soil application. In the absence of free-flowing water, SSF processes involving the development of microorganisms (typically fungi) on moist solid substrates have major economic potential in manufacturing goods for the fruit, feed, medicinal, and agricultural industries [95]. SSF has gained attention as it demonstrates some recovery of goods and lower energy needs. Based on the design of the solid process used, two types of SSF structures can be identified. The first and most widely used approach involves growing on natural substrates. The second method requires cultivation on inert support impregnated with a liquid medium, which is not commonly used. This technique is sufficient for small-scale processing in cottage industries or at the level of individual farmers.

CONSTRAINTS TO COMMERCIALIZATION

The success of biopesticides to suppress pests and diseases depends on the availability of microbes as a product or formulation, which facilitates the technology transfer from lab to land. The constraints to biopesticide development and utilization mirror some of those factors that limit development worldwide. Constraints include:

i. Lack of the right screening protocol for the selection of a promising candidate for *Trichoderma*.
ii. Inadequate knowledge of *Trichoderma* microbial ecology and plant pathogens.
iii. Optimization of fermentation technology and mass production of *Trichoderma*.
iv. Inconsistency in performance and short shelf life.
v. Lack of patent protection.
vi. Expensive registration fees.
vii. Inadequacies in awareness, training, and education.
viii. Inadequate multidisciplinary approach.
ix. Technological limitations.

CONCLUDING REMARKS

Trichoderma spp. can be easily collected from wood using soil washing techniques, and they are also found as isolated colonies, from which isolation and pure culture can be obtained. Differentiation of the characteristics of various *Trichoderma* isolates can be conveniently seen in the growth medium. For the isolation, quantification, and functional analysis of *Trichoderma* organisms, culture media is a more effective and useful tool than non-culturable methods. To distinguish closely related strains, universally premiered PCR (UP-PCR) fingerprinting combined with ITS1 ribotyping is useful. Over the last 35 years, these techniques have assembled *Trichoderma* species into 15 hereditary components, increasing the number of named *Trichoderma* species from 9 to about 80 phylogenetic species. *Trichoderma* species are economically important because they produce industrially important enzymes, antibiotics, and biocontrol agents.

Moreover, using fungi in conjunction with PGPR may be a suitable strategy for sustainable development, although some issues need to be studied further so that the full benefits in terms of improved plant growth can be obtained from this naturally produced population, particularly under stress conditions. Numerous *Trichoderma* species have recently been discovered to have the ability to live as endophytes. *In vitro* and in pod trials, *Trichoderma theobromicola*, an endophytic fungus isolated from cacao in South America, releases a volatile/diffusible antibiotic that prevents the production of *Moniliophthora roreri*, the cause of cacao frosty pod rot. In addition to laccase production for textile industries, cellulases, a significant industrial enzyme from *Trichoderma*, are needed for the breakdown of biomass to produce second-generation biofuels and other value-added products such as fermentable sugars, organic acids, solvents, and softeners. New strains of *Trichoderma* spp. have been developed due to developments in genetic engineering for the production of novel metabolites for a variety of applications through genome sequencing. Development of secondary metabolites and various lytic enzymes generated by *Trichoderma* and their biotechnological and industrial applications in agriculture have been observed. Various *Trichoderma* spp. can be explored for their growth in agriculture.

REFERENCES

[1] Kumar V, Shahid M, Srivastava M, *et al.* Role of secondary metabolites produced by commercial *Trichoderma* species and their effect against soil borne pathogens. Biosens J 2014; 3: 1000108.
[http://dx.doi.org/10.4172/2090-4967.1000108]

[2] Demain AL, Fang A. The Natural Functions of Secondary Metabolites. In: Fiechter A, Eds. History of Modern Biotechnology I. Advances in Biochemical Engineering/Biotechnology, vol 69. Berlin, Heidelberg: Springer 2000; 69: pp. 1-39.
[http://dx.doi.org/10.1007/3-540-44964-7_1]

[3] Cardoza RE, Hermosa MR, Vizcaíno JA, *et al.* Partial silencing of a hydroxy-methylglutaryl-CoA reductase encoding gene in *Trichoderma harzianum* CECT 2413 results in a lower level of resistance to lovastatin and lower antifungal activity. Fungal Gene Biol 2007; 44: 269–283.
[http://dx.doi.org/10.1016/j.fgb.2006.11.013] [PMID: 17218128]

[4] Verma P, Yadav AN, Kumar V, *et al.* Microbes in Termite Management: Potential Role and Strategies. In: Khan M, Ahmad W, Eds. Termites and Sustainable Management: Sustainability in Plant and Crop Protection. Cham: Springer 2017; pp. 197-217.
[http://dx.doi.org/10.1007/978-3-319-68726-1_9]

[5] Yadav AN, Sachan SG, Verma P, *et al.* Bioprospecting of plant growth promoting psychrotrophic bacilli from the cold desert of north western Indian Himalayas. Indian J Exp Biol 2016; 54: 142–50.
[[http://nopr.niscpr.res.in/handle/123456789/33745]
[PMID: 26934782]

[6] Sharma S, Kour D, Rana KL, *et al. Trichoderma*: Biodiversity, Ecological Significances, and Industrial Applications. In: Yadav A, Mishra S, Singh S, *et al.* Eds. Recent Advancement in White Biotechnology Through Fungi. Fungal Biology. Cham: Springer 2019; pp. 85-120.
[http://dx.doi.org/10.1007/978-3-030-10480-1_3]

[7] Tamura A, Kotani H, Naruto S. Trichoviridin and dermadin from *Trichoderma* sp. TK-1. J Antibiot (Tokyo) 1975; 28: 161-2.
[http://dx.doi.org/10.7164/antibiotics.28.161] [PMID: 1089624]

[8] Henrissat B, Driguez H, Viet C, *et al.* Synergism of cellulases from *Trichoderma reesei* in the degradation of cellulose. Nature Biotechnol 1985; 3: 722-6.
[http://dx.doi.org/10.1038/nbt0885-722]

[9] Endo A, Hasumi K, Yamada A, *et al.* The synthesis of compactin (ML-236B) and monacolin K in fungi. J Antibiot 1986; 39: 1609–1610.
[http://dx.doi.org/10.7164/antibiotics.39.1609] [PMID: 3793631]

[10] Cutler HG, Himmelsbach DS, Arrendale RF, *et al.* Koninginin A: A novel plant growth regulator from *Trichoderma koningii*. Agric Biol Chem 1989; 53: 2605–11.
[http://dx.doi.org/10.1080/00021369.1989.10869746]

[11] Andrade R, Ayer WA, Mebe PP. The metabolites of *Trichoderma longibrachiatum*. Part 1. Isolation of the metabolites and the structure of trichodimerol. Can J Chem 1992; 70: 2526-35.
[http://dx.doi.org/10.1139/v92-320]

[12] Howell CR, Stipanovic RD. Gliovirin, a new antibiotic from *Gliocladium virens*, and its role in the biological control of *Pythium ultimum*. Can J Microbiol 1983; 29: 321-4.
[http://dx.doi.org/10.1139/m83-053]

[13] Kontani M, Sakagami Y, Marumo S. First β-1,6-glucan biosynthesis inhibitor, bisvertinolone isolated from fungus, *Acremonium strictum* and its absolute stereochemistry. Tetrahed Lett 1994; 35: 2577-80.
[http://dx.doi.org/10.1016/S0040-4039(00)77175-9]

[14] Qian-Cutrone J, Huang S, Chang LP, *et al.* Harziphilone and fleephilone, two new HIV REV/RRE binding inhibitors produced by *Trichoderma harzianum*. J Antibiot (Tokyo) 1996; 49: 990-7.
[http://dx.doi.org/10.7164/antibiotics.49.990] [PMID: 8968392]

[15] Mazzucco CE, Warr G. Trichodimerol (BMS-182123) inhibits lipopolysaccharide-induced eicosanoid secretion in THP-1 human monocytic cells. J Leukoc Biol 1996; 60: 271-7.
[http://dx.doi.org/10.1002/jlb.60.2.271] [PMID: 8773589]

[16] Parker SR, Cutler HG, Jacyno JM, *et al.* Biological activity of 6-pentyl-2H-pyran-2-one and its analogs. J Agric Food Chem 1997; 45: 2774–2776.
[http://dx.doi.org/10.1021/jf960681a]

[17] Macías FA, Varela RM, Simonet AM, *et al.* Bioactive carotanes from *Trichoderma virens*. J Nat Prod 2000; 63: 1197-200.

[http://dx.doi.org/10.1021/np000121c] [PMID: 11000018]

[18] Evidente A, Cabras A, Maddau L, *et al.* Viridepyronone, a new antifungal 6-substituted 2 H-pyran-2-one produced by *Trichoderma viride*. J Agric Food Chem 2003; 51: 6957-60.
[http://dx.doi.org/10.1021/jf034708j] [PMID: 14611154]

[19] Kawada M, Yoshimoto Y, Kumagai H, *et al.* PP2A inhibitors, harzianic acid and related compounds produced by fungus strain F-1531. J Antibiot (Tokyo) 2004; 57: 235-7.
[http://dx.doi.org/10.7164/antibiotics.57.235] [PMID: 15152810]

[20] Liu P, Yang Q. Identification of genes with a biocontrol function in *Trichoderma harzianum* mycelium using the expressed sequence tag approach. Res Microbiol 2005; 156: 416-23.
[http://dx.doi.org/10.1016/j.resmic.2004.10.007] [PMID: 15808946]

[21] Wu Y-W, Ouyang J, Xiao X-H, *et al.* Antimicrobial properties and toxicity of anthraquinones by microcalorimetric bioassay. Chin J Chem 2006; 24: 45-50.
[http://dx.doi.org/10.1002/cjoc.200690020]

[22] Reino JL, Guerrero RF, Hernández-Galán R, *et al.* Secondary metabolites from species of the biocontrol agent *Trichoderma*. Phytochem Rev 2008; 7: 89-123.
[http://dx.doi.org/10.1007/s11101-006-9032-2]

[23] Contreras-Cornejo HA, Macías-Rodríguez L, Cortés-Penagos C, *et al. Trichoderma virens*, a plant beneficial fungus, enhances biomass production and promotes lateral root growth through an auxin-dependent mechanism in *Arabidopsis*. Plant Physiol 2009; 149: 1579-92.
[http://dx.doi.org/10.1104/pp.108.130369] [PMID: 19176721]

[24] Saxena AK, Yadav AN, Kaushik R, *et al.* Biotechnological applications of microbes isolated from cold environments in agriculture and allied sectors. In: International Conference on Low Temperature Science and Biotechnological Advances, Society of Low Temperature Biology 2015; p. 104.

[25] Suman A, Verma P, Yadav AN, *et al.* Bioprospecting for extracellular hydrolytic enzymes from culturable thermotolerant bacteria isolated from Manikaran thermal springs. Res J Biotechnol 2015; 10: 33-42. [http://krishi.icar.gov.in/jspui/handle/123456789/64691]

[26] Yadav AN. Biodiversity and biotechnological applications of host-specific endophytic fungi for sustainable agriculture and allied sectors. Acta Sci Microbiol 2018; 1: 1-5.
[https://actascientific.com/ASMI/pdf/ASMI-01-0044.pdf]
[http://dx.doi.org/10.31080/ASMI.2018.01.0044]

[27] Weindling R. Studies on a lethal principle effective in the parasitic action of *Trichoderma lignorum* on *Rhizoctonia solani* and other soil fungi. Phytopathology 1934; 24: 1153-79.

[28] Dennis C, Webster J. Antagonistic properties of species-groups of *Trichoderma*. Trans Br Mycol Soc 1971; 57: 41-IN4.
[http://dx.doi.org/10.1016/S0007-1536(71)80078-5]

[29] Brückner H, Graf H, Bokel M. Paracelsin; characterization by NMR spectroscopy and circular dichroism, and hemolytic properties of a peptaibol antibiotic from the cellulolytically active mold *Trichoderma reesei*. Part B. Experientia 1984; 40: 1189-97.
[http://dx.doi.org/10.1007/BF01946646] [PMID: 6500005]

[30] Sivasithamparam K, Ghisalberti E. Secondary Metabolism in *Trichoderma* and *Gliocladium*. In: Kubicek CP, Harman GE, Eds. *Trichoderma* and *Gliocladium* vol 1: Basic Biology, Taxonomy and Genetics. London, UK: Taylor & Francis Ltd 2014; pp. 139-92.

[31] Ghisalberti EL, Narbey MJ, Dewan MM, *et al.* Variability among strains of *Trichoderma harzianum* in their ability to reduce take-all and to produce pyrones. Plant Soil 1990; 121: 287-91.
[http://dx.doi.org/10.1007/BF00012323]

[32] Ruiz N, Roullier C, Petit K, *et al.* Marine-Derived *Trichoderma*: A Source of New Bioactive Metabolites. In: Mukherjee PK, Horwitz BA, Singh US, *et al.*, Eds. *Trichoderma*: Biology and Applications. USA: CAB International 2013; pp. 247-79.

[http://dx.doi.org/10.1079/9781780642475.0247]

[33] Yadav AN, Kumar R, Kumar S, *et al.* Beneficial microbiomes: Biodiversity and potential biotechnological applications for sustainable agriculture and human health. J Appl Biol Biotechnol 2017; 5: 45-57. [https://jabonline.in/abstract.php?article_id=246&sts=2]
[http://dx.doi.org/10.7324/JABB.2017.50607]

[34] Yadav AN, Yadav N. Stress-adaptive microbes for plant growth promotion and alleviation of drought stress in plants. Acta Sci Agri 2018; 2: 85-8. [https://actascientific.com/ASAG/ASAG-02-0104.php]

[35] Djonović S, Vargas WA, Kolomiets MV, *et al.* A proteinaceous elicitor Sm1 from the beneficial fungus *Trichoderma virens* is required for induced systemic resistance in maize. Plant Physiol 2007; 145: 875-89.
[http://dx.doi.org/10.1104/pp.107.103689] [PMID: 17885089]

[36] Gerhardson B. Biological substitutes for pesticides. Trends Biotechnol 2002; 20: 338-43.
[http://dx.doi.org/10.1016/S0167-7799(02)02021-8] [PMID: 12127281]

[37] Bues R, Bussières P, Dadomo M, *et al.* Assessing the environmental impacts of pesticides used on processing tomato crops. Agric Ecosyst Environ 2004; 102: 155-62.
[http://dx.doi.org/10.1016/j.agee.2003.08.007]

[38] Punja ZK, Utkhede RS. Using fungi and yeasts to manage vegetable crop diseases. Trends Biotechnol 2003; 21: 400-7.
[http://dx.doi.org/10.1016/S0167-7799(03)00193-8] [PMID: 12948673]

[39] Schuster A, Schmoll M. Biology and biotechnology of *Trichoderma*. Appl Microbiol Biotechnol 2010; 87: 787-99.
[http://dx.doi.org/10.1007/s00253-010-2632-1] [PMID: 20461510]

[40] Chet I. *Trichoderma*-Application, Mode of Action, and Potential as a Biocontrol Agent of Soilborne Pathogenic Fungi. In: Chet I, Ed. Innovative Approaches to Plant Disease Control. New York: John Wiley & Sons 1987; pp. 137-60.

[41] Marco JLD, Valadares-Inglis MC, Felix CR. Production of hydrolytic enzymes by *Trichoderma* isolates with antagonistic activity against *Crinipellis perniciosa*, the causal agent of witches' broom of cocoa. Braz J Microbiol 2003; 34: 33-8.
[http://dx.doi.org/10.1590/S1517-83822003000100008]

[42] de la Cruz J, Rey M, Lora JM, *et al.* Carbon source control on β-glucanases, chitobiase and chitinase from *Trichoderma harzianum*. Arch Microbiol 1993; 159: 316-22.
[http://dx.doi.org/10.1007/BF00290913]

[43] Elad Y, Chet I, Henis Y. Degradation of plant pathogenic fungi by *Trichoderma harzianum*. Can J Microbiol 1982; 28: 719-25.
[http://dx.doi.org/10.1139/m82-110]

[44] Roldán A, Palacios V, Peñate X, *et al.* Use of *Trichoderma* enzymatic extracts on vinification of *Palomino fino* grapes in the sherry region. J Food Eng 2006; 75: 375-82.
[http://dx.doi.org/10.1016/j.jfoodeng.2005.03.065]

[45] Bai Z, Jin B, Li Y, *et al.* Utilization of winery wastes for *Trichoderma viride* biocontrol agent production by solid state fermentation. J Environ Sci (China) 2008; 20: 353-8.
[http://dx.doi.org/10.1016/S1001-0742(08)60055-8] [PMID: 18595404]

[46] Esposito E, Silva M. Systematics and environmental application of the genus *Trichoderma*. Crit Rev Microbiol 1998; 24: 89-98.
[http://dx.doi.org/10.1080/10408419891294190] [PMID: 9675511]

[47] Kour D, Rana KL, Kumar R, *et al.* Gene Manipulation and Regulation of Catabolic Genes for Biodegradation of Biphenyl Compounds. In: Singh HB, Gupta VK, Jogaiah S, Eds. New and Future Developments in Microbial Biotechnology and Bioengineering: Microbial Genes Biochemistry and Applications. Amsterdam: Elsevier 2019; pp. 1-23.

[http://dx.doi.org/10.1016/B978-0-444-63503-7.00001-2]

[48] Ahamed A, Vermette P. Culture-based strategies to enhance cellulase enzyme production from *Trichoderma reesei* RUT-C30 in bioreactor culture conditions. Biochem Eng J 2008; 40: 399-407.
[http://dx.doi.org/10.1016/j.bej.2007.11.030]

[49] Tripathi P, Singh PC, Mishra A, *et al. Trichoderma*: A potential bioremediator for environmental clean up. Clean Technol Environ Policy 2013; 15: 541-50.
[http://dx.doi.org/10.1007/s10098-012-0553-7]

[50] Khalili E, Javed MA, Huyop F, *et al.* Evaluation of *Trichoderma* isolates as potential biological control agent against soybean charcoal rot disease caused by *Macrophomina phaseolina*. Biotechnol Biotechnol Equip 2016; 30: 479-88.
[http://dx.doi.org/10.1080/13102818.2016.1147334]

[51] Adnan LA, Sathishkumar P, Yusoff ARM, *et al.* Rapid bioremediation of Alizarin Red S and Quinizarine Green SS dyes using *Trichoderma lixii* F21 mediated by biosorption and enzymatic processes. Bioprocess Biosyst Eng 2017; 40: 85-97.
[http://dx.doi.org/10.1007/s00449-016-1677-7] [PMID: 27663440]

[52] Kashyap PL, Rai P, Srivastava AK, *et al. Trichoderma* for climate resilient agriculture. World J Microbiol Biotechnol 2017; 33: 155.
[http://dx.doi.org/10.1007/s11274-017-2319-1] [PMID: 28695465]

[53] Tansengco M, Tejano J, Coronado F, *et al.* Heavy metal tolerance and removal capacity of *Trichoderma* species isolated from mine tailings in Itogon, Benguet. Environ Nat Resour J 2018; 16: 39-57. [https://ph02.tci-thaijo.org/index.php/ennrj/article/view/106514]
[http://dx.doi.org/10.14456/ennrj.2018.5]

[54] Wong KKY, Saddler JN. *Trichoderma xylanases*, their properties and application. Crit Rev Biotechnol 1992; 12: 413-35.
[http://dx.doi.org/10.3109/07388559209114234]

[55] Shanmugaiah V, Balasubramanian N, Gomathinayagam S, *et al.* Effect of single application of *Trichoderma viride* and *Pseudomonas fluorescens* on growth promotion in cotton plants. Afr J Agric Res 2009; 4: 1220-5. [https://academicjournals.org/journal/AJAR/article-full-text-pdf/93EAD4437962/]

[56] Oskiera M, Szczech M, Stępowska A, *et al.* Monitoring of *Trichoderma* species in agricultural soil in response to application of biopreparations. Biol Control 2017; 113: 65-72.
[http://dx.doi.org/10.1016/j.biocontrol.2017.07.005]

[57] Pang G, Cai F, Li R, *et al. Trichoderma*-enriched organic fertilizer can mitigate microbiome degeneration of monocropped soil to maintain better plant growth. Plant Soil 2017; 416: 181-92.
[http://dx.doi.org/10.1007/s11104-017-3178-0]

[58] Jeyarajan R, Nakkeeran S. Exploitation of Microorganisms and Viruses as Biocontrol Agents for Crop Disease Management. In: Upadhyay RK, Mukerji KG, Chamola BP, Eds. Biocontrol Potential and Its Exploitation in Sustainable Agriculture. Boston, MA: Springer 2000; pp. 95-116.
[http://dx.doi.org/10.1007/978-1-4615-4209-4_8]

[59] Jeyarajan R, Ramakrishnan G, Dinakaran D, *et al.* Development of products of *Trichoderma viride* and *Bacillus subtilis* for biocontrol of root rot diseases. In: Dwivedi BK, Ed. Biotechnology in India. Allahabad: Bioved Research Society 1994; pp. 25-36.

[60] Jeyarajan R. Prospects of Indigenous Mass Production and Formulation of *Trichoderma*. In: Rabindra RJ, Ramanujam B, Eds. Current Status of Biological Control of Plant Diseases Using Antagonistic Organisms in India. Bangalore: Project Directorate of Biological Control 2006; p. 445.

[61] Lewis JA. Formulation and Delivery System of Biocontrol Agents with Emphasis on Fungi. In: Keister DL, Cregan PB, Eds. The Rhizosphere and Plant Growth Beltsville Symposia in Agricultural Research. Dordrecht: Springer 1991; vol. 14: pp. 279-87.

[http://dx.doi.org/10.1007/978-94-011-3336-4_55]

[62] Connick WJ Jr, Daigle DJ, Quimby PC Jr. An improved invert emulsion with high water retention for mycoherbicide delivery. Weed Technol 1991; 5: 442-4.
[http://dx.doi.org/10.1017/S0890037X00028402]

[63] Fravel DR, Rhodes DJ, Larkin RP, *et al.* Production and Commercialization of Biocontrol Products. In: Albajes R, Lodovica Gullino M, Van Lenteren JC, *et al.*, Eds. Integrated Pest and Disease Management in Greenhouse Crops Developments in Plant Pathology. Dordrecht: Springer 1999; Vol. 14: pp. 365-76

[64] Sabalpara AN. Mass multiplication of biopesticides at farm level. J Mycol Plant Pathol 2014; 44: 1-5.

[65] Sawant IS, Sawant SD. A simple method for achieving high cfu of *Trichoderma harzianum* on organic wastes for field applications. Indian Phytopathol 1996; 49: 185-7. [https://epubs.icar.org.in/index.php/IPPJ/article/view/20065/10221]

[66] Batta YA. Postharvest biological control of apple gray mold by *Trichoderma harzianum* Rifai formulated in an invert emulsion. Crop Prot 2004; 23: 19-26.
[http://dx.doi.org/10.1016/S0261-2194(03)00163-7]

[67] Balasubramanian C, Udaysoorian P, Prabhu C, *et al.* Enriched compost for yield and quality enhancement in sugarcane. J Ecobiol 2008; 22: 173-6.

[68] Sankar P, Jeyarajan R. Biological control of sesamum root rot by seed treatment with *Trichoderma* spp. and *Bacillus subtilis*. Indian J Mycol Plant Pathol 1996; 26: 148-51.

[69] Jogani V, John P. Evaluation of different application methods of *Trichoderma harzianum* (Rifai) against *Fusarium* wilt of tomato. Crop Res 2014; 48: 76-9.

[70] Tewari AK. Biological control of chickpea wilt complex using different formulations of *Gliocladium virens* through seed treatment. Pantnagar, India: GB Pant University of Agriculture and Technology 1996; p. 167.

[71] Mukhopadhyay AN, Shrestha SM, Mukherjee PK. Biological seed treatment for control of soilborne plant pathogens. FAO Plant Prot Bull 1992; 40: 21-30.

[72] Mukherjee PK, Mukhopadhyay AN. *In situ* mycoparasitism of *Gliocladium virens* on *Rhizoctonia solani*. Indian Phytopathol 1995; 48: 101-2.

[73] Das BC, Hazarika DK. Biological management of sheath blight of rice. Indian Phytopathol 2000; 53: 433-5.

[74] Sriram S, Raguchander T, Babu S, *et al.* Inactivation of phytotoxin produced by the rice sheath blight pathogen *Rhizoctonia solani*. Can J Microbiol 2000; 46: 520-4.
[http://dx.doi.org/10.1139/w00-018] [PMID: 10913973]

[75] Singh D, Maheshwari VK. Biological seed treatment for the control of loose smut of wheat. Indian Phytopathol 2001; 54: 457-60.

[76] Sharma V, Bhandari P, Singh B, *et al.* Chitinase expression due to reduction in fusaric acid level in an antagonistic *Trichoderma harzianum* S17TH. Indian J Microbiol 2013; 53: 214-20.
[http://dx.doi.org/10.1007/s12088-012-0335-2] [PMID: 24426111]

[77] Chavan PG, Apet KT, Wagh SS, *et al.* Integrated management of *Alternaria* leaf spot of cauliflower caused by *Alternaria brassicae* (Berk.) Sacc. Trends Biosci 2015; 8: 1908-13.

[78] Jat JG, Agalave HR. Antagonistic properties of *Trichoderma* species against oilseed-borne fungi. Sci Res Rep 2013; 3: 171-4.

[79] Mishra DS, Singh US, Dwivedi TS. Comparative efficacy of normal seed treatment and seed biopriming with commercial formulations of *Trichoderma* spp. In: 53rd Annual Meeting of Indian Phytopath Soc and National Symposium on Ecofriendly Approaches for *Trichoderma*, Chennai, India 2001; pp. 21-23.

[80] Yadav SK, Dave A, Sarkar A, *et al.* Co-inoculated biopriming with *Trichoderma, Pseudomonas* and *Rhizobium* improves crop growth in *Cicer arietinum* and *Phaseolus vulgaris*. Int J Agric Environ Biotechnol 2013; 6: 255-9. [https://ndpublisher.in/admin/issues/v6i2j.pdf]

[81] Baby UI, Manibhushanrao K. Fungal Antagonists and VA Mycorrhizal Fungi for Biocontrol of *Rhizoctonia Solani*, the Rice Sheath Blight Pathogen. In: Manibhushanrao K, Mahadevan A, Eds. Recent Developments in Biocontrol of Plant Pathogens- Current Trends in Life Sciences vol. XXI. Delhi: Today and Tomorrow's Printers and Publishers, New Society 1996; p. 160.

[82] Kumar S. Integrated management of maydis leaf blight of maize. Ann Plant Prot Sci 2010; 18: 536-7.

[83] Kumar S, Upadhyay JP, Rani A. Evaluation of *Trichoderma* species against *Fusarium udum* Butler causing wilt of pigeon pea. J Biol Control 2009; 23: 329-32.

[84] Reddy K, Narayana PK. Efficacy of *Trichoderma viride* against *Colletotrichum falcatum* in Sugarcane. Indian J Plant Prot 2009; 37: 111-5.

[85] Srivastava RK, Singh RK, Kumar N, *et al.* Management of *Macrophomina* disease complex in jute (*Corchorus olitorius*) by *Trichoderma viride*. J Biol Control 2010; 24: 77-9.

[86] Mustafa A, Khan MA. Inam-ul-Haq M, *et al.* Mass multiplication of *Trichoderma* spp. on organic substrate and their effect in management of seed borne fungi. Pak J Phytopathol 2009; 21: 108-14.

[87] Singh US, Zaidi NW. Current Status of Formulation and Delivery of Fungal and Bacterial Antagonists for Disease Management in India. In: Rabindra RJ, Hussaini SS, Ramanujam B, Eds. Microbial Biopesticide Formulations and Application. Bangalore: Project Directorate of Biological Control 2002; pp. 168-79.

[88] Vasudevan P, Kavitha S, Priyadarisini VB, *et al.* Biological Control of Rice Diseases. In: Gnanamanickam SS, Ed. Biological Control of Crop Diseases. New York: Marcel Decker Inc 2002; pp. 11-32.

[89] Andrews JH. Biological control in the phyllosphere. Annu Rev Phytopathol 1992; 30: 603-35. [http://dx.doi.org/10.1146/annurev.py.30.090192.003131] [PMID: 18647102]

[90] Kumar S, Upadhyay JP, Kumar S. Biocontrol of *Alternaria* leaf spot of *Vicia faba* using antogonistic fungi. J Biol Control 2006; 20: 247-50.

[91] Khan AA, Sinha AP. Influence of different factors on the effectivity of fungal bioagents to manage rice sheath blight in nursery. Indian Phytopathol 2005; 58: 289-93.

[92] Khan AA, Sinha AP. Screening of *Trichoderma* spp. against *Rhizoctonia solani* the causal agent of rice sheath blight. Indian Phytopathol 2007; 60: 450-6.

[93] Singh D, Kapur SP, Singh K. Management of citrus scab caused by *Elsinoë fawcettii*. Indian Phytopathol 2000; 53: 461-7.

[94] Prasad RD, Rangeshwaran R, Anuroop CP, *et al.* Bioefficacy and shelf life of conidial and chlamydospore formulation of *Trichoderma harzianum*. J Biol Control 2002; 16: 145-8.

[95] Mienda BS, Idi A, Umar A. Microbiological features of solid state fermentation and its applications-an overview. Res Biotechnol 2011; 2: 21-6. [https://updatepublishing.com/journal/index.php/rib/article/view/2381]

SUBJECT INDEX

A

Abiotic stresses 117, 120, 145, 153, 160, 162, 193, 198, 201, 208, 210
Acetone-butanol production 9
Acid(s) 7, 8, 9, 30, 40, 56, 57, 60, 63, 71, 72, 73, 74, 75, 76, 77, 80, 86, 108, 109, 132, 135, 150, 200, 208, 234, 244, 258, 259
  caffeic 86
  carbonic 74
  carboxylic 71
  folic 8
  fumaric 74, 75, 80
  fusidic 108
  gluconic 60, 74, 75
  glutamic 63
  harzianic 258
  heptelidic 259
  lysergic 234, 244
  malic 74, 75, 80, 200
  nucleic 109
  organic 7, 9, 56, 57, 71, 72, 73, 75, 77, 80, 132, 135 oxalic 72, 73, 74, 76
  pantothenic 8
  phenolic 71
  phytic 7
  stress hormone abscisic 150
Actinomycetes 30, 31, 61, 66, 67, 102, 103, 104, 112, 181, 182
Activities 50, 209, 261
  defence enzyme 209
  glucanase 261
Agents 79, 226
  antagonistic 226
  anti-bacterial 79
Agricultural 1, 2, 10, 33, 161
  practices, sustainable 33
  1, 2, 10
  system 161
Agriculture, sustainable 1, 155
Amino acids 1, 2, 7, 9, 10, 103, 119, 156, 196, 254

Aminoglycosides 105, 109
Aminopenicillins 108
Amoxicillin 108
Amylases 14, 15, 56, 57, 60, 61, 64, 65, 66, 67, 68, 69
Amyloglucosidase 63
Amyloliquefaciens 35, 37, 61
Amylopectin 70
Antagonistic action 193
Antibacterial 196, 210
  activity 196
  properties 210
Antibiosis 34, 96, 118, 194
Antibiotic 57, 105, 106, 118, 195
  compounds 118
  producers 106
  producing bacteria 105
  resistance 57, 195
Antibiotics, antifungal 108
Antifungal 74, 195, 196, 258, 259, 260
  activity 195, 196, 259
  drug 74
  function 260
  properties 258
  toxins 258
Antimicrobial agents 74, 102, 106, 109
Antineoplastic agent 259
Antioxidant(s) 5, 32, 157, 160, 211
  enzymes 157, 160, 211
  non-enzymatic 157
Applications, synthetic pesticide 85
Arabidopsis 32, 152, 159, 259
Arbuscular mycorrhizal fungi (AMF) 30, 31, 32, 33, 113, 114, 115, 116, 159, 160, 161
Aryl alcohol oxidase (AAO) 184, 186

B

Bacillus 10, 37, 86, 88, 89, 90, 104, 227
  cereus 90, 104
  thuringiensis 10, 37, 86, 88, 89, 227

Bacteria 12, 13, 18, 20, 30, 39, 90, 148, 153
  pathogenic 90
  phosphate solubilizing 30
  phospholipase-producing 90
  phytohormone-producing 153
  soil-inhabitant 12
  soil-inhabiting 13, 18, 20
  sulphur-oxidizing 148
  thermophilic 39
Bioactive secondary metabolites (BSM) 36, 104
Biochemical pesticides (BPs) 10, 86
Biological nitrogen fixation (BNF) 17, 18, 19, 20
Biomass 26, 43, 47, 61, 180, 181, 194
  dry 194
  hydrolyzing hemicellulosic 61
  lignocellulosic 43, 47, 180, 181
  lignocellulosic woody 43
  organic 26
  saccharification 47
Biomolecule inhibitors 109
Biopesticides, fungal 92, 271
Bioprocesses 69, 77
Bioremediation 129, 132, 142
  microbiological 129, 142
  of environmental pollutants and contaminants in soil 129
  of heavy metals by bacteria 132
Bleomycin 105

C

Catalase 5, 6, 61, 63, 157, 184, 199, 203, 209, 211
  activity 203, 211
  defence enzyme 211
  -peroxidases 184
Cellulose 13, 46, 60, 64, 66, 70, 80, 133, 180, 181, 183, 187, 260, 261
  biopolymers 133
  degradation system 187
  -degrading enzymes 66
  fibers 60
  hydrolysis 261
  microfibrils 46, 181
Cellulosic biomass 45
Chemicals, toxic organic 131
Chemotaxonomy 72
Chemotherapy 7

Climate(s) 194, 262
  change 262
  diverse 194
Climatic conditions 103, 146, 226
Conditions 13, 32, 39, 45, 61, 63, 79, 112, 121, 149, 184, 185, 198, 201, 221, 254, 259, 261, 263
  acidic 39
  aerobic 63, 221
  anaerobic 39, 184
  iron-limiting 32
  nitrogen-limiting 45
  nutrient 185
  stressed 149
COVID-19 pandemic 246
Creatininase 62
Crop(s) 1, 32, 43, 49, 113, 145, 146, 147, 148, 161, 162, 194, 219, 262, 267
  peanut 32
  production 49, 113, 145, 146, 147, 161, 194
  productivity 1, 148, 162, 262, 267
  protection systems 219
  sugar 43
Crowded plate technique 109, 110

D

Detergents, enzymatic 70
Diseases 3, 266
  peanut 266
  skin 3
Disorders 3, 79, 158, 161, 178, 245, 246
  dermatological 79
  digestive 3
  nutritional 161
DNA 15, 69, 109, 195
  gyrase 109
  polymerase 69
  recombinant 195
  supercoiled 109
Drosophila melanogaster 91
Drought 32, 33, 119, 145, 146, 154, 156, 159, 162
  stress 119, 154, 156
  tolerance 32

E

Endophytic fungi 194, 239, 240, 272
Endotoxins 10, 87, 88

Subject Index

Energy 28, 43, 176
   renewable 28, 176
   sustainable 43
Environment 15, 17, 26, 27, 28, 38, 79, 89, 97, 115, 129, 130, 133, 139, 140, 142, 222, 227, 229
   chemical 79
   hypoxic 115
   nutrient-rich 222
   radioactive 133
Environmental 129, 141, 142, 261
   degradation 142
   healing 142
   pollutants 129, 141, 261
Enzymatic activity 70, 208
Enzyme(s) 17, 58, 61, 62, 64, 67, 79, 261
   glucose isomerase 62
   hemicellulolytic 261
   industry 58, 61
   invertase 64
   mesophilic 79
   oxygen-sensitive 17
   thermostable restriction 67
Erythromycin 2, 105, 107

F

Fermentation 1, 6, 8, 9, 45, 46, 47, 60, 66, 72, 261, 270, 271
   coffee bean 6, 60
   industrial 9
   industry 1
   solid-state 261, 270
   technology 271
Fermentation process 2, 8, 9, 10, 46, 57, 261, 270
   industrial 2
Fluorescent microscopy 78
Fluorocytosine 108
Foliar application 228, 230, 266, 269
Food 15, 16, 57, 58, 65, 66, 68, 69, 70, 73, 74, 77, 78, 235, 245, 266
   contaminated 235
   inorganic 266
Fossil fuels 26, 28, 43
Fungal 35, 96, 180, 184, 194, 196, 204, 209, 213, 224, 258
   endophytes 96
   inoculation 209
   metabolites 96
   pathogens 35, 194, 196, 204, 209, 258
   peroxidases 180, 184
   phytopathogens 213
   suspension 224
Fungi 11, 133, 184, 267
   anaerobic 184
   antagonistic 267
   biotrophic 133
   edible 11

G

Gene expression 115, 154, 180
   stress-responsive 154
Genetically engineered microorganisms (GEMs) 141
Glucanases 36, 61
Glucoamylase 4, 61, 63, 65, 70
Glucose 5, 60, 62, 63, 184
   isomerase 5, 62
   oxidase 5, 60, 63, 184
   production 63
Glutathione 133, 157
   reductase 157
   transferases 133

I

Immobilized enzyme system 80
Induction of antioxidant enzymes 157
Industrial enzymes 3
Industries 26, 28, 60, 61, 71, 78, 121, 130, 142
   agrochemical 28
   aquaculture 78
   biofuel 60, 61
   candy 60
   emerging 130
   energy 26
   gas 142
   organic acid 71
   pesticide 121
Inhibitors of nucleic acid synthesis 109
Inhibitory activity 258
Insect pest management 223
Insecticidal 87, 91, 92, 95
   toxicity 92
   toxins 87, 91
   virus 95
Insecticides, spore-forming bacterial 87

Integrated pest management (IPM) 85, 86, 97, 231
Iron 19, 29, 32, 39, 41, 43, 115, 118, 196, 201, 257
   chelates 118
   concentrations 257
   deficiency 196, 201
   ferric 39, 41
   ferrous 39, 41
Isozymes 93, 185, 186
   phosphatase 93

K

Klebsiella pneumoniae 5

L

Laccase genes 181
Lesions, necrotic 206
Lignin-modifying enzymes (LMEs) 175, 179, 183, 184
Lignocellulose-degrading system 188
Lipase(s) 3, 57, 60, 62, 63, 64, 65, 66, 67, 68, 70
   and proteinase 60, 64
Liquid state fermentation (LSF) 270
Lytic enzymes 32, 258, 272
   extracellular 258

M

Macrocyclic lactone 107
Maltose syrup production 4
Metabolism, cytokinin 155
Metabolites, toxic 93, 222
Microbial 1, 3, 14, 39, 43, 46, 48, 56, 58, 59, 79, 87, 96, 97, 130, 135, 139, 176, 196, 249, 266, 268
   activity 39, 46, 135, 176, 249
   agents 266
   bacteria 268
   biofuels 48
   biotransformation 1, 3
   degradation process 139
   diseases 196
   enzymes 3, 14, 43, 56, 58, 59
   insecticide preparation 96
   insecticides (MIs) 87, 96, 97
   technology 130
   toxins 87
   transformation 43
   utilization 79
Microbial biocontrol 26, 36, 121
   agents 26, 36
   products 121
Mustard oil 212
Mycobacterium tuberculosis 180

N

Naphthalene acetic acid (NAA) 31
Natural 77, 140, 179
   producers 77
   systems 140, 179
Nausea 236, 239, 241, 244
Necrotrophic behaviour 204
Nitrogen 17, 113, 121, 156
   -containing compounds (NCC) 156
   -fixing bacteria 17, 113, 121
Nitrogenases, oxygen-sensitive 19
Nosema locustae, protozoan pathogen 95
Nucleic acid synthesis 109
Nutrient yeast dextrose broth (NYDB) 198

O

Oil-seed-disease 267
Osmotic 154, 198
   imbalance 154
   shock 198
Oxidative 186
   dehydrogenation 186
Oxidative damage 157, 160
   salinity-induced 160
Oxidative stress 199
   resistance 199
   response 199

P

Pathogenic microorganisms 29, 116
Pathways 77, 85, 115, 130, 135, 141, 157, 161, 181
   antioxidative 157
   chemical signalling 115
   lignin biosynthetic 181
   metabolic 77, 130, 135, 141, 161
   nontoxic 85
Pest 34, 38, 85, 87, 130, 194, 231

control 130, 194, 231
  management 34, 38, 85, 87
Pesticides 10, 13, 27, 86, 87, 88, 94, 97, 113, 116, 121, 131, 133, 141, 150, 220
  baculovirus 94
  halogenated 141
  organic 220
  synthetic 86, 87, 116
Phosphate 10
  solubilization 10
Phosphate solubilizing 30
  bacteria (PSB) 30
  microorganisms (PSMs) 30
Phosphotriesterase-like lactonase (PLL) 67
Plant growth 10, 31, 32, 86, 149, 150, 151, 153, 154, 155, 156, 157, 158, 161, 162, 208, 226, 258
  hormones 153
  promoting (PGP) 10, 31, 149, 150
  promoting rhizobacteria (PGPRs) 10, 31, 32, 149, 150, 154, 155, 156, 157, 158, 161, 162
  regulators (PGRs) 86, 151, 153, 208, 226, 258
Plants 86, 149, 156, 162, 183
  angiospermic 183
  halophytic 149
  -incorporated protectants (PIPs) 86
  inoculated 156, 162
Plasmodium falciparum 259
Project directorate of biological control (PDBC) 266, 267
Pseudomonas fluorescens 31, 32, 37, 75, 76, 132, 152, 156, 182, 262

Q

Quaternary ammonium compounds (QAC) 156

R

Reactive oxygen species (ROS) 157, 199
REV-proteins 258
Rhizosphere 28, 31, 114, 117, 118, 119, 248, 253
  effects 248, 253
  inhabitants 117
  microbiome 28, 114, 119
  microorganisms 31, 117, 118

Rhizospheric/endophytic microorganisms 162
ROS-derived oxidative stress 199

S

Salinity stress conditions 160
SAR gene expression 208
Single-cell protein (SCP) 45
Soil 71, 115, 131, 135, 178, 183
  agricultural 71, 115
  contaminate 131
  grassland 183
  healthy 178
  moisture 135
Spores 13, 220, 222, 223, 229, 230, 250
  fungal 222, 223, 230, 250
  germinating 13
  infectious 229
  infective 220
  microscopic 220
Sporotrichum thermophile 70
Starch 4, 6, 65, 105
  casein agar (SCA) 105
  processing 4, 6, 65
Stress tolerance, oxidative 199
Structure 115, 175
  lobed 115
  material's 175
Systemic 37, 113, 203, 208, 209, 210
  acquired resistance (SAR) 113, 203, 208
  resistance 37, 113, 203, 209, 210

T

Technologies 77, 177
  biodegradable 177
  genetic engineering 77
Thermostable 69, 70
  group II intron reverse transcriptase (TGIRT) 69
  lipases 69
  protease 70
Turmeric plant growth 31

V

Vancomycin 104
Virus, nuclear polyhedrosis 94
Volatile organic compounds (VOCs) 119, 159

Y

Yersinia entomophaga 87, 91

www.ingramcontent.com/pod-product-compliance
Lightning Source LLC
Chambersburg PA
CBHW051143220526
45473CB00003B/644